Behavior

and Ecology

of the

NORTHERN

FUR SEAL

———

Behavior
and Ecology
of the
NORTHERN
FUR SEAL

Roger L. Gentry

PRINCETON UNIVERSITY PRESS

PRINCETON, NEW JERSEY

Copyright 1998 by Princeton University Press
Published by Princeton University Press, 41 William Street,
Princeton, New Jersey 08540
In the United Kingdom: Princeton University Press,
Chichester, West Sussex
All Rights Reserved.

Library of Congress Cataloging-in-Publication Data
Gentry, Roger L.
Behavior and ecology of the northern fur seal / Roger L. Gentry.
p. cm.
Includes bibliographical references (p.   ) and indexes.
ISBN 0–691–03345–5 (cl)
1. Northern fur seal—Behavior. 2. Sexual behavior in animals.
3. Parental behavior in animals. I. Title.
QL737.P63G45   1997   97–8332
599.79'7315—dc21

This book has been composed in Times Roman

Princeton University Press books are printed
on acid-free paper and meet the guidelines
for permanence and durability of the Committee
on Production Guidelines for Book Longevity
of the Council on Library Resources

Printed in the United States of America

1   3   5   7   9   10   8   6   4   2

To Melissa, Erin, and Alison

WITH LOVE AND MUCH REGRET FOR WHAT THIS

BOOK HAS COST THEM: NINETEEN SUMMERS

WITHOUT THEIR DAD

The air on the island is strictly damp, and from the rounding up
of the corpses of a thousand dead fur seals, it was unfavorable
for breathing if not cleaned by an almost continuous wind.

So this time of year ought to be shared by bleakness, when
northern winds prevail for the winter; by dampness, when during
southern winds it rains and during too little wind and dead calm it
is foggy; by springtime, when the grass is green here and there; and
into the illusion of summer, when July and August wholesome light
revives a numbed nature . . . for reassurance of those low spirits,
for whom fate is compelled to carry out there in time.

(Khlebnikova (1830–31), on life at the Pribilof Islands)

Contents

Preface

FEW academics these days get the chance to study, with adequate support and relative research freedom, a species that interests them over a long period of time. Therefore it was very fortuitous when I was approached in 1973 by Dr. George Y. Harry with an invitation to apply for a position that would give me just such a chance. Dr. Harry was the director of a government laboratory in Seattle that was to become the National Oceanic and Atmospheric Administration's (NOAA) National Marine Mammal Laboratory. The position called for long-term behavioral studies on the northern fur seal in Alaska, an animal I knew well from my research on other members of the same seal family. The research was part of U.S. obligations to a treaty and was subject to an international panel of scientists, not the U.S. Marine Mammal Protection Act of 1972. In practical terms, that meant I would not be laboriously applying for marine mammal research permits like all my friends. My debate over whether to leave a soft-money position at University of California at Santa Cruz for this plum lasted many microseconds.

My personal interest in this study was to work on proximate causes of behavior, the observable factors that motivate animal choices or decisions. Proximate causes form the basis of relationships among animals, which in turn determine the character of the observable social order (Hinde, 1976). These interests led the study to be largely empirical and reductionistic. I did not use the deductive method because it was not suited to compiling the detailed species profile that this study required. This project greatly outlived the theoretical questions that were in vogue at its inception (such as group selection).

The problem in writing this book was balancing the presentation for several audiences. I have tried not to limit the book to answering only the practical questions that wildlife managers posed at the outset because that would not serve the theoretical interests of academics. I have tried not to focus just on issues in current theory because future researchers will undoubtedly have different interests (Rubenstein and Wrangham, 1986). Finally, I have tried to present enough detail to satisfy other seal researchers. But I have sequestered these details in the Notes section at the back of the book, where they, I hope, will not muddy the waters for more general readers.

This book deals with aspects of seal behavior that, in my judgment, have a measurable impact on the mating system and on the maternal strategy. The material is organized into five parts under common themes. The first consists of introductory material and a long time series of data to show behavior in the context of time, population size, and sex ratio. The second part deals with individuals and their relationship to the mating system. The third deals with special subjects that are fundamental to the mating system. The fourth deals with the maternal strategy; the chapters here resemble journal articles because, since so much has already been written about this subject, any new contribution will tend to focus on very specific, narrow questions. The final part tries to characterize the species based on the material presented, and summarizes the implications of this work for other studies on this animal family.

My goal here is to present a body of data that will have lasting value as a basis for interspecies comparisons. I hope that the young researchers who try to make these comparisons will have the kind of opportunity that led me to write this book.

Acknowledgments

THROUGHOUT this book, I intentionally use the term "we" to acknowledge the army of people that contributed to it. For example, the following are the people who were involved just in building equipment, catching and marking animals, and collecting field data, the heart of this work: Liam Antrim, Jason Baker, Terrie Brown, John Calambokidis, Sue Carter, Vivian Casañas, David W. Christel, Kathryn Chumbley, Eleanora D'Arms, Darlene Deghetto, Randy Felt, John M. Francis, Camille A. Goebel, Michael E. Goebel, James R. Hartley, Carolyn B. Heath, John R. Holt, Edward C. Jameyson, James H. Johnson (deceased), Andronik Kashavarof, Jr., Dennis Lekanof, Peter R. Lekanof, Alvin Lestenkof, Dimitri A. Lestenkof, Marcus K. Lester, Suzanne K. Macy, Sara Madsen, Maxim Malavansky, Victor F. Malavansky, Jr., Rebecca McGee, Alexis Merculief, Alvin Merculief, Dean Merculief, Lida M. Merculief, Mark Merculief, Philbert D. Merculief, Sarah Merculief, Terenty Merculief, Theodosy Merculief (deceased), Kathlene Newell, Dee Nietert, Alexis Prokopiof, Mark Ryan, Ronald J. Ryel, Isiah Shabolin (deceased), Leslie Slater, Susan Steinacher, E. Dave Thilk, Mark E. Towner, George Zacharof, and Peter Zacharof. Those who have experienced the pale excuse for a summer in the Pribilof Islands know what these people endured to collect data. Those who haven't should try to picture every data point in the antiseptic-looking graphs in this book as being generated by people dressed in a "Pribilof Bikini" (full oilskins) sprawled in the grass in a driving mist, trying in vain to dry a spotting scope lens with a square of tissue. I am indebted to them for their good humor, and above all else, their belief in this work.

I thank the Aleut community on St. George Island for allowing us to live and work among them. Their 200-year-old livelihood of killing seals and processing the skins for market was ended forever in 1972, in part so that our project could proceed. Somehow, the community found the generosity of spirit to tolerate our presence. The community leaders during these years, Nicolai S. Merculief, Victor F. Malavansky, and Maxim Malavansky, along with Anne Prokopiof, Victor Lestenkof, and Alvin Lestenkof, helped us solve the many logistic problems that dominate life on a remote island, from building new roads to finding parts for ancient pickup trucks.

One islander, Gregory P. McGlashan, played a pivotal role in this project. He taught me the Aleut way of working with fur seals, helped invent all the capture and handling techniques we used, captured and moved animals, maintained our captive colonies, cared for our facilities in winter, advised me on community matters, labored beside me to build and repair facilities, taught me to collect Murre eggs, and once saved me from a serious mauling by an adult male fur seal. His steady high spirits helped me through the long years of this project and made him my dear friend for life.

A special thanks goes to the principal research assistants on this project over the years: John R. Holt, John M. Francis, Michael E. Goebel, Wendy E. Roberts, Camille A. Goebel, and Karen Walters. All of them had great dedication to this project. They coordinated the work of other assistants, organized the data, and spent hours arguing with me about fur seal behavior. I especially thank John Francis and Michael Goebel for their dedication, friendship, and care in data collection.

This project benefited greatly from friends and colleagues who worked on it under contract. These included Gerry L. Kooyman, Dave Urquhart, Tom Meyer, Randall Davis, Jack Sarno, Daniel P. Costa, and Steven Feldkamp. Gerry Kooyman and Jim Estes provided much-needed intellectual stimulation through the long years of this project.

The data for this project were originally collected on paper and later entered into a computer database by students from the University of Washington. Camille A. Goebel and Karen Walters oversaw the work of data entry and error checking, a painstaking, two-year process that produced a database that was virtually error-free. All tables were typed by Colleen Lee. Line drawings, maps, and some graphs were made by Katherine Zecca and Wendy Carlson.

This book would never have been written without Joan Dunlap, who conducted all the data analysis and generated the figures for all chapters except chapter 12. Her dedication to this project was remarkable and surprising, given her background in plant genetics. She forced me to frame questions about behavior in ways the computer could answer, and then found ingenious ways to answer them. She also forced me to accept the truth when data analysis failed to support my preconceptions. Her determination in the face of a recalcitrant computer was also remarkable. Anne E. York and Jeff Laake also gave advice on statistics. I thank Cynthia T. Tynan for her help with statistics, the climatological literature, reviews, and for support and encouragement throughout the writing of this book.

My long friendship with Valery A. Vladimirov, VNIRO, Russia, finally culminated in joint research on Medny Island, Russia, in 1990. Many thanks to him for arranging the trip, providing use of the field camp, and his contribution to chapter 13. Thanks to the captain and crew of the Russian Fisheries Inspection Vessel *Merlang* for transportation between Dutch Harbor and Medny, and for instruction in the art of making reindeer piroshki. Many thanks to Alexander Boltnev and his field crew for all living arrangements at the camp, as well as to Victor Nikulin and George A. Antonelis, who helped with all captures and other aspects of the field research. The Russian hospitality and friendship will be long remembered.

Thanks to Dan Costa and Robert Gisiner of the Marine Mammal Project in the Office of Naval Research's Biological Sciences Division, for supporting Mike Goebel while he completed chapter 12. ONR also provided the computer that was used to analyze most of the data and prepare the book manuscript.

Various drafts were reviewed by George Antonelis, Jason Baker, Frank A. Beach, Michael Bigg, Daryl Boness, Ian Boyd, Sue Carter, Doug DeMaster, James A. Estes, Charles Fowler, John Francis, Carolyn Heath, Burney Le Boeuf, Tom Loughlin, Fritz Trillmich, Cynthia Tynan, Anne York, and especially E. H. (Ted) Miller, who has been a reviewer and unflagging supporter throughout this work. Their thoughtful, sometimes challenging comments greatly improved this book. I also thank an anonymous reviewer for Princeton University Press for suggesting the structure of the book as it appears here. I also thank my coauthors for their timely contributions.

This project was supported financially by the National Marine Mammal Laboratory of the National Marine Fisheries Service, NOAA. Three laboratory directors, Drs. George Y. Harry, Michael T. Tillman, and Howard W. Braham, and two center directors, Drs. Dayton Alverson and William Aron, made funds available to conduct the research and complete this book. Dr. Aron supported and encouraged this effort from its inception in 1972. I also thank the National Geographic Society for financial support for the field trip to Medny Island.

Finally, I want to acknowledge the thousands of northern fur seals that I have caught, sexed, herded, tagged, wrestled, pricked, swabbed, anaesthetized, weighed, instrumented, and caged over the years. I hope that what we have learned about the species through them compensates for any indignities they may have suffered.

List of Contributors

John Calambokidis, Cascadia Research Cooperative, P.O. Box 1434, Olympia, WA 98503, USA

Roger L. Gentry, NOAA, National Marine Mammal Laboratory, 7600 Sand Point Way N.E., Seattle, WA 98115, USA

Michael E. Goebel, Institute of Marine Science, University of California, Santa Cruz, CA 95064, USA

Carolyn B. Heath, Division of Natural Sciences, Fullerton College, Fullerton, CA 92663, USA

John R. Holt, 803 18th Ave. W., Kirkland, WA, USA

Mark O. Pierson, Minerals Management Service, Office of Leasing and Environment, 770 Paseo Carillo, Camarillo, CA 93010, USA

Valery A. Vladimirov, VNIRO, 17v Krasnoselskaya, Moscow B-140, Russia

Behavior

and Ecology

of the

NORTHERN

FUR SEAL

PART ONE

Behavior at the Population Level

The chapters of this book are organized under five common themes. The theme of Part One is broad population overviews. The behavior of individuals is the subject of Parts Two (The Mating System), Three (Processes Fundamental to the Mating System), and Four (Maternal Strategies). Part Five presents a comprehensive view of the species and discusses some implications for behavioral studies in the eared seal family.

Chapter 1 introduces the pinnipeds (particularly the eared seal family), describes the northern fur seal, states the problem that led to the present study, and outlines the study and its results. Chapter 2 describes how the behavior of the northern fur seal population changed on a decadal scale as seal numbers and sex ratio changed. Chapter 3 describes northern fur seals on different timescales, from hours to lifetimes, and describes how the behavior of the population differs from that of individuals.

CHAPTER 1

Introduction

THIS BOOK reports on a 19-year study of the northern fur seal, *Callorhinus ursinus* (Linnaeus 1758), a medium-sized eared seal (Pinnipedia, family Otariidae) that inhabits the Pacific Ocean from about 35 to 60 north latitude. The study was designed to answer specific behavioral questions about a long-term decline in fur seal numbers.

The northern fur seal has been of uncommon importance to society for the past 250 years because of the luxurious, durable underfur it produces. To obtain furs, towns were founded, people were enslaved, battles were fought, sealing fleets sailed, and an international treaty was struck. In a 5-year period, the sale of northern fur seal pelts alone repaid the cost of the Alaska purchase.

Because of public interest, support for northern fur seal science has also been unusually good. The species was thoroughly described before 1900 (Steller, 1749; Veniaminov, 1839; Bryant, 1870–71; Scammon, 1874; Allen, 1880; Elliott, 1882; Nutting, 1891; Stejneger, 1896; Jordan, 1898; Townsend, 1899). In 1911, research became a treaty obligation of four nations, organized under an international commission that published extensive comparative data (Lander, 1980a,b). For many years, fur seal research was a line item in the U.S. federal budget. The population database that resulted is the longest and most diverse for any species of marine mammal.

A detailed analysis of the behavior of the northern fur seal is long overdue. This species has had a disproportionate and sometimes inappropriate influence on pinniped research and management. Mainly because of the extensive data on it, the northern fur seal is often used as a representative otariid in compendia of mammalian biology, even though in many ways it is not a typical otariid at all. Similarly, assumptions that are used to model the northern fur seal's population dynamics have been incorrectly applied to other species (York, 1987a). I hope that the details of its behavior will help specialists decide when the use of northern fur seal data is appropriate and when it is not.

This book is intended for general readers and for specialists alike. It discusses general trends and results in the text, and considers the finer details that will be of interest to specialists in the section of notes at the back of the book, referenced in the text by numbers in brackets [ ]. This work is not intended as a review of all northern fur seal literature, most of which relates either to the kill for pelts or to population dynamics, both of which are outside the scope of this work.

In this book I seek to accomplish two goals. The first is to answer specific questions about how northern fur seal behavior changes with population size and sex ratio. Toward that end, in this chapter I will introduce the species, its history of contact with humans, and the problem that led to the creation of the present study. I will outline the study questions and the research program that was put into place to answer them. My second goal is to apply the general principles and approaches of this work to the otariids as a whole.

This chapter introduces the pinnipeds as a group and the otariid family in particular. The relevance of northern fur seals to studies on other otariids will be taken up again in chapter 15, after the details of the northern fur seal work have been presented and synthesized. By describing this species in detail, this book contributes to ongoing attempts to compare the world's otariids (Nutting, 1891; Sivertsen, 1954; Gentry et al., 1986a; Croxall and Gentry, 1987; Boness and Majluf, in prep.).

CHARACTERISTICS OF THE PINNIPEDIA

Antiquity and Taxonomy

The Pinnipedia are amphibious mammals with feet modified as paddles (literally, wing-footed mammals). The suborder includes the true seals (family Phocidae), eared seals (family Otariidae), and walrus (family Odobenidae). The true seals swim with undulating motions of the rear flippers driven by their back muscles. They move caterpillar-like on land on their bellies, and have no external ear pinna. The otariids swim with their foreflippers, using large pectoral muscles. They walk or run on all fours. They can outrun a human on slippery rocks, and can climb nearly vertical cliffs. The otariids are named for their external ear pinna, a short, vestigial, cone-shaped appendage. The walrus has some characteristics of both phocids (swims with rear flippers and lacks an external ear pinna) and otariids (walks on all fours). All pinnipeds bear a single, precocial young on land or ice, and forage at sea (or in some lakes). All pinnipeds, except some of the otariids that have underfur, use blubber as the main means of thermoregulation. On average, pinnipeds are larger than other mammals (range 50–2,000 kg).

Pinnipeds are derived from ursoid (bearlike) stock. About 23 million years ago, the ancestor of all modern pinnipeds was still partly terrestrial, and swam using all four limbs [1]. It had some of the characteristics of all modern pinnipeds, including the walrus.

Amphibious Traits

Pinnipeds are mostly aquatic mammals with some specializations for a terrestrial life, not the reverse. A single example from the visual system will show this difference. The pinniped eye is adapted to low light levels, consistent with feeding at night or at depth (Walls, 1942; Landau and Dawson, 1970). They also have evolved a fishlike eyeball and cornea, shaped to resist water flow (Jameson, 1971). In air, they overcome the strong astigmatism that the distorted cornea causes by closing the pupil to a pinpoint. This technique gives California sea lions similar visual acuity in air and water (Schusterman, 1972). However, it only partly compensates because when the pupil opens as light dims, more of the distorted cornea is exposed (Schusterman, 1972; Lavigne and Ronald, 1975). Thus, aquatic adaptations in vision seem primary, and mechanisms that compensate for them in air seem secondary.

The otariids retain more extensive ties with land than the other pinnipeds. The walrus mates at sea and may suckle at sea or on land (or ice). Most phocids also mate at sea, but all of them suckle on shore. By contrast, the otariids suckle and mate on land. At least four species can also mate at sea [2], but not exclusively and usually to escape high temperatures. Also, otariids hear better in air than phocids do (walrus hearing has not been carefully measured; Richardson, 1995).

Pinniped Mating Systems

The true and eared seals are similar in that females produce a single young and males play no parental role. They differ in whether females feed during lactation, whether mating occurs on land, in the extent of polygyny and sexual dimorphism, in the number of sites used for mating, in time to weaning, and in other factors. Several authors have postulated evolutionary routes by which these differences arose (Nutting, 1891; Bartholomew, 1970; Stirling, 1983; Boness, 1991). A thumbnail sketch of current thinking follows.

Like land mammals, otariid females feed during lactation. Otariid mothers transfer less energy than phocid mothers to their young daily, but they lactate longer and so deliver more total energy (Oftedal et al., 1987). Nevertheless, otariid pups grow slower than phocid pups because of the metabolic cost of fasting during their mothers' feeding absences. Thus, the otariid lactation strategy is energetically more expensive and less efficient than the phocid strategy (Oftedal et al., 1987) [3]. Because of their relatively high energy demands, otariids can rear young only on land sites that are near extremely productive marine areas. Few such sites exist. Therefore, many animals share them; all otariids breed in a few dense colonies unless their numbers are abnormally small. Apparently, concentrations of females have given a few males the opportunity to monopolize mates, with the result that both polygyny and sexual dimorphism for size (males 2–4 times larger than females) have evolved in all the otariids (Bartholomew, 1970).

Most phocid females suspend feeding while lactating [4]. They forage widely before parturition, store energy as fat, and deliver it to the young until weaning without replenishing fat stores by feeding. Limited energy stores in phocid mothers force lactation to end at 4–28 days (Bowen et al., 1985), compared to 4–42 months in otariids (Gentry et al., 1986a; Oftedal et al., 1987). Because phocids are not dependent on concentrated marine production for feeding during lactation, they may mate in many dispersed locations (Stirling, 1983). Apparently, dispersed mating prevents males from monopolizing females. Therefore, most phocids are not nearly as polygynous or sexually dimorphic as the otariids (Stirling, 1983), although three are [5]. Accounting for these exceptions among the phocids has been an ongoing effort in pinniped research.

Litter size in pinnipeds is one. Thermal stress on homeotherms in cold water favors large body size (de Vries and van Eerden, 1995). Therefore, pinnipeds have a single large young, rather than multiple smaller ones. Twins do occur, but because of energy limitations of the mother they are weaned when still too small to survive (Doidge, 1987).

CHARACTERISTICS OF THE OTARIIDAE (GRAY 1859)

Antiquity and Taxonomy

The oldest known otariid was present by 11–12 million years ago (Kellogg, 1925; Repenning and Tedford, 1977; Berta and Deméré, 1986) [6]. It was a fur seal that very closely resembled modern forms, but its dental anatomy shows it was not a direct ancestor (Repenning and Tedford, 1977). The sea lions are the youngest otariids, having arisen in the late Pliocene or more recently (Repenning and Tedford, 1977). The earliest fossils were found in California, suggesting that the otariids evolved around the North Pacific. They apparently spread into the Southern Hemisphere and diversified about 5 million years ago.

Two types of otariids, fur seals and sea lions, were once thought to be distinguishable from each other by the presence (fur seals) or absence of underfur for thermoregulation. This distinction becomes less clear as more is learned about their behavior and molecular biology, although the names are still in wide use. The fur seals tend to be smaller (maximum 250 kg) than the sea lions (maximum 1,000 kg). The scientific and common names of all the seals discussed in this book are given in the Appendix.

Trophic Relations

Otariids are generally high-level consumers, taking fish, cephalopods, and sometimes crustaceans (krill: Boyd et al., 1994). The current thinking is that the fur seals tend to be feeding generalists, taking mostly small, surface-schooling fishes over the continental shelf, or small or juvenile phases of mesopelagic fish and squid associated with the Deep Scattering Layer (DSL) beyond the shelf break (Croxall et al., 1985; Gentry et al., 1986a; Boness and Majluf, in prep.). Sea lions, on the other hand, tend to specialize in large or adult stages of higher-trophic-level prey species found on continental shelves. Sea lions also are known to feed on fur seals (Gentry and Johnson, 1981; Harcourt, 1992, 1993). Both sea lions and fur seals may occasionally prey on marine birds and are themselves preyed upon by sharks and killer whales.

Worldwide Distribution

The otariids' feeding proclivities have allowed them to establish successful populations from about 60 north to about 65 south latitude. They may land on but never breed on ice like the walrus and some phocid seals do. Three species (northern and antarctic fur seals and Steller sea lion) are subpolar breeders. The two fur seals take shallow prey that is highly concentrated but seasonal. They migrate to the high-latitude breeding areas in spring, wean pups at 4 months of age, and migrate back to lower latitudes when winter comes. The Steller sea lion does not feed on the same prey species and migrates in only part of its range (Schusterman, 1981a).

Two species (Galapagos fur seal and Galapagos sea lion) live exclusively near the equator. They exploit prey in areas of local upwelling that are surrounded by vast expanses of unproductive warm water. Neither species migrates. The time to weaning differs by species, as discussed under "Maternal Strategies," below.

All other otariids live in temperate waters, mostly in the Southern Hemisphere. They exploit prey that are usually widespread and dispersed or are concentrated at seamounts, continental shelf breaks, convergences, or at other unique features. No temperate otariid is entirely migratory, but the males of one species (California sea lion) are an exception [7]. Most wean by one year, with several important exceptions, as discussed under "Maternal Strategies," below.

Only three species (northern and Guadalupe fur seals and California sea lion) still inhabit the ancestral waters of the North Pacific Ocean. Otariids probably never invaded the North Atlantic Ocean.

The largest otariid populations (1–2 million animals or more) presently are the northern (Loughlin et al., 1994) and antarctic fur seals (Boyd, 1993) and the Cape fur seal (David, 1987a). Most populations of fur seals are increasing at moderate rates, except the northern fur seal, which is at a plateau. By contrast, most sea lion populations are stable at low levels or decreasing at various rates, except for the California sea lion, which is increasing (Merrick et al., 1996) [8].

General Traits

The eared seals derive almost all their water from the food they consume. Seawater drinking does occur (Gentry, 1981a), but usually among territorial males at the beginning of the fast.

Otariids compare favorably with cats and the higher primates in cognition and learning ability. They can rapidly learn relatively complex tasks (Schusterman, 1966; Schusterman and Thomas, 1966), they have a good memory (Schusterman, 1981b), and show some higher-order abilities (Schusterman and Krieger, 1986; Schusterman and Kastak, 1993), including semantic comprehension (Schusterman et al., 1992). These skills may be associated with their need to find and exploit marine prey that change over time and space.

Unlike other mammals, growth may be indeterminate (Trites, 1991), at least in female fur seals. They stop growing only at old age, after the years of greatest fecundity (Lander, 1981–82; Perez and Mooney, 1985; York, 1987a), probably as a result of delayed epiphyseal closure (Versaggi, 1981). This growth is important because females of prime age (8–13 years) may have an important size advantage over younger females in social interactions, and perhaps can dive to deeper depths (chapter 12). Indeterminate growth has also been reported in the antarctic fur seal (Costa et al., 1988).

Affinity for Sites

Otariids breed on rock, sand, or shingle, often in exposed areas with high prevailing winds and surf, not usually in protected bays or mudflats. The most unusual otariid breeding locations include forest (Crawley and Cameron, 1972), lava tubes (Peterson et al., 1968a; Pierson, 1978), and desert sand beaches (Rand, 1959).

Otariids partition the shore areas they use into central breeding areas, used by adults for parturition and mating, and landing areas used by nonbreeding males. Formerly, these areas were called "rookeries" and "hauling grounds" in sealers' archaic jargon. These terms are neither descriptive nor consistent with modern usage and should be abandoned.

Most species of otariids have strong attachment to their land sites; they do not form new ones or abandon old ones easily. Eared seals persisted in using traditional land sites even when sealers of the last century drove some of them nearly to extinction (Busch, 1985). Steller sea lions in British Columbia persisted in breeding on several sites despite a slaughter of 70,000 of them over 30 years, intended to exterminate local populations (Pike and Maxwell, 1958). Site fidelity has been measured in relatively few species [9].

Use of Islands

Most otariid breeding sites are on islands. Some colonies of the Cape fur seal, South American fur seal, and southern sea lion are found on mainland coasts. One likely explanation for island breeding is that it results from the energy budget for lactation (Costa, 1991a). Female otariids are central place foragers (Orians and Pearson, 1977) that must regularly swim between marine feeding and terrestrial nursing locations and balance their energy reserves to maintain themselves and wean viable young. The distance they can energetically afford to swim depends on the energy available in food, and on the time from birth to weaning (Gentry et al., 1986a; Costa et al., 1989). Oceanic islands, or islands near continental shelf breaks, shorten the commute distance and save energy expended on travel. Not all islands give access to appropriate food resources, so only a few islands with suitable beaches have otariid populations. The avoidance of mainland predators and the availability on islands of cooling winds (Peterson, 1968) were probably secondary benefits of island breeding, and not its primary cause.

Breeding Cycles

All otariids gather once a year to bear young and breed. All species but one do so in the summer months on a strict 12-month cycle [10]. Several adaptations working in concert make a single annual gathering possible. One is that estrus occurs immediately postpartum (within 11 days in all but one species) [11]. Therefore, males gather when females are ashore to bear young, and females need not land a second time to mate. Postpartum estrus is made possible by a bicornuate uterus; one horn prepares for fertilization while the other carries the current fetus to term (Craig, 1964). Another adaptation is embryonic diapause (delayed implantation) by which the development of the embryo is arrested for several months so that gestation always ends during the species' annual gathering (Harrison, 1963) [12].

The breeding season lasts from 5 months at low latitudes to only about 6 weeks at high latitudes [13], but it does not vary clinally at intermediate latitudes (Oftedal et al., 1987; Boness, 1991). In populations with a long season, polygyny is not as extreme as elsewhere because males cannot fast long enough to monopolize all the estrous females. The degree of sexual dimorphism increases with the degree of polygyny (Boness, 1991).

Mating Systems

Given the range of mating systems in mammals (Clutton-Brock, 1989a), the range seen in otariids is very restricted. All otariids are polygynous, sexually dimorphic, and breed in a few dense colonies unless their numbers are atypically low. Boness (1991) classifies eight of eleven species of otariids as following some form of resource defense polygyny (Emlen and Oring, 1977), one as a combination of resource defense and female defense polygyny and two as modified leks. One of the lek species is clearly misclassified [14], and the other may be [15]. Otariid mating systems show relatively minor differences in the duration of the breeding season, extent of polygyny, breeding cycle, and extent of sexual dimorphism. More research should be aimed at processes that link these differences with the environment.

It is important to distinguish between male territorial behavior and the mating system. Males in all species of otariids can and do defend territories. In exceptional circumstances (e.g., in hot climates, when females are sparse, when breeding seasons are long, or a combination of these) some males may forgo fixed territories and move around with single females. But these are exceptions. At most places in most climates, males are territorial. Territories are established mostly on land, a few in water, and some on land that extends into the water [16]. The boundaries of most territories are fixed and are delineated by breaks in topography. All males have a species-specific threat sequence, called a "boundary display" (Gentry, 1975) that they use when defending a territory. Territoriality is a male/male phenomenon only. It is the means of partitioning access to females, and has little to do with how males interact with females within their territory, which varies more across species than with the form of territoriality.

Maternal Strategies

The maternal strategy (see note 3) is the species' means of extracting energy from the environment and converting it into pup growth. Otariid maternal strategies are unusual among mammals in that they involve frequent shuttling between a highly productive marine area and a suitable shore area (Gentry et al., 1986a). The major components of otariid maternal strategies are marine feeding (diving behavior), energetic costs (maintenance metabolism of mother and young and energetic cost of foraging and commuting), milk fat content, the schedule of delivering milk energy to the young (attendance behavior), resultant pup growth rates, and age at weaning. All these components vary somewhat within the otariid family because the environments in which they breed, and the dispersal, quantity, and energy content of the prey, vary. Each species in each location faces a different set of constraints.

Fur seals follow one of three strategies. Northern and antarctic fur seals follow what has been called the "subpolar" maternal strategy (Gentry et al., 1986a). Females are constrained to migrate due to their high-latitude breeding locations. They make foraging trips of 3–9 days, deliver fat-rich milk (> 40% fat), and wean their young at 115–120 days (Payne, 1979; Doidge et al., 1984; McCann and Doidge, 1987). Age at weaning is genetically determined in these species; they cannot exceed 4 months, even where the local environment permits it [17]. The Galapagos fur seal follows a "tropical" maternal strategy that is constrained by food abundance. Females make frequent, brief (0.7 days) foraging trips, deliver milk of highly variable fat content depending on the food, and wean at 1–3 years, again depending on the food supply (Trillmich, 1986, 1990). That is, age at weaning varies inversely with milk fat content. All other fur seals follow a "temperate" maternal strategy that is constrained by the distance to food or its dispersal. Females may make extremely long (12–30 days) trips to sea seeking distant or dispersed food. They deliver fat-rich milk, and wean their young by one year (Kerley, 1985; Gentry et al., 1986a; Croxall and Gentry, 1987; Oftedal et al., 1987; Goldsworthy, 1992; Figueroa Carranza, 1994; Francis et al., 1997; Guinet, pers. comm., 1996).

Sea lions do not follow the same three maternal strategies, even when they are sympatric with the fur seals mentioned above. Subpolar Steller sea lions, tropical Galapagos sea lions, and temperate California, Hooker's, and southern sea lions all wean at approximately one year (Oftedal et al., 1987). However, temperate Australian sea lions wean at 18 months (Higgins and Grass, 1993; Gales et al., 1994). No systematic attempt has been made to tie age at weaning in sea lions with attendance behavior and milk fat content, as it has been for fur seals.

Maternal strategies must accommodate unpredictable changes in the local environment. Otariids in the Pacific Ocean are subject to periodic El Niño/Southern Oscillation (ENSO) events that decrease the abundance of food (see below), and disrupt otariid maternal strategies in several ways (Trillmich et al., 1991). Otariids in other oceans face similar decadal-scale perturbations.

Attendance behavior (the schedule of energy delivery to the young) is central to the success of otariid maternal strategies (Bartholomew, 1970; Stirling, 1983). It gives females the flexibility to compensate for short- and long-term changes in their foraging environment and adjust the flow of energy to fit pup growth to the usual age and size at weaning. Attendance behavior was once thought to vary with latitude and environmental unpredictability (Gentry et al., 1986a). But it is now known that attendance behavior responds to local foraging conditions (see chapter 10 for details).

The extent of parent-offspring conflict at weaning (Trivers, 1974) varies among otariids. The young of northern and antarctic fur seals wean themselves and leave the island before the mother (Kenyon and Wilke, 1953; Macy, 1982; Antonelis, 1990), but the young of Galapagos fur seals and sea lions are driven off by the mother (Gentry et al., 1986a). Too few data are available from the other species to describe a pattern among the otariids.

CHARACTERISTICS OF THE NORTHERN FUR SEAL

Antiquity

Callorhinus is the oldest of the seven living genera of otariids (Repenning and Tedford, 1977; Repenning et al., 1979; Berta and Deméré, 1986) but is not ancestral to them. The genus may have evolved from *Thalassoleon macnallyae*, an otariid of 5–6 million years ago (late Miocene and early Pliocene; Repenning and Tedford, 1977). The extinct *C. gilmorei* inhabited California and Mexico about 2–5 million years ago (Berta and Deméré, 1986). The modern species, *C. ursinus*, probably evolved within at least part of its present geographic range, because fossils have been found in California and Alaska (Repenning and Tedford, 1977).

Physical Appearance

Scheffer (1962) thoroughly described the physical appearance of northern fur seals. Males weigh 200–250 kg and vary in color from black to reddish, with a mane over the shoulders that is often a different color. Females weigh up to 45 kg and are brown to gray. Sexual dimorphism (males up to 370% larger than females; Scheffer and Wilke, 1953) is greater than in most other mammals. Adults (> 5 yr) have white vibrissae, long, slender rear flippers, and a shorter muzzle than other fur seals (fig. 1.1). Juveniles have black vibrissae and a buff-colored belly and neck with light markings on the face. The pups are black. The underfur is brown but cannot be seen in dry animals. It is always dry owing to its great density (ca. 47,000 fibers/

Fig. 1.1. Adult male and female northern fur seals, showing extent of sexual dimorphism for size during the breeding season. (Photograph by V. B. Scheffer.)

cm^2; Scheffer, 1962). The underfur can be seen on wet animals as brown steaks in an otherwise black pelage.

Breeding Islands and Migratory Route

At present, northern fur seals breed on six large islands and on several smaller ones (fig. 1.2). All except animals from San Miguel Island, California, migrate along continental margins from low-latitude winter foraging areas to the breeding islands (fig. 1.3). This migration gives them access to food supplies that are vast (up to 9 metric tons per hectare; Smith, 1981) and predictable but seasonal.

The migratory cycle has been outlined from catch locations of early sealing vessels (Townsend, 1899; Zeusler, 1936) and scientific pelagic collecting (Kenyon and Wilke, 1953; Kajimura, 1980; Bigg, 1990). In November, females and juveniles of both sexes leave the breeding islands and fan out across the North Pacific Ocean. In January and February they concentrate along the continental margins, where some intermixing of stocks occurs [18]. In mid-March, females begin migrating back toward the islands fol-

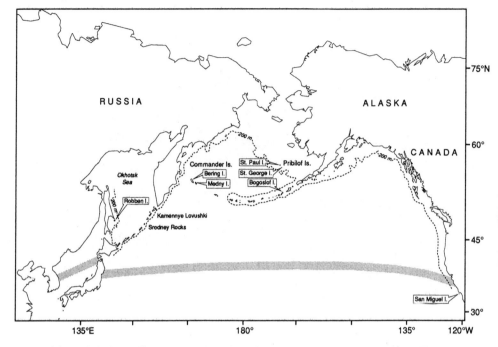

Fig. 1.2. Pelagic distribution of northern fur seals and the location of breeding islands used by the species, including the locations of the continental shelf breaks (dashed line approximates the 200 m contour). The shaded area represents the southern extent of the pelagic distribution.

lowing the continental margins, or across the high seas (Bigg, 1990; fig. 1.3). Juveniles and some nonbreeding females may not return north but instead remain in the Pacific (Kajimura, 1980; Bigg, 1990; Walker and Jones, 1990) to feed in the transition zone between the Oyashio and Kuroshio currents (Wada, 1971). Migratory patterns may have changed as recently as 1900, judging from the presence of adult male skulls in kitchen middens along the coast of Washington state (Gustafson, 1968).

Population Size and Trend

Northern fur seals are the most numerous and widespread otariid in the Northern Hemisphere. Judging from pelts collected by sealers, the population may have numbered 2–3 million when the first breeding island was discovered in 1742. The population reached that size again about 1865, and most recently from 1949 to 1951 (Lander, 1980a; National Marine Fisheries Service, 1993). The present numbers are about 1.32 million worldwide (table 1.1; Loughlin et

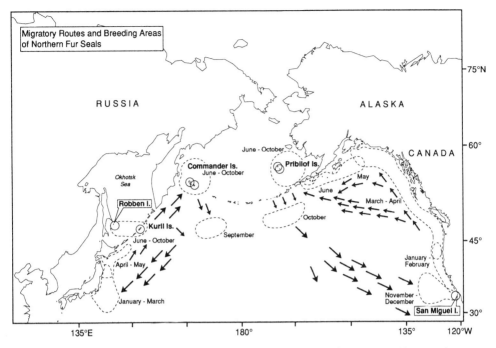

Fig. 1.3. Migratory pathway of northern fur seals based on data from Townsend (1899), Zeusler (1936), and Bigg (1990).

al., 1994), which represents a substantial decline since 1958. Total worldwide numbers have stayed approximately stable since the early 1980s.

The population parameter that presently has the greatest effect on population trend is the survival rate of juveniles. This rate is low and variable (Lander, 1979, 1981–82; York and Hartley, 1981) and difficult to measure because year-lings do not return to land at all (York, 1991a), and others may disperse to non-natal islands. Less than 10% of natural mortality in a cohort (also called a "year class") occurs before weaning at 120 days. Most mortality occurs in the first winter after weaning when animals are pelagic; it may range from about 30% to 80% (Chapman, 1961; Lander, 1979, 1981–82).

Pregnancy rates are thought to have a smaller influence on population trends than survival of juveniles. Major changes in pregnancy rates have not been observed historically (York and Hartley, 1981). Pregnancy rates vary with age and may exceed 85% in the prime 8–13-year age groups (Lander, 1980b, 1981–82).

Prey

About seventy-five species of fish, cephalopods, and crustaceans have been identified in fur seal stomachs across the range (Wada, 1971; Kajimura,

TABLE 1.1

Estimated World Stocks of Northern Fur Seals Just after the Start of the Third
Worldwide Decline (1958) (before Bogoslof and San Miguel Islands
Were Colonized) and in 1992

Island Group	Highest Estimate	
	1958	1992[e]
Pribilof Islands	1,850,000[a]	982,000
Bogoslof Island	0	2,235
San Miguel Island	0	8,211
Commander Islands	152,760[b]	230,000
Robben Island	128,880[c]	50,000
Kuril Islands	5,500[d]	50,000
Totals	2,137,140	1,322,446

Note: Table reports highest estimate instead of ranges.

[a]St. Paul and St. George Islands (North Pacific Fur Seal Commission, 1958–85). The Pribilofs had been at 2.1 million in 1949–51 (National Marine Fisheries Service, 1993).

[b]Bering and Medny islands, estimates based on number of young born in 1958 (reported by Lander, 1981–82) and assuming that young = 25% of total population.

[c]Same data source and assumptions as for Commander Islands.

[d] Lovushki and Srednev Rocks (Kuzin et al., 1973), may be adults only.

[5] Values reported by Loughlin et al., 1994.

1984; Perez and Bigg, 1986; Sinclair et al., 1994; Antonelis et al., 1997).
The diet differs by geographic area and month of the year. During the
breeding season the distribution at sea (Kenyon and Wilke, 1953), diving
pattern (Gentry et al., 1986b), and movements of radio-tagged animals
(Goebel et al., 1991) suggest that many females in the population forage
over deep water in association with the nightly rise of the DSL. Some females
at the Pribilof Islands dive night and day to depths of 175 to 200 m
(Gentry et al., 1986b) on the continental shelf, at or near the bottom (Goebel
et al., 1991). Tracking studies (Goebel et al., 1991; Loughlin et al.,
1993) indicate that females show some site fidelity to foraging locations at
sea. Nothing is known about female diving behavior in winter. More importantly,
little is known about the diet of pups immediately after weaning
(Peterson, 1961).

Environmental Setting

Northern fur seals have experienced at least five glacial cycles that probably
changed their distribution (Davies, 1958). In the interglacial periods,
smaller-scale climate changes probably affected them through their prey.
A change of 1–2 in water temperature would cause relatively small

changes in fur seal metabolic rates (Miller, 1978) but could have profound effects on spawning and larval survival in the food web on which they feed. A climate change that occurred during this study is described here in some detail because it coincided with a change in fur seal behavior (chapter 10).

The northern fur seal's range is affected on the decadal scale by major climate changes referred to as "regime shifts." In one regime, the North Pacific Ocean generally warms while the Bering Sea cools and its winter ice cover increases. In the other regime, the North Pacific cools while the Bering Sea warms and its ice cover decreases (Niebauer and Day, 1989; Trenberth, 1990; Trenberth and Hurrell, 1994). The differences between warming and cooling ultimately come from changes in the pattern of atmospheric pressures over parts of the Northern Hemisphere [19] acting through changes in the wind fields [20]. The strength and position of the Aleutian Low (see note 19) is especially important to the fur seal foraging area. Four interdecadal regimes have affected the North Pacific in this century [21]. They are important to northern fur seals because fish populations are affected by them [22].

A second factor affecting northern fur seal feeding areas is periodic ENSO warm events. In California, ENSO events affect northern fur seals directly by causing the mixed layer to warm and thicken, which drives their prey to depths that are uneconomic for foraging, or which drives prey beyond foraging range with the result that seals may starve (Fahrbach et al., 1991; Trillmich et al., 1991; DeLong and Antonelis, 1991). At high latitudes, ENSO may affect northern fur seals indirectly by altering the strength and position of the Aleutian Low through atmospheric teleconnection [23].

Sometimes the pressure pattern and ENSO events have additive effects on the Aleutian Low, and sometimes they do not (Royer, 1989) [24]. For that reason, oceanic conditions where fur seals feed have changed in a complex way over time [25].

The Bering Sea may be under other influences that do not affect the North Pacific. Sea surface temperatures north of 55 N latitude (Unimak Pass) undergo a 28-year oscillation that does not occur south of the pass (Royer, 1989), and some changes accompany the 18.6-year nodal tidal signal (Royer, 1993).

Breeding Sites

Northern fur seals presently use about thirty-one breeding sites on the islands they inhabit (fig. 1.4). These locations are highly predictable within and between seasons. For example, North breeding area on Bering Island

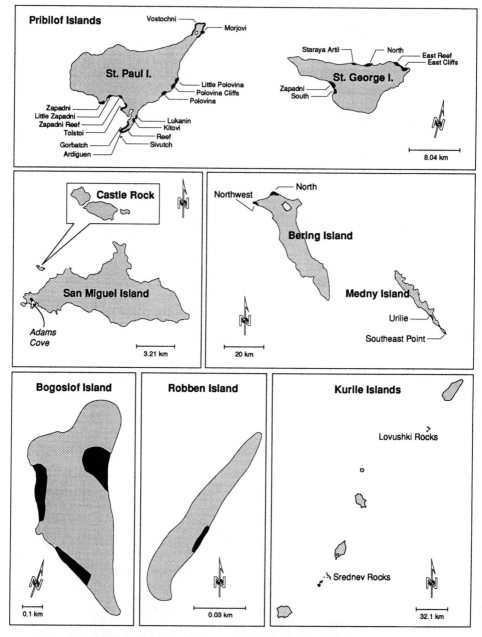

Fig. 1.4. The central breeding areas used by northern fur seals on all islands where breeding occurred as of 1990. Note that scales differ by island.

has existed for at least the past 254 years. Central breeding areas that are abandoned tend not to reform, and new ones are started rarely. Eighteen breeding areas disappeared worldwide since 1786, all of them resulting from a population crash caused by pelagic sealing at the turn of the nineteenth century (Lander, 1981–82; Busch, 1985). The population soon recovered, but only two of the abandoned central breeding areas ever reformed, and only two new ones formed on the same islands [26].

Northern fur seals are conservative at colonizing new islands compared to some other fur seal species (Payne, 1977; Boyd, 1993; Shaughnessy, 1984; Roux, 1987; Shaughnessy and Fletcher, 1987; Bengtson et al., 1990). They have formed only two new ones since 1786. San Miguel Island, California, colonized in 1965 (Peterson et al., 1968b; DeLong, 1982), had not been used by northern fur seals for at least a thousand years (based on archeological evidence, Walker, 1979). Bogoslof Island, Alaska, arose in the 1880s and was colonized by fur seals in 1980 (Loughlin and Miller, 1989) [27].

Sexual Dimorphism and Polygyny

Sexual dimorphism is detectable in embryos two months after implantation (York, 1987a,b; Trites, 1991; York and Scheffer, 1997). Growth rates of males and females are fairly similar until the fifth year of life, when males increase sharply (York, 1987a). Males undergo seasonal fattening (Schusterman and Gentry, 1970) and reach their maximum mass at age 8–11 years, when their copulation frequencies also peak (Vladimirov, 1987). Large size allows males to fast during the breeding season (70 days; Peterson, 1968), and possibly to reach deep winter prey that females cannot reach. The size difference may explain why adult males do not migrate like females and juveniles.

Polygyny can be extreme in northern fur seals, as in other sexually dimorphic mammals (Stearns, 1977). *In utero*, the estimated sex ratio of fetuses is 50:50 (York, 1987b). At birth it has been reported as 50.65% males (Fowler, 1997), and also as 54.7% males (Antonelis et al., 1994). Lack of consensus stems in part from the methods and samples used. After birth, males die at greater rates than same-age females (Lander, 1981–82) and some may be excluded from breeding altogether (Vladimirov, 1987). The result is that females outnumber adult males, sometimes by large amounts (ratios of one male per sixty females were recorded when males were being killed for pelts; Osgood et al., 1915). Most females are at sea on any given day, so the sex ratio on shore (usually 1:5 or 1:10; see chapter 2) is much lower than the sex ratio among all living animals. The large number of females at sea means that the species practices serial polygyny.

Reproduction

The sequence of reproductive events is similar to that of other otariids. Most pregnant females arrive on shore from late June to late July and give birth to highly precocial, single young a day later. Females enter estrus on about day six postpartum, copulate on average 1.2 times (Peterson, 1968), and a day later depart for feeding (Gentry and Holt, 1986; Costa and Gentry, 1986). Embryonic diapause lasts about four months and ends in mid- to late November when lactation ends (Craig, 1964; Daniel, 1981).

Like other sexually dimorphic, polygynous species with high male mortality, reproductive contribution is much more variable in males than in females (Clutton-Brock, 1988). Females can produce a single young yearly from age 5–7 to age 23 (Lander, 1980b; Vladimirov, 1987), so maximum offspring for them varies from zero to perhaps twenty. Many males die between birth and the start of mating at age 8–10 years. The survivors breed on average for 1.5 seasons before they are deposed (Peterson, 1968; Johnson, 1968; Vladimirov, 1987) and usually die before age 16. Therefore, the maximum number of young per male may vary from zero to several hundred. In a species such as this one, according to theory, mothers should invest more in male than in female offspring (Trivers, 1972; Trivers and Willard, 1973).

Reproductive success of northern fur seals is difficult to measure through offspring survival (Clutton-Brock, 1988) because the young do not return to the natal island until age 2, or they disperse among other islands.

In northern fur seals, father/daughter inbreeding may be reduced by a delayed onset of female sexual maturity (Clutton-Brock 1989b). Females usually give birth the first time at age 5–7 years (York, 1983), which is considerably longer than the mean lifetime territorial tenure of males, 1.5 years (Johnson, 1968). This difference means that most fathers pass out of the breeding assemblage before their daughters enter it. No large-scale transfer of either sex is known which would prevent brother/sister or mother/son inbreeding (Clutton-Brock, 1989b).

Behavior

The behavior of northern fur seals has been well described. Steller (1749) and Veniaminov (1839) outlined the general features of the breeding aggregation. Osgood et al. (1915) described the species more thoroughly but presented little quantitative data. Modern research began with Bartholomew's study of marked individuals which considered nearly every important aspect of social behavior. It summarized female reproductive events as a mathematical model (Bartholomew and Hoel, 1953) and accurately described the less quantifiable features of social structure (Bartholomew, 1953, 1959, 1970). Bartholomew's observations on homing in males were later supplemented by Kenyon (1960). In the 1960s, Peterson (1965, 1968) quantified new aspects of behavior, developed new handling methods, and modified Bartholomew and Hoel's model. DeLong (1982) described many aspects of behavior during the formation of a new colony at San Miguel Island. Vladimirov (1987) recorded the behavior of several thousand known-age animals at Medny Island for six years. In addition to these more comprehensive studies, many others have been conducted on specific aspects of behavior (examples: Lisitsyna, 1973; Antonelis, 1976; Griben, 1979; Macy, 1982; Francis, 1987; Insley, 1992; Loughlin et al., 1993; Ragen et al., 1995).

Annual arrival and breeding events are closely timed and always occur on the same dates (Peterson, 1968). The peak of pupping occurs between 4 and 10 July. Over the past 15 years the population on shore has always peaked during the week ending 13 July (see chapter 3). This timing was not affected by the 1982–83 El Niño event in the Pacific Ocean (Gentry, 1991).

Philopatry (repeated returns to the natal site) underlies much of the behavior of northern fur seals. Males are said to defend territories near the site of their own birth, although few data exist to support this contention (Kenyon and Wilke, 1953). Many females give birth and mate near their natal sites (Kenyon and Wilke, 1953; Kenyon, 1960). Females use the same parturition sites in different years, and suckle their young no more than a few meters from these sites until weaning occurs (Peterson, 1968). As a result, the distribution of females on a central breeding area may remain unchanged for at least 60 years (based on photographic evidence; Kenyon, 1960).

On breeding areas, males arrive before females and establish territories on terrain that females later use for parturition. Because females will only use these traditional sites for parturition, they become a resource for which males compete (hence, the mating system is resource defense polygyny; Emlen and Oring, 1977). Males defend identifiable boundaries from mid-May until late July (Peterson, 1968). Depending on population size and rate of arrival, female groups may not conform to the boundaries of male territories. Kenyon (1960) concluded that males defend space irrespective of females.

Juvenile males are excluded from breeding areas until the male territorial structure breaks down in early August. They usually reside on the landing area closest to the breeding area of their birth (Baker et al., 1995). From there they make brief visits to other landing areas (Kenyon, 1960; Griben,

1979; Gentry, 1981b), foraging trips to sea (Gentry, 1981b), and brief forays onto the breeding areas. Extreme site fidelity may explain why juveniles have not abandoned traditional landing areas despite weekly kills there during summers of the past two centuries.

HISTORY OF THE NORTHERN FUR SEAL; CONTACT WITH HUMANS

The islands used for breeding have undoubtedly changed during the northern fur seal's 5-million-year history. The Pribilof Islands (presently home to 72% of the total world northern fur seal population; Loughlin et al., 1994), are only 2.2 million years old (Cox et al., 1966). They have been inundated at least four times by interglacial sea level rises (Hopkins, 1973), and left standing as ice-covered mountains atop a broad plain at least twice by sea level declines (Hopkins and Einarsson, 1966). The Arctic Ocean may have been used by fur seals, as it was used by sea otters (Carter et al., 1986), before it froze 2.4 million years ago (Barry, 1989). No prehistoric central breeding areas have been found, but breeding and feeding sites have probably changed over time. Breeding may have occurred as far south as Baja California (Davies, 1958).

During recorded history, the worldwide population of northern fur seals has declined sharply during three periods. Twice it recovered rapidly when sealing was controlled. Its failure to recover rapidly from the third period of decline led to the present research program.

Although aboriginal peoples hunted northern fur seals for millennia (Gustafson, 1968; Huelsbeck, 1983), the species was numerous when the first breeding colonies were discovered. The Commander Islands (Russia) population was discovered in the spring of 1742 by Vitus Bering's expedition, which wintered on Bering Island after running aground there (Frost and Engel, 1988). The ship's naturalist, G. W. Steller, wrote a first description of the species (Steller, 1749) but did not estimate its population size. Sealing at these islands began slowly (20,000 taken between 1742 and 1760), and later accelerated (100,000 taken between 1760 and 1786) when a European market for skins developed (Stejneger, 1896).

In 1786, Gerassim Pribilof found the Pribilof Islands by following migrating animals northward from the Aleutian Islands. From then until about 1820, many small, competing companies killed fur seals at the Pribilof and Commander Islands, on Robben Island in the Sea of Okhotsk, and on the Kuril Islands (Stejneger, 1896). The kill was large, unregulated, and undocumented. The result was a rapid worldwide decline in numbers from some large but uncounted original population (fig. 1.5).

The herd recovered from this overkilling between 1820 and 1867 at a rate that cannot be calculated because no data were collected. The Russian American Company, which held a concession for sealing on all islands,

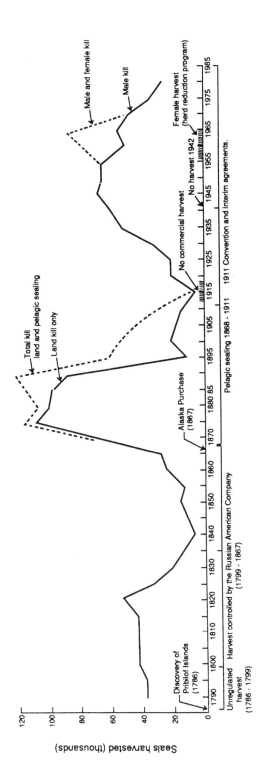

Fig. 1.5. Estimated and actual number of northern fur seal pelts taken at the Pribilof Islands from discovery of the islands to the end of the kill for pelts in 1985. The small numbers taken in 1840, 1915, and 1980 indicate that the herd was seriously reduced. (Reprinted from Lander, 1980a.)

prohibited the killing of females on the Pribilof Islands in about 1834 (Lander, 1980a), and on the Commander Islands, probably in 1843 (Stejneger, 1896). These prohibitions created the first real management plan for fur seals (Stejneger, 1896; Busch, 1985). The kill was euphemistically termed a "harvest."

The success of the management plan depended on the features of the species' mating system. Because of polygyny, more males survived to adulthood than was absolutely necessary for adequate reproduction. Modest numbers could be killed for furs without the adult sex ratio changing enough to jeopardize the pregnancy rate (Roppel and Davey, 1965; Roppel, 1984). Males were killed as juveniles (2–5 years) because they were numerous, their skins were less scarred than those of older animals (hence brought higher prices), and because their segregation onto landing areas helped sealers avoid taking females. Also, juveniles arrived in reverse order of age (youngest last; Bigg, 1986), allowing managers to focus the kill on certain age groups. Throughout recorded history, periods of low population size always coincided with the killing of females (Scheffer et al., 1984; Fowler, 1997).

The management scheme was apparently devised entirely by sealers using practical knowledge of the species. Steller's (1749) was the only scientific account of the animals until Bryant (1870–71) observed them after the Alaska purchase in 1867. If Russian scientists contributed to the sustained management program, their writings have not been found.

Under protective management, the Pribilof herd grew from perhaps 300,000 to about 2.1 million in less than 47 years (fig. 1.5; see also Lander, 1980a; National Marine Fisheries Service, 1993). The Commander Islands population recovered similarly but the numbers are not known. However, the populations did not remain long at high levels. Due to poor management control at both the Pribilof and Commander islands after 1867, and a price increase that encouraged extensive pelagic hunting, the population began another decline (fig. 1.5). Pelagic sealing began about 1868, reached a peak in 1892, and declined to near zero by 1910 (see table 4 in Busch, 1985). The fleet, roughly three hundred boats total, is believed to have killed as many as 75,000 fur seals per year from 1870 to 1910 (see table 5 in Busch, 1985). Most of those killed at sea were pregnant females.

The second decline was detected in the 1880s and was well documented (Allen, 1880; Elliott, 1882; Stejneger, 1896; Jordan, 1898). By 1910 the Pribilof herd had fallen to about 10% of its 1867 level. International negotiations to save the species culminated in the International North Pacific Fur Seal Treaty of 1911. Under its terms, the nations possessing breeding islands (United States and Imperial Russia) shared their land kill with nations lacking such islands (Japan and Britain for Canada) in exchange for a cessation of pelagic killing. The treaty was in effect from 1911 to 1941,

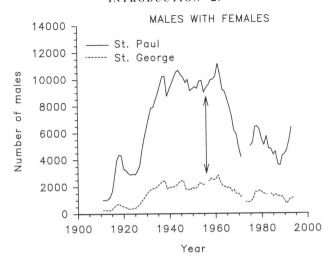

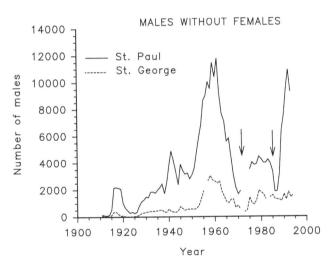

Fig. 1.6. Recovery of northern fur seals from the second (1911) worldwide decline as indicated by the number of adult males counted at the Pribilof Islands with and without access to females. Arrows indicate the 1956 start of intentional herd reduction, and the 1972 and 1985 cessation of commercial killing for pelts at St. George and St. Paul islands, respectively.

when Japan abrogated. An interim treaty was in effect from 1957 to 1985 (U.S. Congress, Senate, 1957).

Between 1912 and 1924, the Pribilof population increased at the rate of 8% per year (based on pup counts; Lander, 1980a). The increase continued, although at a slower pace, through the 1930s to a population high in the 1940s and 1950s (based on counts of adult males; fig. 1.6). Overall, the

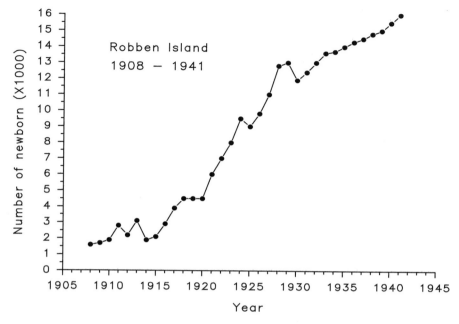

Fig. 1.7. Recovery of northern fur seals from the second (1911) worldwide decline as indicated by the number of young born yearly at Robben Island, Russia. Young at the Pribilof Islands were counted only from 1912 to 1922. (Data from Lander, 1980a.)

Pribilof herd grew from 0.3 to 1.8 million in 30 years. No data are available to trace the recovery of the Commander Island population, but recovery at Robben Island (fig. 1.7) resembled that at the Pribilof Islands.

The third period of decline in fur seal numbers was more complex and difficult to explain than the earlier two. The herd declined from the mid 1950s to the late 1960s (fig. 1.5), increased briefly from 1970 to 1976 (York, 1987a; York and Kozlof, 1987), then declined again from 1976 through 1983. It has shown no trend since 1984 (York, 1990). The decline from 1956 to 1970 directly prompted the research program reported here, and the decline from 1976 to 1983 prevented some of its main goals from being attained.

The extent of the most recent declines can only be estimated. In 1958 the world population may have been about 2.14 million (table 1.1), declining by roughly 57% to 1.25 million by 1984. Most of this decline occurred at St. Paul Island, where the decline rate from 1975 to 1981 was about 7.8% per year (York, 1987a).

At least two factors contributed to the trends from 1956 to 1983. First, the Pribilof population was intentionally reduced in size. For reasons discussed below, managers killed about 315,000 females on land (1956–68; York and Hartley, 1981) and another 16,000 at sea (1958–74; Kajimura,

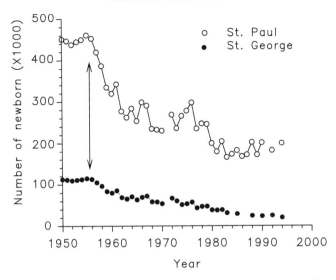

Fig. 1.8. The third worldwide decline in fur seals as seen in the number of pups estimated annually at the Pribilof Islands. Two long periods of decline separated by a brief increase in 1970–76 produced a net loss of about 60% of the population. Arrows indicate the 1956 start of the herd reduction program that accounted for 70% of the decline. (Unpublished data from NOAA, National Marine Mammal Laboratory, and from the North Pacific Fur Seal Commission, 1958–85.)

1984). The number of pups born per year started declining immediately after the first female kills in 1956 (fig. 1.8). After a brief lag time, the number of adult males also began to decline (fig. 1.6) because of the smaller number of male pups being produced.

The kill of females accounted for about 70% of the decline in fur seals between 1956 and 1980; methodological problems in estimating pup numbers, or some unspecified change in the marine ecosystem, accounted for the remaining 30% (York and Hartley, 1981). During the decline, the pregnancy rates of prime-age females (8–13 years old) declined, and the mean age at first birth increased (Trites and York, 1993).

The suggestion that the ecosystem changed finds some support in that the Robben Island population spontaneously started to decline in 1964 (fig. 1.9). This decline began before females there were killed (1972–76; Lander, 1981–82) and was too large to have been accounted for by the number taken (3,100). Estimation errors were not a factor because census was by direct count.

Two facts about the third worldwide decline affect the search for its causes. First, not all the populations declined between 1956 and 1983; only the Pribilof and Robben Island populations did so (figs. 1.8, 1.9). The others increased during this period (fig. 1.10). Immigration may have accounted for some of population increases but emigration alone cannot ex-

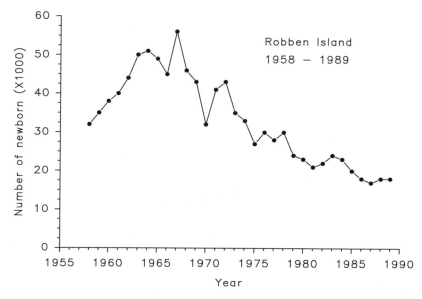

Fig. 1.9. Third (1956–80) worldwide decline in fur seals as seen in the number of pups born annually at Robben Island, Russia. The trend cannot have been caused by a kill of females, but may have been caused by an unspecified change in the marine ecosystem.

plain the decreases. The sharp drop at San Miguel Island in 1983 was a local response to the 1982–83 El Niño event, not part of the worldwide decline (DeLong and Antonelis, 1991).

Second, fur seals declined simultaneously with seabirds (unpublished data, U.S. Fish and Wildlife Service; Murphy et al., 1986; Roby and Brink, 1986), harbor seals (Hoover, 1988a; Pitcher, 1990), and Steller sea lions (Merrick et al., 1987; Hoover, 1988b) in the Bering Sea. At present it is not known whether these declines were merely congruent, or were linked through some common cause, such as climate-driven changes in prey abundance.

Parties to the fur seal convention began an organized search for the cause of the decline in 1982. However, the convention was ended before definitive research could be done. In 1985, repugnance by U.S. citizens toward the killing of animals for fur ended U.S. involvement in the convention because a kill for pelts was one of its objectives. Without U.S. participation, the convention expired and formal international cooperation ended. Research budgets disappeared due to the lack of a convention, and field activities became focused only on population assessment.

As of 1996, the St. George Island population appears to have finally stopped declining [28]. Robben Island may still be declining (Vladimirov, 1993). But the San Miguel (DeLong, 1990) and Bogoslof Island popula-

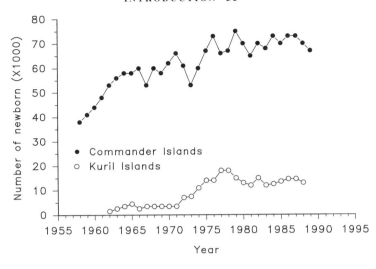

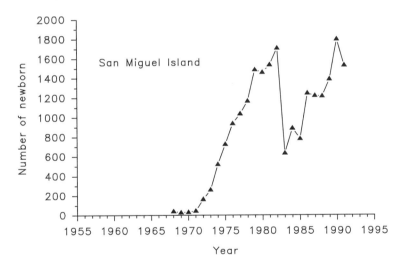

Fig. 1.10. Population trends of northern fur seals at three island groups that did not decline in 1956–80. (*Data sources.* Commander Islands: Lander, 1980a; North Pacific Fur Seal Commission, 1958–85; Kuril Islands: Kuzin et al., 1973; North Pacific Fur Seal Commission, 1958–85; San Miguel Island: National Marine Fisheries Service, 1963–91.) The sharp decline in the San Miguel population represents the 1982–83 El Niño event (DeLong and Antonelis, 1991).

tions (Merrick, 1990) have increased to the extent that worldwide numbers show no overall decrease (Loughlin et al., 1994). The cause of the 1976–81 decline remains an important question because future population trends could be affected by it.

STATEMENT OF THE PROBLEM

The problem that started the present research project began in the 1950s, when managers noted that the number of pups being born annually equaled the number born per year in the productive period 1932–37, but that fewer juvenile males were being taken for pelts despite an equal effort (Anon., 1973). They surmised that density-dependent mortality had changed the relationship between the abundance of neonates and the survival of juveniles compared to the 1930s. Postulating that smaller pup cohorts would reduce this mortality and allow greater survival to the juvenile years (hence, more pelts), and that pregnancy rates would increase (Anon., 1973), they killed adult females for seven years (1956–62) until the number of young born annually equaled that in the period 1932–37 (400,000). For six more years, they killed additional females equal to the number estimated to have been annually added to the herd by recruitment.

Based on the recovery rate from 1912 to 1924, managers expected the population to recover from intentional reduction at the rate of 8% per year. Instead, it continued to decline, coincidentally at about that rate (York, 1987a). By 1972, 10 years after the desired cohort size had been reached, the expected increase in pelts had still not occurred. Managers then concluded that either they had incomplete information about factors that regulate population size, or that some known regulatory factor had changed.

In 1972, with approval of the North Pacific Fur Seal Commission, managers began an ambitious 15-year program (Anon., 1973) to study the relationship between survival and abundance (the density-dependence question), the relationship of fur seals to other marine resources, and the effects of management and research activities on reproduction. This program was commonly called the St. George Island Program [29]. It included coordinated studies on behavior, population dynamics, and pelagic investigations. This book is the final report for the behavioral portion of this program only.

The behavioral part of this program was instructed to investigate two questions. The first was whether the changed relationship between survival and abundance was the result of behavioral changes brought about by artificial selection (killing males), human disturbance, or density-dependent behavioral processes on breeding areas. The second was whether behavioral studies could show the potential effects on fur seals of the fishery for walleye pollock (see Appendix for scientific names), which greatly increased at the same time as fur seals were declining (fig. 1.11) [30].

The behavioral project required a study site where males were not being killed for pelts. No such areas had existed since 1786, so a research preserve was created by stopping the kill on all parts of St. George Island after

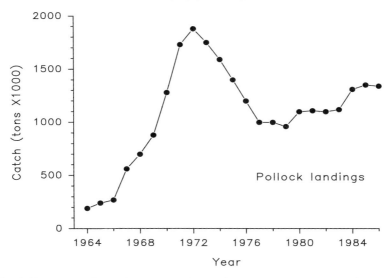

Fig. 1.11. Total landings of walleye pollock, *Theragra calcogramma*, in the Aleutian Islands and Bering Sea areas, 1964–1986. (Reprinted from Wespestad and Traynor, 1988.)

the 1972 season. Because 60–80% of males breed on their natal island (Anon., 1973), the adult sex ratio was expected to change on St. George Island but not on St. Paul Island, where the kill was to continue unchanged. The changed sex ratio on St. George Island was expected to improve the pregnancy rate and thereby cause the population to grow.

At the 1973 start of this program, scientists were unaware that climate regime shifts affect the fur seal's range. Therefore, researchers were not prepared to collect relevant data when a major regime shift occurred in 1977 [31]. It was only through luck that some of the routine data reflected the change (chapter 10). In hindsight, this program would have asked very different questions had the existence of regime shifts been known at the outset. Unfortunately, the effects of these shifts on the Bering Sea ecosystem were not compiled until just before this book went to press (National Research Council, 1996).

THE BEHAVIORAL RESEARCH PROJECT

This project measured behavior as a function of changing sex ratio and population size as the St. George Island population slowly approached a no-kill equilibrium. The project was to have continually compared behavior at St. Paul Island where the kill was ongoing, but the lack of funds made this comparison sporadic. Virtually any aspect of behavior that could affect population size was open to study. The island had comfortable accommodations, vehicles, an abundant labor pool, and a laboratory (actually the former skin processing plant) which included water tanks and outdoor cages for holding animals captive. Also, the animals could be easily approached and handled.

For the first 3 years, we collected descriptive data on many subjects to establish a baseline of behavior at the high sex ratios the former kill of males had produced (Gentry and Johnson, 1975). After a hiatus of 8 years, we repeated these baseline measures to record behavior at low sex ratios, typical of more pristine conditions. These comparisons appear mostly in Parts One and Two of this book. Throughout the study we made counts and recorded the reproductive histories of known animals (reported in Part Two). During the 8-year hiatus we conducted field and captive experiments on subjects that were fundamental to the fur seal mating system (Part Three) and maternal strategy (Part Four). In the final years of the study we focused on comparing diving behavior (Part Four). We collected the last data from St. George Island in 1992.

The behavioral project measured the behavior of known individuals on small, selected plots, compared this behavior over space and time, and from this behavior inferred the processes that control herd size. The population dynamics project used another approach. Each year it assessed the number of adult males as well as the number of pups that were born and died on shore, and it combined these into an index of population size (Fowler, 1990). Yearly indices were used to infer the behavior of individuals in an attempt to explain the processes that control herd size. These two approaches are obviously complimentary.

We established behavioral study sites at East Reef and Zapadni on St. George Island (fig. 1.4) in 1974 because they differed from each other in distance from the sea, number of animals, and terrain [32]. We established our St. Paul site at Kitovi (fig. 1.4), where several previous behavioral studies had been conducted (Bartholomew, 1953, 1959; Bartholomew and Hoel, 1953; Peterson, 1965, 1968). Each study site had a permanent grid painted on the rocks for mapping purposes, and during the breeding season each site had an observation blind. About 10,000 observer hours were spent in these blinds collecting the data reported here.

Virtually all land areas of St. George Island that fur seals used were included in this study. We documented the behavior of juvenile males on the landing areas (especially at Zapadni; fig. 1.4), and we compared pup growth rates at Staraya Artil (St. George Island) and Reef (St. Paul Island; fig. 1.4) [33]. We made direct observations at sea in areas where fur seal pups were being preyed upon by Steller sea lions (Gentry and Johnson,

1981), and we made indirect observations at sea (via instruments) in areas were females foraged.

We marked thousands of individuals using permanent tags and sometimes brand marks (Gentry and Johnson, 1978; Gentry and Holt, 1982) to help build long-term reproductive records.

GENERAL RESULTS OF THE PROGRAM

The behavioral project answered most of the questions that the St. George Island Program posed about behavior and the effects of human disturbance on it. It was particularly successful in answering questions about the pelagic activities of fur seals. In 1976 we helped devise a new instrument, a Time-Depth Recorder (TDR), which recorded all the dives that foraging females made on a trip to sea (Kooyman et al., 1976, 1983a). This instrument gave us the first quantitative measures of fur seal foraging behavior. It also made possible a comparison of northern fur seal maternal strategies with those of other fur seals worldwide (Gentry et al., 1986a). In some ways, these instruments substituted for the Pelagic Investigations portion of the St. George Island Program, which was terminated after 1975.

Contrary to predictions, the female population at St. George Island continued to decline after 1972. The number of adult males in the population initially increased as expected, which caused an expansion of landing areas and changes in male behavior on central breeding areas. However, after 1979 the male population began to decline in parallel with the female population. For that reason we could address only limited questions about density dependence in behavior. We were able to compare behavior as a function of changing sex ratio, but only in a declining herd, not an increasing one.

Wear and breakage of the plastic tags that we applied to females, juveniles, and pups precluded any major analysis of adult survival rates, kinship relations, and direct measures of reproductive success on breeding areas. This part of the program produced far fewer results than expected.

The key questions posed by the St. George Island Program, and some answers from the standpoint of behavioral studies, are presented in chapter 14.

SUMMARY

The mating system of this species is inextricably interwoven with the species' history of contact with humans. For most of the past 250 years, humans took advantage of polygyny, site fidelity, male exclusion from breeding, and other features of the mating system to kill northern fur seals for their fur. In so doing, they changed the size and sex ratio of the herd. The present research program sought to reverse these changes, and in restoring the herd to its former state to reveal the behavioral processes that contribute to the regulation of population size.

The northern fur seal is a top-level consumer, taking many species of small prey, usually at shallow depths. In most parts of the range they take prey that migrate vertically in deep, offshore waters and approach the surface at night. Females and juveniles migrate between subpolar waters that have abundant, predictable prey in summer, and lower latitude areas in winter where they intermix with animals from other populations. Adult males remain in the northern end of the range in winter, which separates the sexes and reduces intraspecific food competition.

Males give no assistance in rearing the offspring. The young are precocial and require only initial maternal protection and the input of energy (milk) to achieve nutritional independence at about 120 days of age, just prior to the southward migration. Weaning occurs without parent-offspring conflict.

Mothers alternate between feeding themselves at sea and feeding their dependent young on shore. The need of lactating females to remain within commuting distance from deep-water foraging areas limits the number of islands that are suitable for breeding. This limitation concentrates females onto a few land areas, allowing a small number of males to monopolize access to them. Thus, only part of a female's range is defendable by males. Sexual dimorphism for size combines with male fasting to increase the ability of males to monopolize mates. Fasting has important implications in several parts of the mating system.

Females mate within male territories and then rear their young there with energy they import from sea. Females are indifferent to male territorial boundaries with a few exceptions. From the male standpoint, territories are for mating only.

Female parturition sites act as resources for which males compete. The oldest females are the largest due to indeterminate growth.

The mating season fits the brief summers at the latitude of breeding islands by being highly synchronized, brief, and stable as to date. A postpartum estrus allows impregnation soon after parturition. Estrus is brief and usually includes a single copulation, thus contributing to high serial polygyny. Embryonic diapause forestalls implantation until after lactation ends and contributes to the timing of arrival in the following year.

Variability in lifetime reproductive contribution is greater for males than for females. The sex ratio at birth may slightly favor males. Lower survival rates and reproductive exclusion of males result in adult sex ratios of between five and ten females per male when no kill of males biases the ratio. Males delay breeding longer, breed for a shorter period, and die at an earlier age than females. Females can bear a single young per year from sexual maturity until death. First breeding in females is delayed long enough to prevent inbreeding with fathers holding territory.

The mating system permits efficient reproduction (high pregnancy rates) over a wide range of sex ratios (up to 1:60 when males are being killed) and population densities. Populations have been artificially reduced to as little as 10% of their equilibrium levels and have fully recovered in 30–45 years with no shift in the mating system.

Site fidelity insures long-term stability in the location of breeding and landing areas. Sufficient numbers of females breed on non-natal islands so that major differences in gene frequencies among sites are not likely to develop. Such wandering does not occur on a scale sufficient to insure that brother/sister or mother/son inbreeding does not occur. Site fidelity inhibits the colonization of new islands.

Population size has fluctuated during three broad periods of recorded history. The declines from 1968 to 1970, and from 1976 to 1983 differed from previous declines in that only about 70% of the declines could be explained by the kill of females. The present project was started to determine whether behavioral changes contributed to these declines.

Not all population centers declined during these periods; some of them increased. But populations of other vertebrates did decline simultaneously, which suggests the fur seal decline was related to the food web. The foraging area used by the northern fur seal is affected by long- and short-term atmospheric and oceanographic changes, some of which are known to affect fisheries landings.

The organized search for causes of the decline ended in 1985 when the convention that supported such research lapsed. Public opinion was instrumental in starting the collection of a long-term database on fur seals when furs were desirable, and it significantly reduced fur seal research when furs became undesirable. Worldwide northern fur seal numbers have been stable since the early 1980s.

CHAPTER 2

Population Changes, 1974 to 1986

THIS CHAPTER describes in detail how the St. George Island population of northern fur seals changed in the first 14 years after the kill of males ended in 1972 (chapter 1). Our data cover the period 1974 through 1988. During these 14 years, the male population increased as predicted, but the female population unexpectedly continued to decline. These changes had profound effects on the size and shape of breeding aggregations on central breeding areas, the sex ratio, and the size and distribution of nonbreeding males on their traditional landing areas.

Unlike the remainder of this book, this chapter focuses on changes that occurred in groups of animals. Loose groups of various size constitute a central breeding area. Individuals affect their immediate group, and in turn are affected by it. The factors that affect group formation and density are essential for understanding the context in which the behavior in the following chapters was measured.

By starting at the population level I do not mean to imply that the herd has unique behavioral traits that individuals lack. It probably doesn't. However, changes at the population level, especially changes in the size, density, shape, or composition of groups, imply certain behavioral tendencies of group members toward each other, or toward the terrain, that are difficult to measure except in groups. Behavioral data at the population level are important because of what they imply about behavioral tendencies of individuals. These tendencies can be considered intrinsic factors that affect the distribution of animals on shore. Extrinsic factors, such as light cycle, environmental temperatures, and the like, have long been believed to control this distribution (Temte, 1985; Fowler, 1990).

The main data in this chapter came from maps that were drawn daily at two study sites, East Reef and Zapadni on St. George Island (fig. 1.4), throughout each season of the study. The maps show the locations of individual males and the size, shape, and location of female groups. Terrain differences at the two sites (chapter 1) provide another dimension for comparison. These maps give the spatial details of the population decline that occurred throughout this study.

Between 1914 and the present, East Reef has declined more than most other breeding areas on St. George Island, but Zapadni has not (Ream et al.,

1994). However, during this study the part of East Reef that we used changed much less than the part of Zapadni we used. Zapadni is composed of an uphill, inland portion (called Upper Zapadni), and a larger beach area (Lower Zapadni). Upper Zapadni, which was used for this behavioral study, declined much more than Lower Zapadni and certainly more than East Reef. All the results reported for "Zapadni" refer to the study site on Upper Zapadni.

The increase in juvenile males that resulted from stopping the kill for pelts was measured by making weekly counts on traditional male landing areas. However, the counts there are only a rough index to the total number of juvenile males. Counts of landing areas reflect the number of male pups born 2–7 years before the count, the survival rate each cohort has experienced, and emigration/immigration rates. (Juvenile males arrive on landing areas in reverse order of their age [Lander, 1980a; Bigg, 1986; Fowler et al., 1993], with yearlings and 2-year-olds landing in August.) Males of age 8 or 9 years are usually not included because such males move off to reside around central breeding areas. On average, only 19% of known juveniles are on shore daily due to absences at sea for feeding (Gentry, 1981b). The number on shore fluctuates seasonally, in a circadian manner, and in response to storms. As a result, total counts of seals on landing areas include the effects of age structure and foraging needs of the males, month of year, time of day, and weather. As imprecise as they are, counts of landing areas are the only empirical measure available for the size of the juvenile male population. We collected no data on juvenile females because they are inseparable from the adult females.

This chapter addresses broad questions that are answerable only when a population undergoes major changes in animal numbers and sex ratio. The most obvious is: How do these changes proceed in terms of the number of females, number of males, adult sex ratio, and number of peripheral and juvenile males? How do female groups change location as the population declines? Do territorial males make the same kind of location changes as female groups? Does the natality rate change when the sex ratio changes? Does the density within female groups change when the population declines (that is, as population density declines)? How do density and territory size change for adult males when more males are added to the population? The latter question is important because of the agonistic behavior that accompanies territorial defense. Is there a seasonal signal in either group location or group density that is independent of population size? Are any of the above properties of groups affected by differences in terrain? Many of these questions pertain to otariid seals in general. They have not been answered to date because detailed studies were not made while populations declined.

METHODS

Study Grids

The East Reef site was on the north side of St. George Island in the middle of a long, narrow breeding beach. The Zapadni site was on the south side of the island, inland and uphill from a broad breeding beach. At each study site we built one or more permanent observation blinds, and painted a grid on the rocks with intersections every 10 m. Each 10 ∞ 10 m grid sector was numbered, and each sector was subdivided into imaginary quarters (5 ∞ 5 m), hereafter called "subsectors." Sector and subsector numbers were used to record animal locations. The grids at East Reef and Zapadni were 100 meters long by 30 and 40 meters wide, respectively. The same grid boundaries were maintained at each site throughout the study.

Census and Distribution Maps

Observers drew daily maps at each site throughout each season showing the distributions of groups [1], location of all known males, number of females per group, and the population totals. The outline and size of each female group was rendered as precisely as possible using the grid intersections as landmarks. Male locations were also written on history cards for each individual (chapter 5). Only animals inside the study grid were included in this analysis.

In most years, a single observer drew the maps at each site to reduce observer bias. Nevertheless, observers still differed in their judgment of group discreteness, size, and shape. The maps were most consistent in quality at East Reef, where for 10 years only two different observers worked (John M. Francis and Michael E. Goebel). Observers changed almost yearly at Zapadni.

Not all the available maps were analyzed [2]. Instead, three periods were selected to represent the breeding population in its early, middle, and late weeks [3]. The middle week always produced the peak counts of births, copulations, and females on shore. Data from the middle, or peak, week of the season are used here as the main index of population trends.

Population Data and Sex Ratios

The daily census and distribution maps were digitized and produced data on the total area occupied by the female population, parts of the study sites being used, female and male densities, contact between males and female groups, and average male territory size [4].

Data from the census and distribution maps were also used to calculate two different sex ratios. One was the "in-group sex ratio," which included all animals within breeding groups [5]. It was the ratio at which animals found mates. It changed hourly as females moved to sea and back. The in-group sex ratio was calculated only for the peak week to show the ratio at which most copulations occur.

The "population sex ratio" included all adults present on an entire breeding area, in and out of groups [6]. Its importance was that, since it included all nonbreeding males, it indexed male/male competition for breeding sites and all the social interactions that this competition caused. This ratio changed over years as the total abundance of the two sexes fluctuated (males increased after killing of males for pelts stopped). The population sex ratio was computed for the early, peak, and late weeks of the season. The sex ratio among all living northern fur seals is never observable at once because some females are at sea on any given day (for example, only 34% of known females were ashore per day during the middle week of the season; see chapter 4).

Natality Rate

Each year we calculated natality rates (the proportion of known females supporting young) for each study site. Natality rate is important because it substitutes for pregnancy rate (Vladimirov, 1987), which cannot be determined directly without shooting females at sea (Trites, 1992a). In calculating natality rate we classified each female as parous, nonparous, or undetermined [7], then calculated the proportion of each year's population in each category. Nonparous females visited shore only briefly and on an irregular schedule compared to parous females (Gentry and Holt, 1986).

Juvenile Males

Changes in the population of nonbreeding, juvenile males were measured through whole-island censuses made from 1978 through 1985. At approximately weekly intervals from early June through early August, observers visited every landing area on St. George Island and counted all males that lacked a mane. Counts made after 5 August included unknown numbers of females and were excluded from analysis. No counts were made on stormy days when landing areas were normally vacated.

In all statistical tests, significance is taken as $p \leq 0.05$.

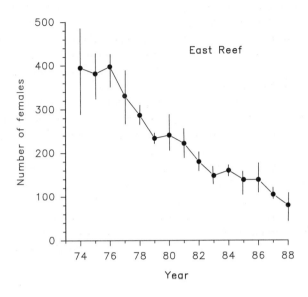

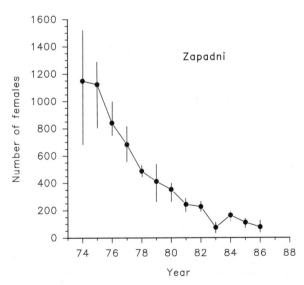

Fig. 2.1. Female populations at the East Reef and Zapadni study sites during 1974–88. Each point is a mean based on counts for 2–8 observation days between 7 and 14 July, inclusive. The regression line, using all observations, is *ln(population size) = 13.502 − 0.1007(year)* for East Reef, and *ln(population size) = 25.933 − 0.2528(year)* for Zapadni. Vertical lines give the minimum and maximum counts in the sampling period.

TABLE 2.1

Annual Increment of Population Change in Northern Fur Seal Females Counted
on Shore during the Period 7–14 July at Three Study Sites

Year	East Reef % Change	Zapadni* % Change	Kitovi % Change
1975	−3.3	−2.2	
1976	4.2	−25.3	
1977	−16.8	−18.8	−19.7
1978	−13.3	−28.3	11.4
1979	−18.5	−15.7	
1980	3.0	−14.3	
1981	−7.9	−31.2	
1982	−18.9	−7.0	
1983	−17.8	−67.3	34.9
1984	8.1	121.6	−23.3
1985	−13.1	−32.3	
1986	0.0	−30.6	
1987	−24.5		
1988	−23.8		

Note: Values represent the change from the previous year's count.

*Refers to study site on Upper Zapadni only, not to entire Zapadni breeding area.

RESULTS

Female Numbers and Area Occupied

The female populations on our study sites declined drastically during the
study (fig. 2.1). The East Reef and Zapadni study sites declined by 65% and
93%, respectively, between 1974 and 1986. Statistical analyses showed
that the population decline at each site was significant, and that Zapadni
declined about 2.5 times faster than East Reef [8]. East Reef continued to
decrease until at least 1996. The East Reef study site changed like the rest
of the East Reef breeding area. But the Zapadni study site changed much
faster than the rest of the Zapadni breeding area. In fact, by the end of this
study the Zapadni study site was almost uninhabited.

The female populations at East Reef and Zapadni did not decline in a
uniform or coordinated pattern from year to year (table 2.1). Each site had
a variable decline rate from year to year, and the rate at one site was usually
not the same as at the other. For example, one site might decline four times
faster than the other in one year, and in other years one site might increase
while the other decreased.

As the number of females declined, the area they occupied also declined.
The total area occupied during the peak week of the breeding season decreased from 1974 to 1986 by 87% at East Reef and by 95% at Zapadni.

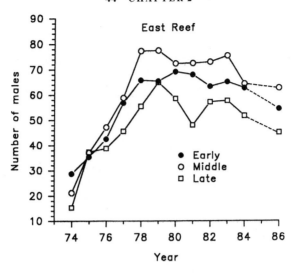

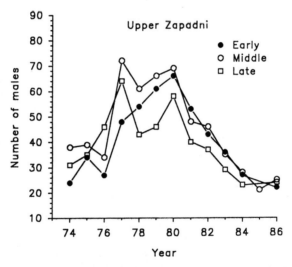

Fig. 2.2. The number of adult territorial males at East Reef and Zapadni breeding areas in the early, middle, and late weeks (see note 3 for dates) of each season from 1974 to 1986. Each data point averages counts from five to eight maps. Variances (small) were excluded due to lack of space. Dotted lines indicate a year of missing data.

Male Numbers

Unlike the females, which showed a consistent decline from the start of the study until the end, males at first increased and then decreased (fig. 2.2). The initial increase (1974–78) was predicted before this study began (Anon., 1973) on the basis that stopping the kill of males would allow more

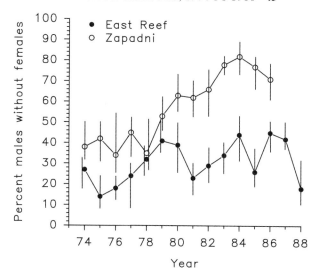

Fig. 2.3. The mean percentage of adult territorial males at two sites that lacked direct contact with females. Each data point is a mean for the peak week of each breeding season and is based on 3–8 daily counts. Vertical bars indicate maximum and minimum values for the week.

of them to survive to adulthood. However, the decline following the brief plateau in male numbers was not predicted. For most years of this study, adult males were most numerous during the middle week of the season and second most numerous during the early week.

The changes in the male populations (abundance) were highly statistically different (all $p \leq 0.001$) at the two sites, across years, and among weeks, as was the site-by-year combination [9]. None of the other two- or three-way interactions showed any significant differences.

The male population changed not only in total numbers, but in the proportion that failed to get access to females. This proportion increased as the male population increased and the female population decreased. Males without females in the middle week of the season peaked at about 45% for East Reef and 80% for Zapadni (fig. 2.3). The proportion of males at East Reef that lacked contact with females was actually greater than this analysis shows because many of them were outside the study grid and were therefore excluded from analysis.

Sex Ratio

The population sex ratio (females to all males present) began at about 20–30 females per male, declined rapidly for 4 years, then leveled off at about 4–6 females per male for the ensuing 9 years (fig. 2.4). In the first 4 years,

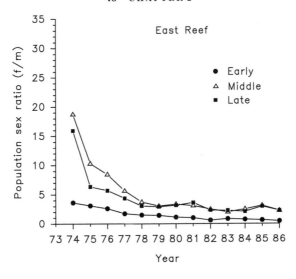

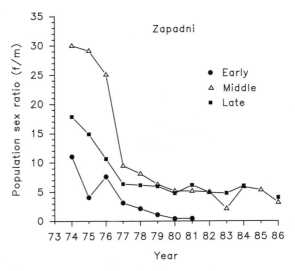

Fig. 2.4. The "population" sex ratio (females to all males present) in the early, middle, and late weeks of the season (see note 3 for dates) at two study sites throughout the study. Each data point is a mean of 3–8 daily counts. The population sex ratio reflects the ability of the male population to replace males that fatigue and abandon their territories.

when the male population was small and growing, the population sex ratio showed more of a seasonal change (early, middle, and late weeks of the season differed from one another; fig. 2.4) than it did in the later years [10]. The early week of the season always had the lowest population sex ratios because most females had not arrived by then [11].

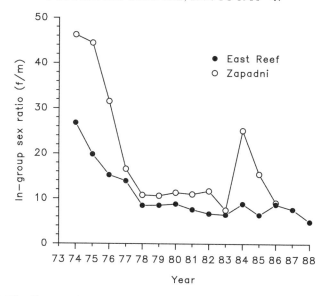

Fig. 2.5. The "in-group" sex ratio (females to males that have direct contact with them) at two sites throughout the study period. Data are for the middle week (7–14 July) of each season when births and copulations reach a peak. The in-group sex ratio reflects the ratio at which males detect estrous females.

The in-group sex ratios (females to males they contacted) declined like the population sex ratios, but did not so closely approach 1:1. After the initial decline, the in-group sex ratio leveled off at about nine females per male at both study sites (fig. 2.5) [12]. The relatively constant population and in-group sex ratios from about 1978 onward (figs. 2.4, 2.5) were a surprise, given that the male and female populations were behaving very differently (compare figs. 2.1 and 2.2).

Natality Rate

We classified 650 females as parous, nonparous, or undetermined at East Reef and Zapadni from 1974 to 1989. Many were seen in more than one year, so the data set included 1,429 female-years of data. Only twenty-five females (East Reef = 14) were classed as being of undetermined natality. These were mostly marked, parous females from other breeding areas that were making brief visits to our study sites during a foraging trip. Such visits were especially common from August onward.

Parous females comprised 65.7% of the remaining 1,404 female-years of data. The proportion that was parous was not different between the two study sites, and it was not different for females that were present for one year compared to those present for several years [13]. Methodological problems (small sample sizes due to tag loss) caused at least as much variation in the results as natural variation would have, so we can conclude nothing about the effects of yearly environmental change on the natality rate [14].

The proportion of parous females in the sample did not change with relative age of the females, as might be expected. We analyzed the records of our most consistent females by year of tenure (up to 13 years for some females) and found that although the proportion of parous females fluctuated (from 70.5% to 79.5%) it did not change consistently with age [15].

Group Density

The density of females within groups changed little if any during this study (fig. 2.6), compared to the massive decline in number of females (fig. 2.1) and the decline in sex ratios (figs. 2.4, 2.5). Females usually maintained mean densities of from 1.5 to 2.5 animals per square meter within groups, no matter how many males or females occupied the breeding area. This finding is very important because it implies that the densities at which animals find mates stays fairly constant as the total population size fluctuates.

The density that females maintained within their groups was not different at the East Reef and Zapadni study sites [16]. This implies that terrain had little effect on group density because the two sites differed in surface roughness.

Group density varied by a statistically significant amount over many years [17]. Density increased at East Reef in the last year of this study, and it fluctuated without trend at Zapadni. In both cases the changes in density, although statistically different, were so slight that field observers could not perceive a change. In all likelihood, these slight density fluctuations resulted from errors in drawing the daily maps. These significant changes in density did not reflect the larger changes that occurred in population size.

Group density showed a strong seasonal decline each year. Density was greatest in June when females first arrived, and it declined progressively thereafter [18]. The same seasonal pattern was seen at both study sites despite the fact that the terrain differed greatly. The density of females was not different between the early and peak weeks, but density in the late week of the season was significantly lower [19].

The greater group density early in the season was not likely to have been a thermoregulatory response to low temperatures (Gentry, 1973). Dense groups formed in springtime when air temperatures were moderate, but females spaced about a meter apart in November despite occasional light dustings of snow.

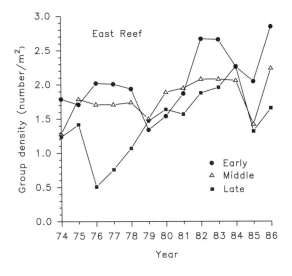

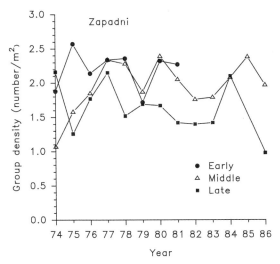

Fig. 2.6. Intragroup density of females (number/m^2) at East Reef and Zapadni study sites based on census and distribution maps drawn daily each year. Each point is a mean based on 2–93 groups in the early week, 15–332 groups in the middle week, and 22–211 groups in the late week of breeding seasons (total n = 3,401 groups) (see note 3 for dates). The number of observation days per week varied from one to eight.

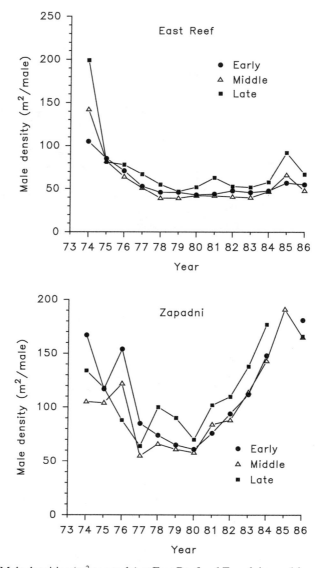

Fig. 2.7. Male densities (m² per male) at East Reef and Zapadni over 14 years. Each point is a mean of daily counts for the early, middle, and late weeks of the breeding season (see note 3 for dates). At Zapadni in 1985, data were collected only during the middle week. The 1985 values for East Reef are based on an area that was smaller than in other years.

Average and Minimum Territory Sizes

As the male populations increased during the first 4 years of the study, male territory sizes rapidly decreased to a third (East Reef) or a half (Zapadni) of their original sizes (fig. 2.7). Thereafter, males at East Reef maintained average territory sizes of about 50 square meters for 9 years with little change. This estimate is somewhat inflated because the method of measure [20] included some grid sectors that no males used. Territory size increased at Zapadni late in the study (fig. 2.7) because no new, young males were moving to this site (fig. 2.2), which left more space for the few males that remained. This was a local effect that was not seen elsewhere.

Male densities, measured in number of males per grid sector [21], increased from 1–2 males per grid sector in years when the male population was small, to 3–4 per sector or higher (up to ten males per sector; fig. 2.8) when it was large. The type of terrain available strongly influenced male densities. Densities of seven males per sector were rare at Zapadni, where the flat terrain provided males with few natural boundaries along which they could delineate their territorial boundaries. However, such densities were common at East Reef, where large rocks provided large natural boundaries (see also chapter 4).

A minimum territory size can be estimated from fig. 2.8 by dividing sector size (100 m^2) by male density. Using eight males per sector as the highest densities that males routinely reached, each male would have about 12.5 square meters, or a circle of radius 1.99 meters. This was almost exactly the mean body length of adult males.

NUMBER OF FEMALE GROUPS AND FEMALES PER GROUP

The number of groups into which females divided themselves on land correlated closely with the number of males present. Females formed into few groups (fig. 2.9) when males were sparse (fig. 2.2), and into many groups when males were abundant. The correlation between the number of female groups and the number of males was positive and linear, and the correlation coefficients were high [22] (fig. 2.10).

Even though the number of groups that females formed correlated well with the number of males present, the number of females per group declined with the number of females in the population. The average number of females per group in the peak week of the season declined from sixty to eight between 1974 and 1986 [23]. Zapadni declined faster than East Reef in part because the flat terrain there allowed groups of hundreds of females to form there in the early years, and in part because females eventually almost totally abandoned the site.

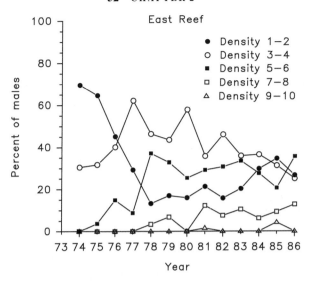

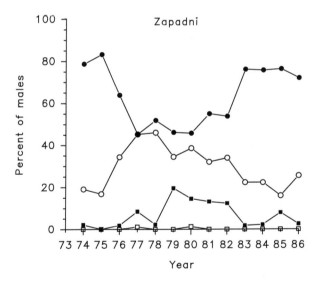

Fig. 2.8. Interannual variation in the density of males at East Reef and Zapadni study sites in 1974–86. Each year's data show the proportion of 10 ∞ 10 m grid sectors that fell into each of the five density categories (units are males per sector). Annual results may not total 100% due to slight rounding errors.

The number of females per group changed in a clear pattern at East Reef, but not at Zapadni. Groups of from two to ten females dominated the East Reef population from 1976 onward (fig. 2.11). No such trend was evident at Zapadni, where group sizes fluctuated without apparent trend.

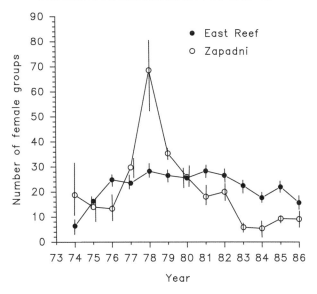

Fig. 2.9. The number of female groups at East Reef and Zapadni, 1974–86. Each point is a mean based on counts for 3–8 days during the period of 7–14 July each year. Vertical lines represent 50% of the minimum and maximum number of groups counted in each period (plotted as 50% to prevent overlap of lines). The 1978 Zapadni point is probably spurious (see note 22).

Abandonment of Breeding Sites

As the population declined, females abandoned parts of breeding areas that did not give them direct access to the sea, and persisted on parts that did. Females had access to the sea along the entire length of the East Reef site. As the population there fell, females abandoned very little of the study site. Groups formed in the same locations over years (seen as unchanging X–Y positions in fig. 2.12), but the number per group declined (height on the Z axis), and the total area they occupied also declined. The east and west ends of the entire East Reef breeding area contracted toward the middle, but the study grid did not include these two areas.

At Zapadni, females had access to the sea (100 m away) only down an inclined rocky ramp situated at the northwest corner of the study grid. As the population there fell, females abandoned all but two sectors (5% of the study grid) immediately adjacent to the access ramp. By 1980, the number of groups had increased, group size had decreased, and groups had shifted northwest (to the right in fig. 2.13). By 1986 the number of groups had decreased and the shift to the north was more pronounced.

The locations of males on the breeding sites matched well the locations of female groups at East Reef but not at Zapadni. Males at East Reef used

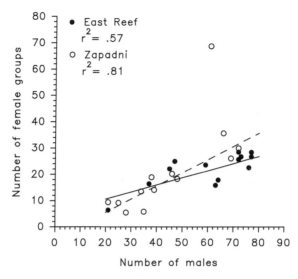

Fig. 2.10. Linear relationship between the number of female groups and number of males at East Reef and Zapadni (see note 21). Each point is a mean of 3–8 daily counts taken during the middle week (7–14 July) of the breeding season. The outlier in the Zapadni data is plotted but was not included in calculating the regression line. The two slopes are significantly different from each other and from zero ($p < 0.05$).

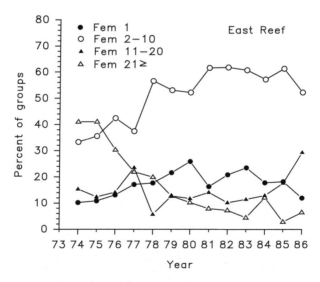

Fig. 2.11. Interannual variation in the number of females per group at the East Reef study site, 1974–86. The plot shows the proportion of female groups during the peak week of each season that was assignable to each of the four size categories shown (units are females per group). Yearly totals may not equal 100% due to slight rounding errors.

all parts of the grid throughout the study in greater or lesser densities (fig. 2.14). Late in the study, males became sparse immediately in front of the observation blind (at the 50 m mark).

Males at Zapadni did not shift to the northwest when females shifted there between 1976 and 1980, and they moved only a little in response to the pronounced shift that females made between 1980 and 1986 (compare figs. 2.13 and 2.15). That is, as females declined, males persisted at their previously used sites. The implications of male site fidelity for reproductive success will be discussed in chapter 4.

Number of Peripheral and Juvenile Males

Most males that lacked contact with females occupied the periphery (hence the name) of areas used by females. The number of peripheral males increased as the female population declined (example in fig. 2.3), but the total increase in their numbers was not documented. Their territories were often outside the observation grids, so they do not appear in figs. 2.3, 2.14, and 2.15. Furthermore, they were not included in the yearly whole-island counts of "idle" males mandated by the fur seal convention, because they were somewhat smaller than adult males (fig. 5 in York, 1987a). Peripheral males occurred at all central breeding areas in varying numbers, depending on terrain.

During this study, all landing areas used by juvenile males increased in size by unmeasured amounts. The expansion was first observed in 1977, when a new landing area developed at East Reef and when the one at Zapadni expanded about fivefold, the single largest such increase on the island.

The population of juvenile males initially increased (fig. 2.16), apparently in response to the cessation of killing juveniles from 1973 onward. After 1980 their numbers declined again, most likely because fewer pups were being born yearly as the female population declined (fig. 2.1).

DISCUSSION

From a management standpoint, the single most important result of this study was that it showed that the kill of males for pelts was not a cause of the herd decline. This was the key question that started this program (Anon., 1973). Unfavorable sex ratios, resulting from killing some males for their pelts, cannot have caused the decline, and the more favorable ratios created by ending the kill did not reverse it. The in-group sex ratio (females to breeding males) fell from twenty-seven or forty-six females per

East Reef

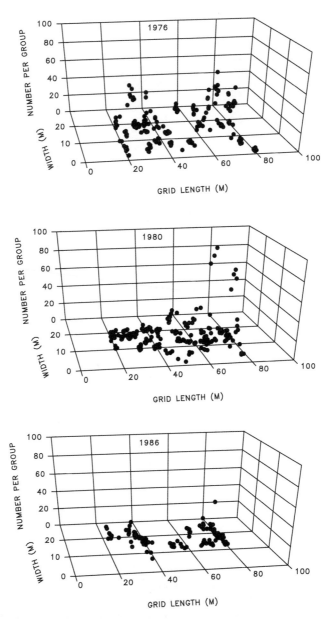

Fig. 2.12. Spatial distribution of all female groups and solitary females at East Reef during the peak week at large (1976), medium (1980), and small (1986) female populations. Graph is plotted as a north-facing observer would view the study grid with the observation blind at 50 on the X axis, and the sea on the 0 baseline where the Y and Z coordinates meet. Group location is indicated by X–Y position, group size by Z height. Clustered points are usually the same group on different days.

Zapadni

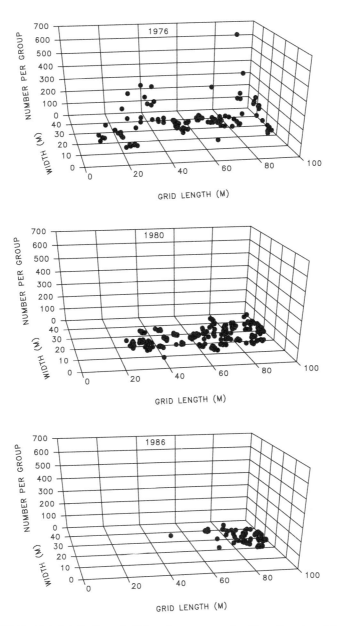

Fig. 2.13. Spatial distribution of all female groups and solitary females at Zapadni during the peak week at large (1976), medium (1980), and small (1986) population sizes. Graph is plotted as a west-facing observer would view the study grid with the zero baseline between the Y and Z coordinates representing the cliff edge, and the intersection of X, Y, and Z representing the narrow corridor that gave access to sea. Group location is indicated by X–Y position, and group size by Z height. Clustered points are usually the same group on different days.

East Reef

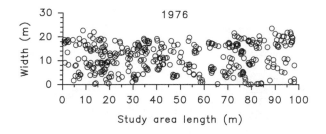

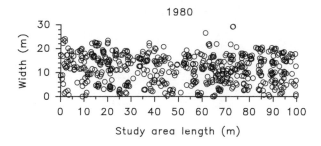

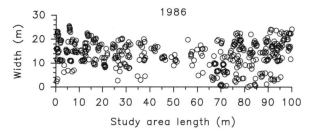

Fig. 2.14. Spatial distribution of all males on the East Reef grid during the period 7–14 July in 3 years. North is at the top and the observation blind is at 50 on the X axis. Each circle gives the location of a given male once a day for a week. Closely clustered circles represent the same male on different days.

male (depending on the site) to about nine, and remained there for 10 years without reversing the downward population trend (fig. 2.5). Furthermore, the St. George Island population continued declining for 22 years after the kill there ended in 1972, whereas the St. Paul Island population stopped declining 4 years before the kill there ended in 1985. Therefore, the population declines from 1956 onward (York and Hartley, 1981; York and Kozloff, 1987) were independent of the kill for pelts.

Reducing the sex ratio probably failed to reverse the population trend

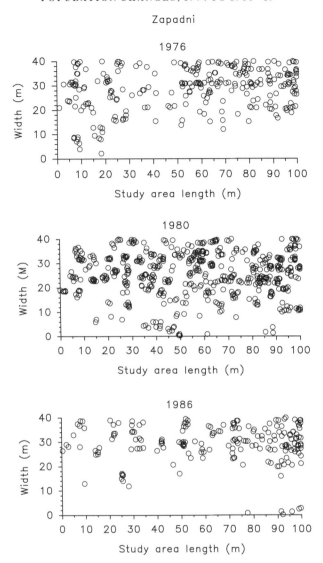

Fig. 2.15. Spatial distribution of all males on the Zapadni grid during the period 7–14 July in 3 years. Each circle gives the location of a given male once a day for a week. Closely clustered circles represent the same male on different days.

because pregnancy rates were so high initially that the addition of more males could not have increased it sufficiently to offset whatever caused the decline. Pregnancy rates during the decline were 86–89% for prime-age females (8–13 yr) (York and Hartley, 1981; Lander, 1981–82; York, 1983) and the total sex ratio (pups born/adult males) was about 33:1 (York,

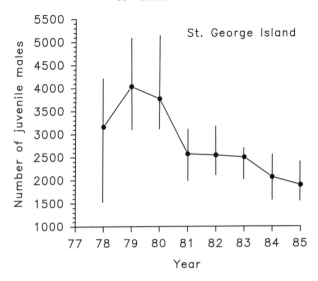

Fig. 2.16. The mean number of juvenile males counted on all landing areas of St. George Island by year. Each data point is a mean of 3–6 counts made between 24 June and 31 July each year. Vertical bars show minimum and maximum counts involved in the mean. The number counted ashore on a given day represents on average 19% of the total juvenile males in the population (Gentry, 1981b).

1987a). Pregnancy rates were not measured after 1974, so it is not known to what extent, if any, they responded to the declining sex ratio.

The present data show how the decline proceeded. The rate of decline at a single study site was not stable over time, and the rates at two study sites fluctuated out of synchrony with each other. Therefore, the factors that caused the decline were not acting uniformly over time or in parallel on all breeding sites in the population at once. That is, the overall population decline comprised considerable local variation. Shore counts alone cannot reveal the causes of the decline because total numbers only reflect the outcome of annual recruitment minus losses, both of which fluctuate independently (Lander, 1979, 1981–82).

At St. George Island, the annual timing of arrival seems insensitive to long-and short-term environmental changes. The female population peaked from 7 to 14 July over a 15-year period without fail. It did not change with the 9-year cycle in Bering Sea surface temperature (Niebauer and Day, 1989), nor did it respond to the massive 1982–83 El Niño warm event (Niebauer and Day, 1989; Trillmich, 1990; Gentry, 1991). El Niño events have a greater effect on fur seals at San Miguel Island than at the Pribilof Islands (DeLong and Antonelis, 1991; Gentry, 1991).

Perhaps the most important new finding in the data is that intragroup density in females is independent of both population size and sex ratio.

Group density in the peak week of the season did not change by four- to elevenfold as the numbers at East Reef and Zapadni changed by that amount (fig. 2.1), nor by three- to fourfold as the sex ratio decreased (figs. 2.4, 2.5). Instead, it remained the same at Zapadni and underwent a slight but statistically significant increase at East Reef, which was likely an artifact of the map drawing method (see above).

The importance of this finding is that it suggests that behavioral processes on land are likely not the causes of density-dependent changes in reproduction. For example, if pregnancy rates decline at high population levels (Fowler, 1990, citing unpublished data by Baba and by Bigg and Fawcet), it is not due to crowding on shore because animals mate at the same intragroup density regardless of population density. Any such changes in pregnancy rates must be caused by changes at sea, such as intraspecific food competition. Similarly, density-dependent pup mortality (Fowler, 1990) is not likely to result from increased contact with adults because these contacts may not increase as numbers increase. Because of intragroup density, an individual's immediate surroundings may not change much as the population changes size.

Intragroup density does not correlate with the number of males, but the number of groups does. Intragroup density changed seasonally every year, but it did not change as the number of males increased. It would have if intragroup density were determined by the herding actions of males (Francis, 1987). However, the results suggest that the greater number of males caused the female aggregation to break into a greater number of smaller groups within which the densities remained the same. There are no quantitative measures of female density from other northern fur seal populations. From the present results, density appears to be very similar among sites while group size and number fluctuate with population size.

The sizes of territories vary inversely with the number of males present and sometimes affect male reproductive success. Individuals defended up to 400 square meters when the population contained few males, and as little as 12.5 square meters, a circle about one body length in diameter, when males were numerous. Such small territories were only defended on rocky, irregular terrain, not on flat ground. Few females resided in these small territories because space was limiting, and because the proximity of neighbors forced male territory holders into frequent agonistic encounters that females avoided (chapter 5).

These results suggest that during a population decline, breeding areas that afford direct access to the sea will change less than those with limited access (compare figs. 2.12 and 2.13). Shoreline areas change only in group size, but inland areas change in both group size and group location. As a result, inland extensions of breeding areas disappear fastest during population declines. Like Zapadni, the inland extensions of East Cliffs, North, and

Staraya Artil (fig. 1.4) declined more than their shoreline portions. The largest inland central breeding area occupied by northern fur seals in historic times (Palata Point on Medny Island; Stejneger, 1896) completely disappeared in the population crash of 1911 while other breeding areas on the shoreline survived.

Males were more site-faithful than females as the population declined. Males (the same individuals) persisted in using the Zapadni study site long after females had abandoned it. Eventually nearly 80% of males there lacked contact with females. Such rigid adherence to a site during population declines seems maladaptive (see chapter 4).

By extension of present results, northern fur seal populations that are not subjected to a kill of males can soon be expected to develop a large population of sexually mature males that lacks contact with females during the main breeding period. Management refers to such animals as "idle bulls," an inaccurate inference of their reproductive status (Standing Scientific Committee, 1963; Lander, 1980a). The term "peripheral males" is more appropriate because it refers to their locations relative to breeding rather than to their reproductive role. The concern that increases in young males would jeopardize pup survival led to a call for the resumption of the kill at both islands. The present results suggest that this fear was groundless; males form a self-regulating society (chapter 9). Peripheral males play a valuable role in replacing breeding males, inseminating nulliparous females (Vladimirov, 1987), and reducing disturbance (chapter 9).

The northern fur seal can successfully breed at a wide range of adult sex ratios. This study suggests that a daily population ratio of nine females per male may be typical (fig. 2.5). This ratio affects pregnancy rates because it represents the number of females from among which males must detect those in estrus. If peripheral males are included, the daily ratio is about 4:1 (fig. 2.4). Assuming that the females on shore daily during the peak week of the season make up 34% of all females in the population (chapter 3), then sex ratio in the entire herd is usually 26.5 females per breeding male, or 11.7 adult females per adult male (breeding or not). The ratio of 1:9 did not develop until 8 years after the last kill of males, but thereafter it did not fluctuate for 9 years.

SUMMARY

Breeding females at St. George Island declined markedly throughout this study (Zapadni, −93% in 1974–86; East Reef, −80% in 1974–88). The decline was not uniform but occurred in a mosaic pattern, with inland areas falling nearly twice as fast as shoreline areas. The decline at a given site progressed irregularly across years and out of synchrony with other sites.

The male population initially increased and then declined. Juvenile males increased in numbers and area occupied because they were not killed commercially after 1972. They began to decrease after 1978, apparently because adult females had declined and fewer pups were being born. The adult male population followed the same pattern but lagged several years behind due to the time required to reach maturity.

The decrease in females and increase in males resulted in the decline of daily population sex ratios from twenty-seven or forty-six females per male (depending on the site) to about nine, a ratio the herd then maintained for 9 years. The total adult sex ratio, including females at sea, was estimated at 11.7 females per male during this time. Many of the males, here termed peripheral males, defended territories that contained no females. Peripheral males were important because they mated with young females.

The density of females within groups was independent of the population size and sex ratio. Density within groups, which was determined by females, underwent a seasonal decrease of from 2.5 to 1.5 females per square meter, but was not affected by increasing numbers of males over years. The number of female groups was linearly related to the number of males present. As males increased and females decreased, females formed a larger number of smaller groups over time, always maintaining their typical density within groups.

Male densities increased greatly as their numbers increased. The minimum area that males seemed able to tolerate was 12.5 square meters per male, or a circle about one male body length in radius. There was no apparent upper limit to the area they would defend when males were sparse. Males were more site-faithful than females when the population declined.

On flat terrain, female groups had slightly higher densities, males had lower densities, and the sex ratio was higher than on broken terrain. On flat terrain, females formed a smaller number of discrete groups and maintained a wider variety of group sizes (females per group) than on broken terrain.

Where access to the sea was unlimited, the declining female population kept breeding in the same locations over years. But where access was limited, the declining female population rapidly abandoned the site. Access to sea did not affect territorial males that way.

The numbers on shore peaked in the same week of the year for 15 years despite known short- and long-term fluctuations in oceanographic and meteorological conditions in the Bering Sea. Therefore, if the timing of spring arrival is governed by extrinsic factors, these factors vary over a longer period than measured here.

The kill of juvenile males for pelts cannot have caused the decline in females that was documented here. The St. George Island herd continued to decline for at least 24 years after the kill there ended, and the St. Paul Island population stopped declining 4 years before the kill for pelts ended there.

CHAPTER 3

Temporal Factors in Behavior

THE PREVIOUS chapter showed that some behavioral tendencies of individuals can be inferred from the behavior of the herd at large, specifically that groups reflect the spacing tendencies of their members. There are other cases in which the behavior of individuals cannot be predicted well from the behavior of the population. As this chapter will show, individuals are sometimes more narrow and specific in their behavior than is the population they form. The difference between individuals and the population becomes important whenever humans manipulate seal populations for some specific purpose.

One way to differentiate between the behavior of the population and of individuals is to compare them on the basis of time. The present chapter considers northern fur seals on three timescales: yearly cycles, daily/weekly cycles, and the scale of individual lifetimes. This treatment also broadly defines how the species relates to its environment and provides a basis of comparison with other species.

The yearly cycle for females includes migrating north to the Pribilof Islands, bearing young, mating, weaning young, and starting the southward migration, all in a 5-month window bounded by freezing weather. Suckling may not extend beyond mid-November, when pups depart on their southward migration (Ragen et al., 1995). Therefore, the timing of spring arrival is critical to growth and survival of the young and growth of the population. This chapter addresses several important questions about the yearly cycle, beginning with spring arrival. When do males arrive relative to females? Do females arrive more synchronously than males, as they do in the antarctic fur seal which has a similar annual cycle (Duck, 1990)? Does the arrival of females vary over the years, suggesting that it is under the influence of weather or climate? Antarctic fur seals tend to arrive late after unusually severe winters (Duck, 1990). Luckily, we were collecting data when a major climate regime shift occurred in 1977 (chapter 1). How much do arrival dates of individuals vary over their lifetimes, compared to the arrival dates of the population? Are nonbreeders more or less flexible in their arrival dates than breeders?

This chapter also addresses whether the time of mating within the annual cycle is stable, specifically whether the median date of copulation can shift over the years. Changes in the median copulation date might suggest that either the duration of embryonic diapause or of gestation varies. Gestation begins with implantation the previous autumn, coincident with the end of suckling. The median date on which the implantation chamber forms is 14 November (95% CI = 10–18 November; York and Scheffer, 1997); the mean weaning date (measured as the mean date that radio-tagged pups migrate away from the island) is 13–17 November (Ragen et al., 1995). Implantation is preceded by embryonic diapause, which begins at fertilization the previous summer. Changes in the median copulation date, or in the date of arrival, might coincide with changes in population size, and thereby imply that human disturbance has affected behavior (Anon., 1973).

In terms of the daily/weekly timescale, the most important question is, what is the flux of animals on and off shore? What proportion of the female population is on shore on any given day? Without that information, managers cannot estimate herd size from a simple head count. A fixed estimate of the portion at sea has been used in population models (Bartholomew and Hoel, 1953; Peterson, 1965), but the variation is not well known. In California, northern fur seals are nocturnal while on shore (Antonelis, 1976). Has this nocturnal habit been altered by two centuries of killing for pelts at the Pribilof Islands? Does the activity cycle change with the population size, thereby showing a human influence on behavior (Anon., 1973)? Northern fur seals are mostly solitary at sea (Kajimura, 1980; Kajimura and Loughlin, 1988) [1]. When they are on shore, do pairs rest together or move to sea and back in a way that would suggest social bonds? How does the shore aggregation respond behaviorally as females come and go?

This chapter also examines behavior on the scale of individual life spans. We followed many females and males through their reproductive lives at two study sites. The most important question about life span is, how many years do the two sexes devote to reproduction? This question is important because it gets at the average lifetime reproductive contribution by sex. Does the longevity of males (number of years of breeding) change as the number of males increases? What can the longitudinal study of individuals show about survival rates? Past research, based on anonymous animals shot at sea, showed that average female survival exceeds average male survival (Lander, 1981–82; York, 1985) by a considerable margin.

METHODS

Timing of Arrival

The seasonal arrival of the fur seal population and its subsequent history were analyzed using data from daily census and distribution maps [2], normalizing the data to compensate for the population decline. The arrivals of known individuals were compared against the arrival of the population at large using data from female history cards. Appropriate statistical tests were used to look for yearly differences in arrival date within and between our two study sites [3].

The seasonal arrival of juvenile males was documented through weekly censuses of all landing areas on St. George Island [4]. Counts were made in early afternoon hours when peak numbers were on shore, except during stormy weather when the animals went to sea.

Timing of Reproduction

The annual timing of reproduction was analyzed from copulations that were recorded daily at each site every year [5]. These records did not include many virgin females that mate later (mid-August to mid-September; Vladimirov, 1987) or at sea (Baker, 1989). For each copulation we recorded appropriate descriptive data [6]. We did not use parturition date to measure the timing of reproduction because births were too inconspicuous to see reliably in the dense masses of females.

Observer effort was recorded daily so that behavioral events could be reported as rates. Observer effort for copulations was calculated between the dates of the first- and last-observed mating of each year.

We looked for yearly shifts in the median date of mating by focusing on the period 1–28 July because most of the copulations occurred then [7]. The analysis involved calculating a median copulation date for every year and comparing days before and after the median.

History Cards

The reproductive lifetimes of known males (n = 1,542) and females (n = 1,063) were recorded on history cards throughout the study. We identified males at Zapadni (n = 673) by tags we applied, but we identified males at East Reef (n = 869) by natural marks because the terrain there was too rocky for our capture methods (Gentry and Holt, 1982). Entries were made on male history cards once each day the male was ashore, noting time, location, and behavior.

We applied several restrictions to the final data set to ensure that only fully mature breeding males were included [8]. These restrictions excluded the records of most of the young, wandering males which were not making concerted attempts to breed but spent time on the periphery of breeding areas.

We identified most females (East Reef, 477; Zapadni, 586) from brands or tags we applied to them as adults, although we tagged some of them as pups. History card entries of time, location, and behavior were made once each day the female was ashore and each time certain special events occurred [9].

One group of females at Zapadni, called "angle-branded" females, was analyzed separately. They had been marked before this study began using marks that were difficult to read [10]. Because of this difficulty, we followed them only in certain years and ignored them in others. The records of these animals show maximum longevity, but cannot be used to calculate survival rates because of our irregular sighting effort.

We could not restrict the analysis of female history cards to ensure that only residents were included. Tagged adults from other breeding areas occasionally appeared on our study sites, and the cards that were made for them could not be removed from the database by a simple numeric restriction as for males [11]. As a result, the estimated lifetime survival and annual mortality rates of females reported here are inflated to an unknown but probably small degree. This inflation mainly applies to females seen in a single year.

Arrivals and Departures

The movements of females on and off the breeding area imply a certain timing between female movements and their offshore food base. To determine whether that timing changed as the population changed size and sex ratio, we quantified hourly arrivals and departures of females at East Reef when the population was large (1974, 1975) and when it was smaller (1979) [12].

When this study began, little was known of the daily movements, activity cycles, or behavior of nonbreeding males on landing areas. To determine whether they moved like adults on the breeding areas, we collected data at a special study site at the Zapadni landing area in 1978 [13]. The data showed that, like adult females, juvenile males arrived in the mornings and departed in the evenings (Gentry, 1981b). Given that pattern, we collected hourly data on the activity cycle and behavioral interactions among these young males.

Daily Activity Pattern of Adults

To test the hypothesis that the activity cycle of adults may change when the population size or sex ratio change (Anon., 1973), we quantified the amount and timing of activity on breeding areas when the population was large (1974) and again when it was small (1981–83), using a scan sampling

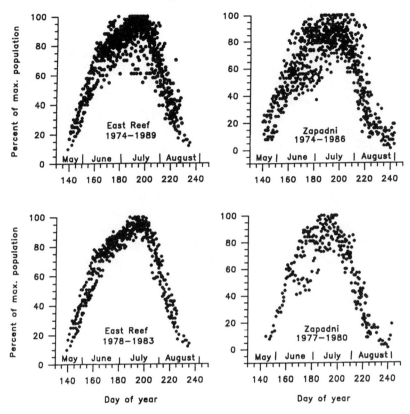

Fig. 3.1. Counts of adult male fur seals on shore at two sites between 20 May and 23 August over several seasons. Each count is expressed as a percent of the highest count made that year to compensate for yearly changes in population size. *Top left*: East Reef in 1974–89 excluding 1987 (911 counts; 18–94 counts per year); *Bottom left*: East Reef for 6 years when males were at peak numbers (457 counts; 67–94 counts per year); *Top right*: Zapadni in 1974–86 (817 counts; 6–86 counts per year); *Bottom right*: Zapadni for 4 years when males were at peak numbers (277 counts; 53–85 counts per year).

procedure [14]. Sampling was not suspended during highly active periods because each sample required only minutes to complete.

RESULTS

Timing on the Yearly Scale: Population Arrival

The population arrived in the same general pattern each year of this study. Males arrived well before the first females, established territories [15], mated, and abandoned these territories before the last females arrived. Males

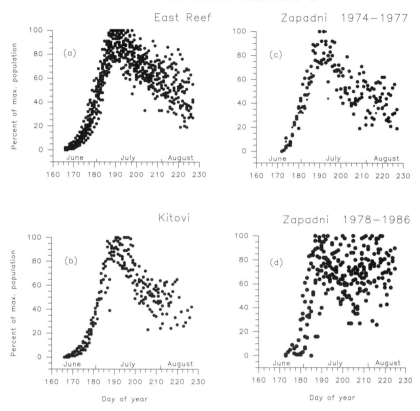

Fig. 3.2. Yearly timing of adult female fur seals on shore at three sites between 15 June and 15 August over the seasons indicated. Each count is expressed as a percent of the highest count made in the same year to compensate for the declining population size. (a) East Reef, 1974–89, excluding 1987 (709 counts; 18–62 counts per year); (b) Kitovi, 1976–78, 1983, and 1985 (239 counts; 42–51 counts per year); (c) Zapadni in four seasons when the population was large; and (d) Zapadni over nine seasons when the population was small (448 counts; 3–59 counts per year).

began arriving about 20 May (fig. 3.1), sometimes while snow still covered the rocks. After establishing a territory, males fasted and rested but rarely fought. About 80% of the male population arrived before 15 June when the first females arrived (fig. 3.2). Male numbers on shore peaked in mid-July, when female numbers also peaked (fig. 3.2). On about 21 July males began a rapid decline as they reached the end of their fasting ability and abandoned territories. This decline coincided with a drop in the number of estrous females (i.e., a decline in copulation rate). By 23 August all adult males had been replaced by smaller males, which were not included in the graphs.

We found that adult males arrived at the island in a fairly narrow range of dates over the years (scatter in fig. 3.1). The years that were least variable

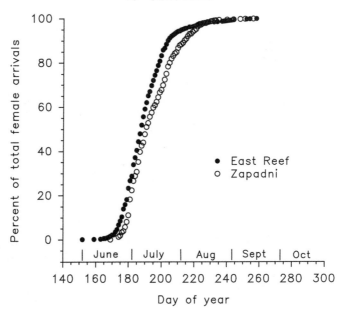

Fig. 3.3. The arrival dates of known females at East Reef (n = 477) and Zapadni (n = 586; total 2,292 female-years of data) from 1974 to 1986 plotted as cumulative percent of new arrivals by date.

were those that had the largest male populations (the bottom two panels of fig. 3.1). A testable hypothesis is that competition for space accounts for male arrival in a narrow time frame.

The female population arrived on even more predictable dates than the male population (fig. 3.2), growing from zero animals on 15 June to 100% of their annual peak between 7 and 14 July. Females at Zapadni became irregular in their arrival dates after 1978 (fig. 3.2d) because of the combined effects of a declining population and female avoidance of males [16]. The arrival dates of known individuals varied within a much more narrow range than for the female population at large (fig. 3.3).

The suggestion that females arrive more precisely by date than males was confirmed by comparing the records of known individuals over their lifetimes. For about 55% of marked females the range from earliest to latest arrival dates was 10 days or less (table 3.1). By comparison, only 29% of marked males arrived within a 10-day span over their lifetimes.

The female population remained at its peak for only a few days. After 19 July it declined because fewer new females were arriving and because previously arrived females had begun making foraging trips to sea.

Even when the population of females on shore was at its peak, many females were still at sea. In one typical year (1982; fig. 3.4), only 34% of marked females were ashore daily during the peak week. Thereafter the frac-

TABLE 3.1

Precision in Arrival of Male and Female Fur Seals at Two Study Sites

Range (Days)	East Reef		Zapadni	
	Males	Females	Males	Females
1–10	29.9	59.3	28.1	55.6
11–20	29.4	24.9	27.5	23.6
21–30	14.2	10.4	18.3	12.5
31–40	14.7	2.3	7.8	6.9
41–50	8.8	3.6	5.6	1.4
51–60	2.5	2.3	3.3	0
61–100	0.5	2.2	9.4	0
Sample size	204	221	153	74

Note: Entries represent the percentage of animals observed for which arrivals over their recorded lifetimes fell within the indicated range.

tion on shore daily declined and varied widely, especially in the second half of July. Because of this variability it is not likely that a close estimate of total female numbers can be derived from using a single value to represent absent females. It could if that value remained stable over the years, but it did not. The portion ashore daily during the peak week varied from 23% to 42% over a decade (table 3.2). Nevertheless, total numbers can be estimated if shore counts are conducted while females at sea are simultaneously estimated. We used this procedure to estimate the East Reef population for a decade (table 3.2) [17], showing that it declined from about 1,500 to about 400 animals. This is a more empirical way of estimating population size than earlier procedures (Bartholomew and Hoel, 1953; Peterson, 1965).

The arrival of juvenile males was not as precipitous as for adult females. They began arriving in early May when the first adult males arrived, and peaked in August after adult females had mated (fig. 3.5). The apparent double peak in their population reflected the arrivals of two different age groups (3+ years old in early July, and 2 years old in early August). Scatter in the figure was caused by responses to bad weather.

Annual Variation in Arrival

Females were very constant in their arrival dates over a long period (East Reef, 14 years; Zapadni, 13 years). Statistical tests showed that the median date of arrival did not change over the years in either data set [18]. This result was unexpected because the environment was not constant. For example, arrival dates were virtually identical before and after the long-term change in the climate regime that had its greatest effect in the Bering Sea in 1979 (1977 in the North Pacific; Niebauer and Day, 1989) [19]. Also,

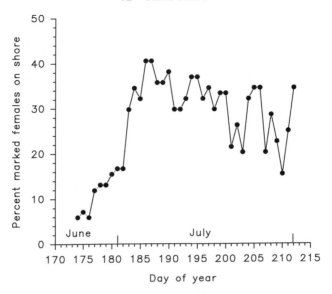

Fig. 3.4. The percentage of eighty-four marked females that was on shore daily throughout the 1982 season at East Reef study site.

arrivals differed by only 5 days between the coldest and warmest years [20] of the study period. It is doubtful that this 5-day difference had much impact on rearing and weaning of young.

Timing on the Annual Scale: Mating

The precision in the dates of arrivals of individuals (fig. 3.3) was matched by the precision of events after they arrived. On average, parturition occurred 1.3 days after arrival, copulation occurred 5.3 days after parturition, and departure on the first foraging trip occurred 1.2 days after copulation. The most striking feature of these means was their low variance among individuals [21]. Field observers could anticipate to the nearest half-day when these events would occur for a given female.

The precision of arrival and the events following meant that copulations were highly predictable by date. They occurred in an almost logistic curve (fig. 3.6), with the steepest part from about 30 June to 21 July. Interestingly, 21 July was the date on which adult males began to abandon their territories (fig. 3.1). That is, males began to leave land when about 20% of the year's copulations still had not occurred. The copulation curve slowly approached 100% by mid-August when our observations ended. From mid-August until September, young females mated for the first time, usually with younger males. We did not record these copulations.

Since the arrival dates did not differ over the years, it follows that the

TABLE 3.2

Percentage of the Northern Fur Seal Population Seen Copulating
during Daylight Hours at the East Reef Study Site (by year)

Year	Marked Pop.[a]	% Marked per Day[b]	Mean Daily Count[c]	Est. Pop.[d]	No. Cops.[e]	% Seen Copulating[f]
1974	14	0.36	395	1,097	411	37.5
1975	39	0.26	382	1,469	343	23.3
1976	31	0.27	398	1,474	287	19.5
1977	46	0.32	331	1,032	294	28.4
1978	75	0.29	287	990	225	22.7
1979	44	0.26	234	900	294	32.7
1980	55	0.23	241	1,048	192	18.3
1981	86	0.23	222	965	270	28.0
1982	84	0.34	180	529	275	51.9
1983	69	0.36	148	411	280	68.1
1984	53	0.42	160	381	222	58.3

[a] Number of marked animals in the population.

[b] Average daily fraction of the marked female population ashore during peak week of season. Assuming unmarked females move to sea and back like marked females, this number represents the portion of the entire female population that was ashore daily during this period.

[c] Mean size of the daily population during the peak week from counts.

[d] Number of females estimated to have used the study site by year. Calculated as Mean daily count divided by the proportion of the population ashore at peak week [i.e., divided (c) by (b)].

[e] Number of copulations observed during the entire season at this site.

[f] Percentage of the estimated population that observed copulations represent [i.e., divided (e) by (d) and multipied by 100].

median date of copulations would also not differ. This was the case. The median copulation date varied over the years, but was not statistically significant [22]. It occurred on 12 July at East Reef (9 years of data) and 10 July at Zapadni (5 years of data), a nonsignificant difference. Individuals showed remarkable precision in estrus dates over years (chapter 6).

Timing on the Scale of Days

FEMALE ARRIVAL AND DEPARTURE

Females moved on and off shore at all hours, although we quantified movement only during daylight (fig. 3.7). During daylight, most of the animals we scored were arriving, so departures must have predominated at night. Movements in both directions peaked in the early evening at about thirty animals per observation hour. On some evenings, most of the females on shore moved in one direction or the other. Judging from the movements of tagged females, both arrivals and departures continued at night.

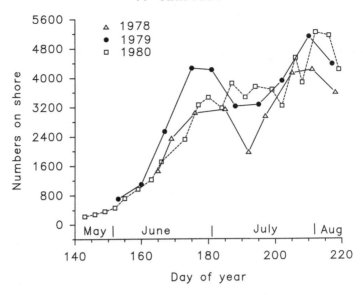

Fig. 3.5. Arrival pattern of juvenile males on all landing areas of St. George Island in 3 years. Only males lacking secondary sexual characteristics were counted. Scatter resulted partly from animals vacating land during stormy weather.

ADULT ACTIVITY PATTERNS

Most of the adult fur seals were inactive for most of the daylight hours. At any hour, on average no more than 25% of them were active (fig. 3.8) [23]. But activity was highly variable (see average SD bars in figure). It varied daily because of the weather; animals tended to lay flat during high winds, stand upright during rain, and actively move around during high surf. It varied hourly because of behavioral disturbances—especially fights when new males arrived—and forays by nonterritorial males (brief dashes onto the breeding area; see chapter 9). Nevertheless, the three categories were different from one another (Inactive Down > Inactive Upright > Active) [24]. The activity "cycle" consisted merely of a peak in activity from 0600 to 0900 hours daily [25]. Analysis showed that breeding site, sex of the seals, and year of data collection all affected the number of animals in each behavioral category [26]. Most animals in the Inactive Down category were males, and females predominated in the other two.

The general activity pattern shown in figure 3.8 did not change as the population underwent large changes in size and sex ratio over nearly a decade [27]. It was especially important that the proportion of animals in the Active category did not change, as it would if the amount of activity had been affected. This finding answered a key question posed in the establishment of this program.

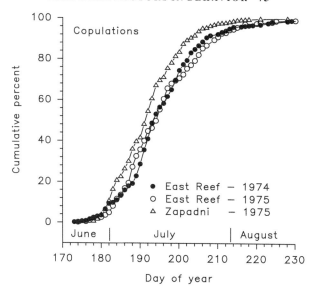

Fig. 3.6. The cumulative percentage of copulations recorded by date for the three longest seasons (East Reef, 1974 [n = 406] and 1975 [n = 343]; Zapadni 1975 [n = 683]).

JUVENILE MALE ARRIVALS, DEPARTURES, AND ACTIVITY PATTERNS

Juvenile males moved on and off shore in a crepuscular pattern, much as adult females did. They arrived on shore in the morning hours, reached a peak in numbers and a low in activity between 1300 and 1900 hours, and became active and departed for sea in the evening (fig. 3.9). Activity levels at dawn and dusk were not different from each other, but both were significantly greater than activity in the afternoon hours [28]. Just as for adults, bursts of activity were likely at any hour (error bars in the figure). As with adults, this activity sometimes resulted from weather and sometimes from behavioral disturbances.

COPULATION

Animals mated throughout the daylight hours at a relatively low and steady rate (fig. 3.10). Slight peaks in mating activity were suggested at dawn and midday [29], with a major peak at dusk. Unfortunately, the dusk peak was based on few observation hours (12 at East Reef, 20 at Zapadni). It undoubtedly occurred, but whether its magnitude was greater than six copulations per hour, as indicated, is uncertain.

Clearly, most of the females in the population copulated at night. We usually saw copulations for fewer than a third of all females present (8 of

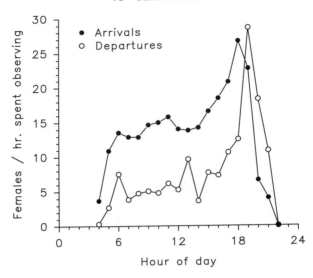

Fig. 3.7. Female arrivals and departures per hour of observer effort for each daylight hour (based on 649 hours of observation occurring on 32 days in 1974, 24 days in 1975, and 8 days in 1979). The mean observation effort per hour from 0400 through 2200 hours was 34.2 hours (SD = 18.0, max = 54.3 h at 1400, min = 2.0 h at 2200).

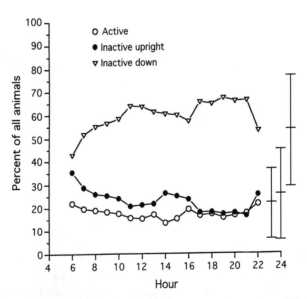

Fig. 3.8. Activity patterns of male and female northern fur seals on 218 days at East Reef (1974 and 1981–82) and Zapadni (1981–83) study sites (total = 2,817 hourly scan samples divided equally between sexes; 147,736 individuals). Average standard deviations are indicated. Dawn and dusk are approximately at 0530 and 2130 hours, respectively, at this time of year.

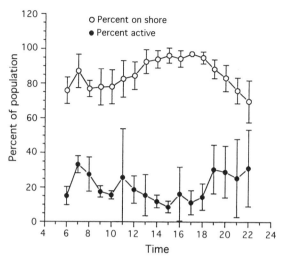

Fig. 3.9. Presence on shore and activity cycle of juvenile males at Zapadni landing area on 4 days in 1978 (n = 66 hourly samples). On shore = the percent of the daily maximum count observed at each hour of the day. Active = the percent of each hour's count that was moving (see note 28).

11 years; table 3.2). Even when we recorded copulations continually from dawn to dusk (1974–75) we did not see copulations for more than 37% of females present. In fact, the number we saw in daytime was usually similar to the average number of females ashore daily during the peak week of the season [30].

Timing on the Scale of Individual Lifetimes: Longevity

MALES

Males spent only a small fraction of their lives attempting to breed. Most of them (about 75%) appeared on the breeding areas for only a single season before they disappeared permanently (fig. 3.11). Two exceptional males reappeared for eight and ten seasons, respectively, and all others spent 2–7 years on territory. The average for all males at both sites was 1.45 years of breeding, very close to the value of 1.5 years calculated previously (Peterson, 1968; Johnson, 1968; Vladimirov, 1987). Disappearance from our study sites was tantamount to disappearing from the pool of breeding males [31].

Our records included few records of wandering, nonbreeding males, and these had little impact on the results. Only the copulation restriction (males had to be seen copulating to be included in the analysis) removed

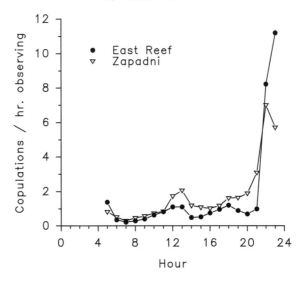

Fig. 3.10. The hourly copulation rate among northern fur seals at two sites (East Reef, 1974–86; Zapadni, 1975–83). The plot used 5,771 copulations recorded in 6,510 hours of observation (see note 29).

more than 1–2% of the records [32]. Removing the records of males that were not seen copulating did not change the shape of the plot of lifetime territorial occupancy (fig. 3.11). Also, the mean number of years that males were present was not changed when the records of wandering males were excluded (wandering males excluded: East Reef, 1.46 years; Zapadni, 1.38 years; wandering males included, 1.45 years at both) [33]. In other words, our history cards gave a fair representation of the adult male population.

Given that most males spent only a single year of their lives breeding, the duration of their territorial tenure may have been very important to their reproductive success. Did this tenure change throughout the study as male densities increased and females decreased? The variability in male tenure clearly decreased as male density increased (fig. 3.12) [34]. Perhaps competition for sites forced males to this more uniform tenure pattern, just as it seems to have forced males to arrive in a narrower time span. However, the average length of tenure was no different in the early years when males were scarce compared to later years when they were numerous [35]. Also, territorial tenure was about the same at the two study sites despite large differences in terrain (East Reef, 32.7 days; Zapadni, 30.3 days—a nonsignificant difference) [36]. However, tenure was 8–12 days longer for males that were seen mating than for the population at large [37].

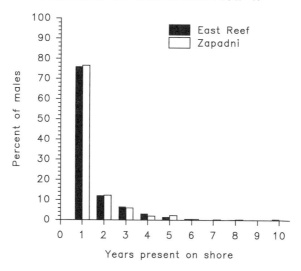

Fig. 3.11. Lifetime territorial occupancy of identifiable male fur seals at East Reef (1974–88, n = 869), and Zapadni (1974–86, n = 673). The data are unrestricted as to date, location, and whether males copulated (see text).

FEMALES

Females had a significantly longer reproductive life than males (fig. 3.13) at both of our sites (table 3.3). Many more females than males returned for 5 or more years. The angle-branded females at Zapadni showed the greatest longevity (fig. 3.13, bottom), with some individuals returning for 14 years (minimum age 18, assuming marking at age 4 years). Our records indicated the average female reproductive life was only 2.3 or 1.7 years long, depending on the site (table 3.3). These numbers were probably not realistic because of tag loss and wear [38]; actual longevity is probably considerably longer. From table 3.3 it appears that females at Zapadni had a shorter reproductive life than females at East Reef. However, they may not have; many Zapadni females abandoned our study site and moved to Zapadni beach, where they could not be seen.

When females disappeared from our records it did not signify that they had stopped breeding, as it did for males. Some females returned for years having lost, broken, or worn tags that prevented our identifying them (see note 38). In addition, we found two or three females with substantial records at East Reef that were suckling pups at another breeding area 2 km away (see East Cliffs in fig. 1.4) a year or two after their disappearances from East Reef. The number of females that move among breeding areas is not known but is probably small (a few percent of the females).

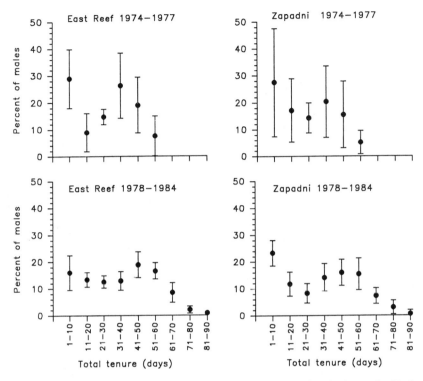

Fig. 3.12. The maximum within-season territorial tenure of male fur seals. Each male on territory was assigned to the 10-day category that most closely matched his maximum tenure, and the percent of males in all categories was calculated annually. Yearly percentages were used to calculate the means and SDs plotted here. The sample represents 10 years at each study site (East Reef = 869 males, Zapadni = 673 males).

DISCUSSION

Most males arrive before the first females, with the earliest of them preceding females by 4 or 5 weeks. They spend the time before females arrive firmly establishing boundaries. The male population arrives less synchronously than the female population, as in antarctic fur seals (Duck, 1990). Over years, individual males vary more in arrival date than individual females. The difference may reflect the fact that female arrival is driven by impending parturition, whereas male arrival is based on past experience. Males may have less latitude in timing their arrival when the male population is large than when it is small.

Early arrival gives males the advantage of something like a prior residence effect (Braddock, 1949) in that once they are on territory it is difficult

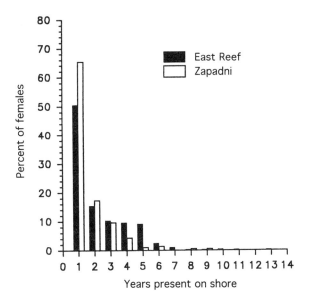

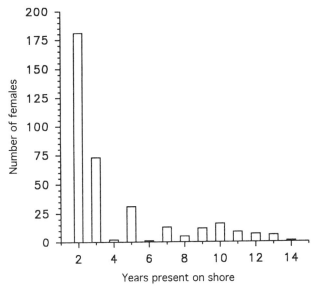

Fig. 3.13. Reproductive lifetimes of marked, adult female fur seals. *Top*: East Reef (1974–92, n = 373); Zapadni (1974–86, n = 354). *Bottom*: Angle-branded females at Zapadni, 1973–85. Population means were 2.33 years at East Reef (SE = 0.09), 1.68 years at Zapadni (SE = 0.09), and 2.09 years for non-angle-branded females (SE = 0.09) at Zapadni.

TABLE 3.3

Reproductive Lives of Male and Female Fur Seals at
Two Pribilof Island Study Sites (in years)

	Mean Years	SE	n
East females	2.33	0.09	447
East males	1.45	0.03	849
Zapadni females	1.68	0.09	208
Zapadni males	1.45	0.04	653
Zapadni angle- brand females	2.09	0.09	348

χ^2 analysis of 3-way frequency table with 6 df (compare to tabled $\chi^2_{(.05,6)} = 12.59$ and $\chi^2_{(.10,6)} = 10.65$)
χ^2 for site = 15.7, sex = 63.7, site-by-sex = 16.2

χ^2 analysis of site-by-sex interaction (same df and tabled values)
χ^2 for site differences among females = 21.8
χ^2 for site differences among males = 3.5 (NS)
χ^2 for sex differences at East Reef = 105.4
χ^2 for sex differences at Zapadni = 15.7

One-way ANOVA
   East Reef females > males F = 127.1, p < .05
   Zapadni females > males F = 7.3, p < .05
   East females > Zapadni females F = 21.8, p < .05
   East males = Zapadni males F = 0, p = .98
   East females = angle-brand females > Zapadni females $F_{(2df)}$ = 30.2, p < .05

for subsequent arrivals to replace them, although they can be displaced. The disadvantage of early arrival is that it increases the time males must fast while awaiting estrous females. The average time on territory for all males is 33 days per year, and about 43 days for males that do most of the mating. For early arrivals, this period elapses before copulations wane in mid-August. In fact, males begin to abandon their territories in late July, leaving approximately 20% of the copulations to later-arriving, younger males. The disadvantage of early arrival is apparently preferable to the disadvantage of late arrival—namely, wresting from established males territory space among females. The former strategy merely risks reproductive success without jeopardizing future chances of mating, while the latter risks substantial injury.

As the male population grew, males changed their behavior in at least two ways. Male arrival dates became less variable, and the pattern of tenure among males stabilized (figs. 3.1 and 3.12). These changes can be interpreted as resulting from increased male-male competition. As competition increased, males may have been pushed to the earliest arrival dates and maximum tenures possible (hence reduced variability in both measures).

The female population arrived with remarkable precision over the years, given the length of the migration. Despite an El Niño event and a climate regime shift in 1979 (Niebauer and Day, 1989), no statistical differences were found in the median date of arrival at St. George Island. This date remained within a 10-day span (28 June–7 July) over 14 years. Unlike antarctic fur seals (Duck, 1990), arrivals did not occur later following unusually cold winters, but they did shift 4 days earlier in an unusually warm spring (the year the regime shift occurred). Thus, timing of arrival showed only a slight nonsignificant shift coincident with a major change in climate.

The median date of copulation also varied over the years at St. George Island, but not by a statistically significant amount. The median varied from 9 to 14 July at our two study sites over a 9-year period. This difference was not large given that copulations for adult females may occur for about 40 days (18 June–28 July). The annual shifts had no trend over years. The mean parturition date for the population can be inferred from the mean copulation dates because of the low variance in birth-to-copulation interval (see note 21). These suggest that the mean parturition dates varied from 3 to 11 July with a 9-year mean of 4 or 6 July, depending on site. Pupping occurs about 11 days earlier on San Miguel Island, covers a comparable range (21–29 June with a 10-year median on 24 June; DeLong, 1982), and may be more affected by severe El Niño events than at the Pribilof Islands (DeLong and Antonelis, 1991). Trites and Antonelis (1994) concluded that mean pupping date at the Pribilof Islands coincides with long-term annual peaks in air temperature and relative humidity.

Individual females tend to arrive and give birth on or near the same date each year (4 days either way of a date particular to each one). Antarctic fur seal females also have a narrow preference for parturition date (Boyd, 1996).

The low annual variation in mean arrival and copulation dates supports the hypothesis that the annual reproductive cycle is driven by the seasonal light cycle. Temte (1985) compared the mean parturition dates of northern fur seals at San Miguel and St. George islands and concluded that the time difference in their peaks could be explained by differences in the date on which a mean photoperiod of 12.5 h/day occurred at the two sites. This day length, which occurs 62 days after the mean date of parturition (11 September for the Pribilof Islands), may trigger the implantation sequence and thereby establish the date of subsequent parturition.

Ichihara and Yoshida (1972) concluded from pelagic observations that northern fur seals are nocturnal at sea before the breeding season. By observing animals at night on land using a night-vision device, Antonelis (1976) concluded that adult male fur seals are nocturnal, females are crepuscular, and that copulations occur in a nocturnal pattern. Graphs of activity level by hour from both these studies closely resembled the curve for active animals in figure 3.8 (active at dawn and dusk, inactive throughout the day, and copulate mostly at night). That is, human killing of seals at the Pribilof Islands has not altered their basic activity cycle (Anon., 1973). Antonelis (1976) attributed the nocturnal pattern to the avoidance of high daytime temperatures at his California study site. However, animals that are not exposed to high temperatures (those at sea and those on land in Alaska) have the same cycle.

Nocturnal habits may result from feeding at night on vertically migrating prey. Northern fur seals off Japan (Taylor et al., 1955) and in the mid-Pacific (Walker and Jones, 1990) feed mostly on myctophid fish and squid that rise to the surface at night. Most females at the Pribilof Islands (Gentry et al., 1986b) and all females at Medny Island (chapter 13) follow prey that ascend at dusk and descend at dawn. In California, fur seals feed at night in oceanic water even though the same prey species are available in the nearby neritic zone (Antonelis et al., 1990a). Foraging animals sleep at the surface for extended periods during daytime (Gentry et al., 1986b). In short, daily activity cycles at sea seem geared to the nightly rise of prey.

During the few days per year that fur seals are on land they maintain the same nocturnal activity cycle. Most sleep in the afternoon when pelagic animals also sleep, have bursts of activity at dusk and dawn when animals at sea are making their deepest dives, copulate at night, move into the sea at dusk, and move onto land in the morning. The potential to mate appears not to affect the daily activity cycle because nonbreeding juveniles sleep by day and move to sea at dusk (Gentry, 1981b). Weather and social factors that fur seals encounter on shore create wide variances in activity level at any hour of the day, but these do not mask the underlying tendency for nocturnal activity.

The activity cycle on shore did not change as the population declined and the male/female ratio increased over nearly a decade. Density-dependent change in activity levels was postulated as one possible result of terminating the commercial kill of fur seals on St. George Island (Anon., 1973). However, the present results suggest that the influence of nightly feeding on the activity cycle of fur seals is so fundamental that it is not affected by changes in numbers or sex ratio on shore.

Sex differences in longevity and lifetime reproductive output cannot be closely compared from the present data. Males appeared in our records for less time than females (mean 1.45 vs. 2.33 years; maximum 10 vs. 14 years). However, males were usually identified from scars which did not change over time, but females were identified from tags that failed in various ways. Our estimate of male longevity agrees closely with Johnson's (1968) and seems reasonable. However, our method underestimates female longevity by an unknown amount and is not close to that predicted by Lander (1981–82).

Males devoted a smaller portion of their lifetimes to breeding than females. Their 1.45 seasons of breeding occurred at least 4–5 years after the onset of sexual maturity. In the interim, males attained the size and experience required to defend territory. Females had no hiatus between sexual maturity and the onset of mating. They mate for the first time at age 3 or 4, and thereafter live and mate longer than males. A closer estimate of the proportion of life devoted to breeding is not possible because of the tag problem for females. The same problem also prevents a close estimate of total lifetime reproductive output by the sexes, but a value for males is given in chapter 4.

SUMMARY

The annual cycle of this species, measured by the dates when breeding islands are occupied in spring and the dates when mating occurs, varied little over 14 years. Males began to arrive well before the first females, and their numbers peaked when female numbers peaked. Adult males began to abandon their territories before the final 20% of females had arrived.

Individual males arrived less precisely by date than individual females (55% of females arrived in a 10-day span over their lifetimes, compared to only 29% for males). Female arrival is driven by impending parturition, whereas male arrivals were related to age and fasting ability. Males arrived within narrower dates when competition for space was high (the male population was large) than when it was low.

Despite a change in climate regime and a strong ENSO event during this study, the median arrival date of the female population did not change by a significant amount over 13–14 years; female numbers peaked from 7 to 14 July every year. Some relatively unchanging environmental cue, such as light cycle, may control the date of arrival. This would explain why arrival differed by 11 days between the extreme northern and southern ends of the range.

Events in the female's first visit to shore were closely timed and had low variance. Because there was little variation in either arrival dates or the arrival-to-parturition interval, copulations peaked on either 10 or 12 July yearly, depending on the study site.

On average, only 34% of the female population was ashore on a given day. The remainder was either at sea foraging or had not yet arrived at the island for the season. The proportion on shore varied greatly on a daily, seasonal, and yearly basis. This variation means that a close estimate of population size cannot be made using a single value to represent absent females.

During the brief period that northern fur seals were ashore they retained the same nocturnal habits they followed at sea, suggesting that the daily cycle was most likely related to feeding on the nightly rise of prey to surface waters. This cycle has apparently not been affected by human disturbance because it was the same at St. George Island (disturbed) as at San Miguel Island (relatively undisturbed).

Shore activity for adults and nonbreeding juveniles tended to peak at dawn, dusk, and briefly at noon. Active animals seldom exceeded 25% of those on shore. Most animals, especially females, slept during daylight hours. Site, sex, and year all affected the number that was active, but these effects were partly masked by responses to weather and especially to behavioral disturbances. Females moved on and off shore at all hours. Movement in both directions peaked at dusk, and movements after dark were mostly departures. About two-thirds of the females copulated at night. Trends in activity cycle, movement patterns, and timing of copulation remained the same throughout this study, despite large changes in sex ratio and population size.

Males dedicated a smaller proportion of their lives to actual mating than did females. Males delayed mating longer than females, mated fewer years, and died at an earlier age. Males spent on average only 1.45 seasons on territory (range 1–10 years); few exceed 5 years. Males that mated averaged about 43 days per year on territory, or about 62 days in an average lifetime. Tag failure thwarted a realistic measure of female longevity. Our estimate—that females mated for an average of only 2.3 seasons before disappearing—was undoubtedly a low value. Many females returned for more than 5 years, and some returned for 14 years or more.

PART TWO

The Mating System

In the previous section the northern fur seal was characterized as a widely distributed, highly pelagic, mostly solitary nocturnal species in which most females make a yearly migration of some 5,000 km to rear young in a few dense aggregations located near seasonally abundant prey at the cost of compressing mating and pup rearing into a brief summer/autumn period. In this section we will consider the kind of mating system that evolved to accommodate the brief period that females and males are ashore each year (35 and 70 days, respectively). The mating system is defined as the species' general behavioral strategy for obtaining mates (Emlen and Oring, 1977).

Various authors have tried to interpret the northern fur seal's mating system for the past 250 years, often with conflicting results. Since the species has had a disproportionate influence on how vertebrate biologists view the otariids, these various interpretations need to be thoroughly reexamined in the light of new data.

In this section we will examine the mating system largely from the behavior of individuals, not from the structure of social groups or from the sex ratio. The chapters are presented as male/male, male/female, and female/female associations to represent all the possible kinds of social interactions individuals have. This division is somewhat artificial because each individual must contend with at least two of these simultaneously. The male contribution to the mating system is presented first because males precede females onto breeding areas.

Behavior of Adult Males

IF MALE northern fur seals are to mate, they must defend space among females. A few mate at sea (Baker, 1989) or on landing areas but so infrequently that they make little genetic contribution to the population. Males are territorial. They use aggressive behavior and topography to delineate borders around nonoverlapping domains within which they have exclusive access to females (Steller, 1749; Veniaminov, 1839; Elliott, 1882; Jordan, 1898; Osgood et al., 1915; Bartholomew, 1953; Bartholomew and Hoel, 1953; Kenyon, 1960; Bychkov and Dorofeev, 1962; Peterson, 1965, 1968; Johnson, 1968; DeLong, 1982). Peterson (1965) described territoriality in northern fur seals in their typical setting (at high latitude on rock substrate in dense aggregations) and DeLong (1982) described it in an unusual setting (at low latitude, on sand, at a small population size, during space competition with sea lions). No one has suggested that northern fur seal males maintain dominance relations, share territories or home ranges, or form leks or other forms of social organization.

At the start of this study, northern fur seal territories were believed to be highly volatile because of "uncompromising aggressiveness" of males (Bartholomew, 1953), and because a pool of nonbreeding rivals apparently waited to replace breeders that faltered (Johnson, 1968). Fighting was believed to be an important cause of death among territorial males (Johnson, 1968). In all, pervasive and unstructured male competition was believed to make the male territorial system susceptible to human disturbance.

The importance of territory to the mating system suggests several questions that this chapter will try to answer. First, how robust, resilient, or volatile is the territorial system? Is the turnover rate among territory holders so high that neighbors are essentially strangers, or are they more like dear enemies? When and how are territories established relative to the arrival of females? How labile are boundaries? Do males change location to increase their mating opportunities? How does territory defense change with male age, or with size of the male population? What behavior do males use to defend boundaries? What effect does the presence of females have on male territorial defense?

The length of territorial defense is limited by the male's ability to fast, which in turn is limited by his activity level (mating, defending the territory, interacting with females, chasing juveniles). Clearly, fasting in otariids is more subject to behavioral factors than in terrestrial mammals that fast during hibernation. Given that fasting limits territorial defense and thus curtails mating opportunity, how long do males fast each year? How do males time their fast relative to the arrival of females? Do males that fast longer mate more? Does the duration of fasting change with age or with the number of males in the population?

Male sexual selection begins at the time of territorial establishment. In a stable population, some males (estimated 30%; Vladimirov, 1987) never win a territory and are excluded from mating for life. Even those that defend a territory are not assured of copulation. Factors such as size, fighting ability, size and location of the territory, skill at interacting with females, ability to detect estrus and copulate, aggressiveness, fasting ability, and timing of fasting and territory defense affect the number of times males mate (Bartholomew, 1953; Miller, 1974, 1975). Given this wide array of factors, is there a typical "profile" of behavioral and morphological traits that typifies males of high reproductive success? This chapter addresses that question.

Adult males accidentally cause some mortality among pups. Pups sleep in dense groups in male territories, are twenty times smaller (average mass in July = 10 kg and 205 kg, respectively), and cannot always avoid adult territorial conflicts, with the result that some die from trauma. Organ rupture and suffocation caused by adult males accounted for 2.1% of deaths among 2,500 pups (North Pacific Fur Seal Commission, 1980). However, this measure was made when adult males were sparse. Does the rate increase when adult males increase, just as overall pup deaths increase when the population increases (York, 1985; Fowler, 1990)? The possibility that the rate could increase has twice been used to argue against ending the kill of males for pelts (at St. George Island in 1972, and at St. Paul Island in 1985). This chapter contains a partial answer to that question.

METHODS

We estimated the turnover rate, or annual loss rate, of males by calculating the percentage of known animals from one year that failed to return the next [1]. These values are high because a few males returned that could not be identified due to tag wear. However, this was an infrequent problem because males were identified by color, location, and other characters in addition to tag numbers.

We analyzed male use of space from entries on their history cards (chapter 3), using subsector numbers to record their locations. Subsectors were quarters of the 10 ∞ 10 m grid that was painted on the rocks. Use of subsectors had some drawbacks [2], but they were the smallest unit of land that was practical given our study methods.

The history cards also provided one measure of male behavior. Once a day, we recorded one of eighteen different types of male behavior on the cards. Observers scored the behavior that was underway during a random glance at the male. Thousands of such scores over the years produced a behavioral time budget. These data were used to determine whether behavioral patterns changed with changes in population size and sex ratio.

Male behavior was also recorded on a subject-by-subject basis. Observers continually scanned the population and recorded information about selected behavioral events as they occurred. Data on male-male aggression, copulations, female stealing, and other types were recorded on separate forms (protocols), which were eventually entered into a database. The types of behavior that were recorded varied by year, as discussed below.

The purpose of scoring aggressive acts among males was to test the hypotheses that territorial defense behavior increases with male population size and copulation frequency. Male aggressive acts were recorded only in years before the male population reached its peak [3], and again after it had peaked. As a control, we recorded male aggressive acts at Kitovi on St. Paul Island (fig. 1.4), where the male population was being kept small by a continued commercial kill of juveniles. We defined several types of aggressive acts [4] and used these to construct a protocol.

Males used various vocalizations as territorial advertisement (Peterson, 1968). However, we did not record them because they occurred too frequently (almost continuously), for us to record given the other behavior we were scoring.

We recorded male-pup interactions in 1984 to measure how often adult males hesitated or stopped threatening each other if their movements would involve trampling pups [5].

History cards gave a good measure of male fasting because entries were made each day the male was on territory. The evidence for fasting and breaks in fasting was in seeing males in the same locations each day, and noting when their pelage had been washed clean of dirt and feces by a trip to sea. We used the single longest sequence from each history card as an index of that male's fasting ability [6]. This index was probably conservative.

In order to correlate various aspects of male behavior with body weight, we had to weigh territorial males alive. We devised live-capture and weighing methods [7] for handling males (Gentry and Johnson, 1978; Gentry and Holt, 1982). Captures were so strenuous and dangerous that a limited number of them (175) was made. Three of the males that were weighted at the start of a breeding season were reweighed the day they abandoned territories as a measure of weight loss during fasting.

The question of the extent to which the location of a male's territory contributed to his reproductive success required a comparison of two rankings; both involved the copulations that were recorded throughout the study. One ranking was the observed copulation frequency, and the other combined the reproductive potential of a site (number of copulations seen there over years) with the number of days the male spent there [8].

We used the relative age of males to determine the effects of age on reproductive success [9]. At Medny Island, Russia, males known to be 9 years old reportedly performed more copulations (37.4%) than known younger or older males (Vladimirov, 1987). Since our males' ages were unknown, we compared their copulation frequencies by year of territorial tenure.

RESULTS I: FIELD EXPERIMENTS

Experiments on Territorial Defense

We conducted field trials on the stability of the male territorial system at the start of the study in 1974. We wanted to know how fragile the male territorial system was before we risked disturbing our study sites. Our method was to drive all males off their territories and measure the success each had in reclaiming his site, the behavioral means of doing so, and the extent to which competition for these sites was open.

Experimental Methods

Two weeks before the first females arrived in 1974, we made distribution maps and history cards for all scarred males present at the East Reef and Zapadni study sites. Then we systematically drove all males off from territory and away from the periphery [10]. Immediately thereafter, we began to record data on the returns of the displaced animals [11]. Observations continued until dark, or until 75% of the original number of males had established territory.

Experimental Results

Males at the two sites differed greatly in their readiness to abandon territory. At East Reef, a single worker gently prodding with a bamboo pole could drive a male from his territory into the sea. At Zapadni, 4–6 people prodding the animal hard and using noisemakers were required to drive them off. Some males had to be driven 100 meters or more before their persistent attempts to return were thwarted.

Displaced males returned to East Reef faster than to Zapadni. Only 2.5 hours after being driven off, 82% of the seventeen East Reef males had reestablished their territories. All occupied their former sites. No fights or biting and only twelve boundary displays occurred (0.86 aggressive acts per returning male). These males did not wander, but arrived at their formerly held sites by the shortest routes possible while submitting to and avoiding males that had returned before them.

Most of the Zapadni males (95% of twenty) were back in their places within 7 hours. Most of them (65%) arrived during the third and fourth hours. We recorded three fights of less than 4 minutes' duration (brief compared to the population average; see below). We also recorded forty-four boundary displays (two with biting) and nine chases. In all, we observed 2.95 aggressive acts per returning male. About 80% of the males reoccupied the site they used before being driven off. Again, males reached their sites by the most direct routes and showed submission to males on all sites except their own.

To determine whether distance to the sea caused the difference between East Reef and Zapadni, we drove to sea three territorial males on Zapadni beach that were as close to the water as those at East Reef. They resisted little and returned quickly, like East Reef males.

Experimental Conclusions

Territorial organization is highly resilient to disturbance. Even if it is completely disrupted by humans it will return to almost its original form within about 7 hours, with far fewer (average 1–3) interactions than males use to first establish their territory (compare below). Where males had ready access to sea they abandoned territory readily and reoccupied it quickly; where sea access was distant, males resisted being driven off and returned slowly. There was no open competition for sites made available by disturbance. Instead, males preferred very specific sites and went to them directly after landing. They performed displays and fought only on their own site, and showed submission behavior to others they had to pass to reach their own sites.

RESULTS II: LONG-TERM OBSERVATIONS

Turnover Rate

From 1974 to 1984, our study sites lost males at a seemingly high yearly rate (East Reef, 66%; Zapadni, 67%) [12]. The rates did not change much even in those years when the male population was largest and male compe-

TABLE 4.1

Area Used by Adult Male Fur Seals in the First and Subsequent Years of Tenure, and for Males That Were Seen Copulating and Those That Were Not

Location	1 Year[b]			2–10 Years[b]		
	No. Males	No. Sectors	% Area	No. Males	No. Sectors	% Area
East Reef	646	3.4	2.85	223	4.12	3.43
Zapadni	500	4.3	2.67	172	6.09	3.81

Location	Copulating[c]			Noncopulating[c]		
	No. Males	No. Sectors	% Area	No. Males	No. Sectors	% Area
East Reef[a]	419	3.94	3.28	450	3.28	2.73
Zapadni[a]	232	5.37	3.36	440	4.31	2.69

Notes: Sectors were 5 x 5 m in size. Compare z-values against $t_{05,2,\infty} = 1.96$.
Wilcoxon rank sum test on the number of sectors occupied by a male.
Sources of variation:
[a] Site (East Reef vs. Zapadni): z = 4.80.
[b] Year (1 yr vs. multi-yr): z = 21.94.   [c] Copulation (cop. vs. noncop.): z = 11.33.

tition was greatest (1978–83). Our rates were about twice as great as the 38% rate calculated by Johnson (1968) for adult territorial males. When our data set was restricted to adults only, the rate was very similar to Johnson's (East Reef, 34%; Zapadni, 40%). That is, turnover rates among large, adult territory holders may be similar over time, but the turnover rates among all males that are trying to breed may be greater (about double in our case), depending on the number of young males present.

Establishment of First Territory

Most males gained a breeding territory by defending a peripheral area during the breeding season, and later (late July or early August) moving among females on an adjacent site that had been held by a larger territorial male since May. Only a few males gained their first territory by appearing at the breeding areas as a full adult and defeating an established male in a fight. The proportion of males gaining territory by these two routes was not analyzed.

Males established territories in locations that seemed highly specific, not random. Peripheral males moved onto breeding areas adjacent to their usual sites, not on any area that became available. Males that won a territory by defeating an established male usually attacked only that male and acted submissive to all others. If unsuccessful, they would return later and fight the same male again. One new male fought the same established rival four times before prevailing, and fought none of the other ninety males present.

Table 4.2

Frequency of Sighting Known Adult Male Fur Seals at Numbered Locations

Study Site		*1**	*2*	*3*	*4*	*5*	*6-11*
				Location			
East Reef	Mean	66.7	19.0	7.6	3.7	1.6	0.3
	SE	2.8	1.4	0.8	0.3	0.2	0.04
Zapadni	Mean	56.8	19.7	9.1	5.2	3.1	1.1
	SE	2.3	0.6	0.5	0.4	0.3	0.1

Notes: Each location 5 ∞ 5 meters in size. Sample sizes: East Reef, 419 males; Zapadni, 232 males.

*Mean percent of all sightings at the first through eleventh most commonly used locations on the study site.

The success that males had at obtaining a particular site depended on the size and vigor of males already there. Newcomers could almost never drive a robust adult male from an established territory. At best they could force such a male to yield a piece of his territory, which they would then occupy. However, they could defeat and replace old, debilitated males outright.

Male Use of Space

Each territorial male was seen in relatively few of the numbered subsectors on the grid (usually 3–6 subsectors; table 4.1). The number of sites a male used depended on several factors. Males that were present for more than one year (multiyear males) used more locations than males seen for only a single year; males that copulated used more sites than those that did not; Zapadni males used more sites than males at East Reef (see table 4.1 for statistical results). In all cases the differences between groups were small but significant because of the large sample size.

Males favored some rest sites within their territories over others [13]. Our analysis (table 4.2) showed that males at East Reef were at a single location specific to them on 50–66% of the sightings. They were at a second location on 20% of the sightings, and all other sightings were distributed among five or fewer rest sites.

Males changed the way they used space as they aged, gained experience, and declined in physical prowess. We analyzed the records of multiyear males for age-related changes in the number of sectors they occupied, and in their tendency to reoccupy previously used sites (here called "site fidelity"). Site fidelity was measured using a ratio method [14]. We calculated population averages only through the fifth year of tenure because sample sizes decreased thereafter.

Table 4.3

Average Number of 5 ∞ 5 meter Sectors Defended per Tenure Year by Territorial
Northern Fur Seal Males and the Ratio of New to Previously Used Sectors

Year of Tenure	Sample Size (n)	Avg. No. Sectors Occupied	SE Sectors Occupied	Avg. Ratio New/Old Sectors	SE New/Old Sectors
East Reef					
First	204	4.40	0.16	1.00	0
Second	204	4.21	0.16	0.75	0.07
Third	101	3.84	0.21	0.27	0.03
Fourth	45	3.98	0.39	0.16	0.03
Fifth	19	3.84	0.43	0.15	0.05
Zapadni					
First	153	7.55	0.44	1.00	0
Second	153	5.92	0.32	0.90	0.11
Third	73	5.60	0.34	0.26	0.03
Fourth	34	5.62	0.56	0.16	0.04
Fifth	21	6.67	0.73	0.21	0.05

As a generalization, males defended more space when they were young than when they were older (table 4.3), no matter whether they were on the periphery or in the center of the breeding area. Assuming that they occupied the entire subsector in which they were seen, each would have occupied, on average, a maximum area of 110 square meters at East Reef and 189 square meters at Zapadni (table 4.3). However, most males did not defend the entire subsector at each location they occupied. Therefore, the actual defended space was smaller than these maxima suggest.

The usual history of territorial change was that males occupied whatever sites they could in year 1, returned to some of these same sites in year 2 while expanding to a few adjacent new sites, and in years 3 and later they used mostly previously held sites. Males that returned from four to eleven years used almost exclusively the sites they had used before. These sites were often delimited by terrain breaks, such as cliffs or boulders, which made them easy to defend. But these breaks also confined them to a small part of the breeding area, most often where few females resided.

The ratio method confirmed that few males made a large expansion in territory size after the first year. Only ninety-six males out of 2,307 male-years of data showed a great expansion (e.g., had a ratio 1; table 4.3). In 69% of these cases, the male moved into areas vacated by or won from neighbors, and in 22% of the cases a peripheral male shifted to territorial status by moving among the females. In 3% of cases the male was defeated in a territorial fight in year 1 and wandered the periphery in year 2.

Very few males changed the location of their territory entirely, despite the large number of territorial locations available, and the many territories that were abandoned by females when their numbers dwindled. Only 10 of the 1,542 males in this study established a second territory more than 10 meters distant from their first. Males that tried to do so but lost their new sites within 2 days were not counted as successes, nor were peripheral males that were establishing their first territory. Apparently, changing the location of the territory was not a common response to the disappearance of females.

We found a low but significant correlation between the location of a male's territory and the number of copulations he performed [15]. More of the variation was caused by males that had access to many females but mated infrequently than by males that had access to few females but mated often. Poor mating success resulted from the territory being too small to hold many females, females avoiding it because of frequent male fights, or the males harassing females so constantly that the females moved to other areas. Residence on good territories (among females) was a prerequisite for high mating success, but did not insure it.

Persistence in Territory Location

Adult males usually maintained a territorial site for life, even if it was not heavily used by females. After females abandoned all but two sectors of the Zapadni observation grid (fig. 2.13), peripheral males disappeared soon thereafter but adult males persisted for several years more. From 1983 to 1987, about 80% of Zapadni males had no contact with females during the main breeding season (fig. 2.3). Had these males been situating their territories to maximize reproductive success, they would have moved by substantial distances, but few did so. Males had no reproductive motive for defending these territories, except that a few females moved there in August after most mating had ended. The same individuals returned for several years but were not replaced by peripheral males, so that their numbers eventually dwindled.

Date of First Arrival for the Season

As males aged they arrived to reoccupy their territories within an increasingly narrow time frame. Early in their years, males arrived as late as 5 September, but by their third year of tenure almost all of them were arriving before 30 June (fig. 4.1) [16]. This narrowing could be seen in the records of long-lived individuals; it was not an artifact that resulted from the inclusion of young, highly variable males [17].

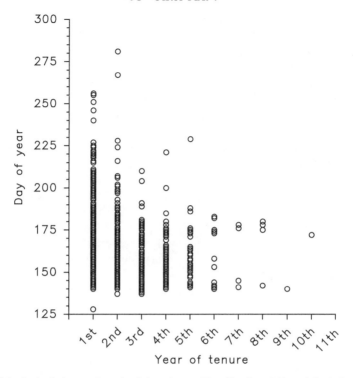

Fig. 4.1. Arrival dates of territorial males at East Reef and Zapadni study sites plotted by year of tenure in which the territory was defended. Plot includes 1,030 male-years of data from 357 multiyear males that had no years of missing data (see note 16).

Weight and Fasting

Males that were weighed in June as territorial tenure began averaged 208.7 kg [18]. The smaller animals came from peripheral areas where contact with females was unlikely. Males that were later seen copulating averaged 16 kg more than those that were never seen copulating, a significant difference. Nevertheless, there was no statistical correlation between the individual's initial weight and the number of copulations subsequently performed.

Based on three animals weighed at the start and end of territorial occupancy, males appeared to lose on average 32% of their initial body mass while fasting [19]. Their average mass loss over an average fast of 46 days was 1.2 kg per day, or 0.7% of initial body mass per day of fasting. Clearly a larger sample size is needed to characterize the species.

Males with the longest fasts remained from May through July. About 50% of them returned to their territories after an absence of 2–4 weeks (termed "post-territorial" behavior by Peterson, 1965). This second period was not included in fasting durations reported here.

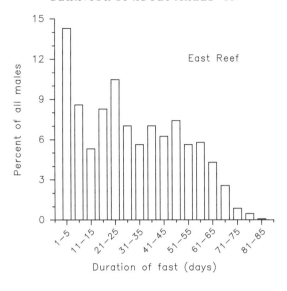

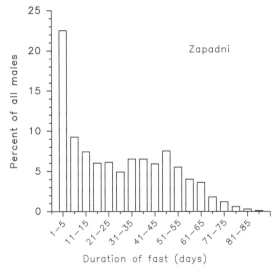

Fig. 4.2. The duration of fasting in 5-day increments for 869 males at East Reef (1,280 male-years of data including single- and multiyear males) and 672 males at Zapadni (995 male-years of data) from 1974 to 1986. Fasting was calculated as the single longest sequence of days each male was recorded on shore. Sequences could include any number of one-day absences but no absences of 2 or more days.

The average fast, considering all territorial males, lasted about a month (East Reef, 30.1 days; Zapadni, 27.4 days) [20]. The longest fasts we recorded were 81 days at East Reef and 87 days at Zapadni, but few males fasted for more than 60 days (fig. 4.2). The shortest fasts recorded, 1–5

Table 4.4

Duration of Fasting for Adult Male Fur Seals as a Function of (a) Whether They Were Seen Copulating, and (b) Whether They Spent More than One Year on Territory

	East Reef [a]			Zapadni [a]		
	n	Mean	SE	n	Mean	SE
Copulation[b]						
1 year[c]	239	32.8	1.0	129	31.9	1.3
≥ 2 years[c]	180	42.2	1.0	103	38.1	1.3
No copulation[b]						
1 year[c]	407	15.8	0.7	372	16.6	0.9
≥ 2 years[c]	43	19.3	1.6	68	23.3	2.5

Notes: Mean = mean duration of fasting in days.
Results of 3-way Kruskal-Wallis test:
[a] Site: H = 0.26, p > .05
[b] Copulation: H = 218.8, p < .05
[c] Years: H = 29.7, p < .05
Interactions: NS.

days, were also the most frequent class (fig. 4.2). These represented wandering, peripheral animals that spent few days at the study sites before disappearing, and contributed little to reproduction. The most successful breeding males in the population averaged 41–43 days on territory, and an average fast of 39 days [21]. Because of brief male absences, the days spent fasting represented on average 73–86% of the days of tenure.

Most of the mating was done by males that returned for several years and spent relatively long periods fasting each year. Males that copulated fasted significantly longer (by 21–25 days; table 4.4) than those that did not, and males present for several years fasted significantly longer (by 12–15 days; table 4.4) than those present for one year [22].

Unexpectedly, the duration of fasting seemed to change very little as males aged [23]. Males fasted for a shorter time during their first year on territory than in any other year. But after the second year, the length of the fast remained the same, about 41 days, until the male disappeared from the population. Therefore, there was no evidence that fasting duration increased with age, as Versaggi (1981) had concluded with size. Also, the duration of fasting did not increase to a maximum (at age 9) and then decrease again as Vladimirov (1987) had concluded with time on territory. Therefore, fasting duration did not seem to be linearly related to size in this species.

The length of time males fasted per year seemed unrelated to the size of the male population, that is, to the degree of male-male competition. The duration of fasting showed no significant shifts as the male population increased and then decreased in size [24]. It varied but showed no trend by calendar year.

Despite the fact that males that copulated fasted longer than males that did not, the number of a male's copulations could not be predicted from his fasting duration [25]. In fact, males that copulated the most fasted for about 30 days, whereas those that fasted the longest (80 days) averaged only one copulation [26]. Therefore, although fasting was necessary for reproduction, other factors (such as location), were necessary to insure it.

Behavior

We found no changes in male behavior that were profound enough to have caused the population changes that occurred during this study. During daylight hours over all years and sites, males spent most of their time (ca. 93%) "Inactive," less than 3% "Interacting with Females," and less than 2% defending their territories [27]. A few weak trends were found (table 4.5). Males at East Reef interacted with females more often than males at Zapadni, but the frequency of male/female interactions did not change with male population size. Also, males were most active ("Inactivity" was lowest) in years when the population was largest, and least active when the population was smallest, as would be expected. Finally, more male "Moving" was scored when the female population was large than when the male population was large. These were the only significant trends found.

The most frequent behavior that males used to defend their territories was the two-male display (82% of all acts recorded; table 4.6) [28]. The behavior related to territorial defense was virtually identical at East Reef and Zapadni. This similarity suggests that terrain differences did not affect the kind of behavior that males used to defend their territories.

True fights, the ferocity of which is often used to characterize the species, were infrequent (two hundred total, or 1.4% of recorded encounters). In all but 3 years of the study [29], males averaged fewer than one fight each. The most fights scored for an individual in a single year was eight (Zapadni, 1981). I saw one male die within an hour of losing a fight; no other similar deaths were reported to me or recorded on the history cards. The importance of fighting as a means of territorial defense, and as a cause of male mortality, seems to have been greatly exaggerated for this species, at least during daylight hours.

Fighting ability, measured as the ability to injure and push the opponent, did not assure fighting success. Males that won fights were often those that lost pushing contests and received more bites than they delivered. In these cases a resolute, unyielding bearing seemed more important to winning than the ability to inflict injury. Large size and great speed were also important to fighting ability. But both could be nullified by a small opponent making skillful use of terrain to keep an uphill advantage.

Table 4.5

Time Budget for Adult Male Behavior Taken from the History Cards of 1,542 Adult Male Fur Seals from East Reef and Zapadni Study Sites from 1974 through 1988

Component Behavior	Frequency	Percent	Behavioral Class	Percent	$H(per)^a$	$H(Site)^b$	$H(Per*Site)^c$
Prone	49,705	69.9					
Upright	15,026	21.1	Inactive	92.9	13.73*	2.71	1.31
Groom	1,373	1.9					
Interact w/ female	1,606	2.3	Interact /w Female	2.7	5.97	11.40*	2.73
Copulate	294	0.4					
Moving, general	969	1.4					
Moving, female	377	0.5	Moving	2.2	8.00*	0	0.16
Moving, male	202	0.3					
Boundary display	770	1.1					
Fight	202	0.3	Terr. defense	1.5	3.14	0.39	0.42
Chase	105	0.1					
Return	277	0.4	Depart/Return	0.6	1.31	0.10	0.61
Departing	159	0.2					
Drink	42	0.1					
Interact w/ pup	8	0	Miscellaneous	0.1			
Swim	26	0					
Disturbance	16	0					
Total	71,157						

Notes: Frequency represents all behavioral entries on male history cards. H values are the results of a two-way Kruskal-Wallis test. *Indicates differences significant at p < 0.05.

Significant comparisons: inactive, period 2a, 3ab, 1b; interact with female, East > Zapadni; moving, period 1a, 3ab, 2b.

[a]Per compares the data for time periods; 1 = 1974–77, 2 = 1978–83, 3 = 1983–88. These periods coincide with low, high, and intermediate male populations respectively (compare to fig. 2.2).

[b]Site compares the data from the East Reef and Zapadni study sites.

[c]Per*Site is an interaction between time period and site.

Table 4.6

The Behavioral Means Used by Adult Male Northern Fur Seals to Defend Their Territories

Site	Two-Male Display	One-Male Display	Multi-Male Display	Fight	Chase	Unreturned Bite
East Reef	82.1	12.2	1.1	1.0	2.4	1.2
Zapadni	81.5	13.9	1.0	1.1	2.1	0.4

Notes: Entries represent the percentage of recorded aggressive acts at East Reef (n = 5,759) and Zapadni (n = 9,049) study sites. The table pools 11 years of data collected before and after the male populations peaked (see text for dates and behavioral definitions).

The rate of territorial defense (measured as acts per male per 10 hours of observation) varied with male population size (fig. 2.2). Twice when the male population increased in size the rate of territorial defense increased significantly [30]. Once when the male population decreased, the rate of territorial defense decreased, but not by a significant amount [31]. Together these results suggest that males defend their territories appropriate to the size of the male population. The key variable may be the proximity of neighbors; males are closer together when the population is large and therefore territories are defended more often. That is, the frequency of territorial defense is determined by a male's immediate surroundings.

The rate of territorial defense did not change as males aged. We analyzed territorial defense relative to the number of years spent on territory because absolute ages were not known [32]. This result was unexpected because males defended fewer sites and stopped moving into new sites as they got older. This more conservative behavior might be expected to lead to less frequent territorial defense. Age differences may have occurred but may have been masked by the changes in male density that these males experienced.

The rate of territorial defense waned in the presence of females. The rate peaked before the first females arrived and declined thereafter (fig. 4.3). Statistical tests [33] showed that the arrival of only 50% of the females was associated with a significant decline in the rate of male territorial defense.

The reducing effect of female presence on male territorial defense could be measured in individuals. Our analysis showed that territorial defense decreased by about three displays per 10 hours when one or more females were in the territory (table 4.7) [34]. The effect of females was not additive; a regression analysis of male defense rate on number of females present showed a poor correlation. The difference between no females and a few was greater than the difference between a few females and many.

The effect of females seemed to result from a shift in attention by the territorial males. When no females were present the male attentively watched the movements of neighbors and displayed at them frequently. When females were present the males watched them (chapter 5) and largely

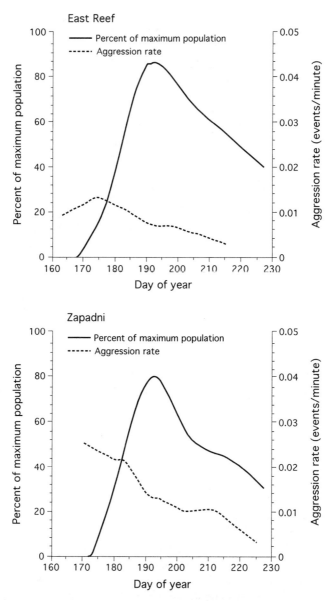

Fig. 4.3. The reducing effect of female arrivals on the rate of territorial defense among males. Curves for the female populations result from a Lowess smoothing procedure on the data presented in figures 3.2 top left and right. Curves for male territorial defense rate result from Lowess smoothing of daily rates for four seasons each (East Reef, 1981–84; Zapadni, 1977, 1981, 1982, 1984).

Table 4.7

The Reducing Effect of the Presence of Females on Male Aggression Rate
at East Reef (1981–84) and Zapadni (1977, 1981–84)

Number of Females	East Reef			Zapadni		
	n	Mean	SE	n	Mean	SE
0	274	9.0	0.54	527	12.0	0.54
≥ 1	505	6.6	0.3	365	9.6	0.72

Notes: Mean = mean number of aggressive acts per individual male per 10 hours of observation. n = number of separate observation periods involving known focal males. Wilcoxon Mann-Whitney two-sample rank test; East Reef, $Z = -3.56$, $p \leq .0004$; Zapadni, $Z = -4.37$, $p \leq .0001$.

ignored neighboring males except when they approached the mutual boundary. Males that were interacting with females near the territorial boundary vacillated between threatening neighbors and herding females, but usually did the latter. The decline in male territorial defense occurred too early in the season to be caused by fatigue. Some process such as distraction or selective attention seemed a more likely cause.

We tested whether males that defended their territory more often also copulated more. We found that in years when many females were present there was a correlation between the frequency of territorial defense and the frequency of copulation, but that this correlation was not present when there were few females in the population [35]. These results suggest that frequent territorial defense was of marginal value. It gave males a slight reproductive advantage when the female-male ratio was high or falling, but none when the ratio was low and stable.

Male/Pup Associations

To a greater extent than was previously recognized, males actively avoided injuring pups. They rarely stepped directly on pups with their full weight, but stepped on the ground and lifted their bellies so that pups were rolled, not crushed, by their passing. When two adults came together in a threat interaction and pups blocked their path, males stopped or hesitated before interacting in 22.5% of cases [36]. They never did so in interactions where no pups were present. Hesitation was a clear response to the presence of pups. It measurably reduced the number of pups contacted (rolled or stepped on) during adult interactions. Males contacted a pup in 29% of the interactions at which pups were present. Of these, 72 pups were contacted when males hesitated, and 351 were contacted when they did not, a significant difference [37]. The 423 pups contacted amounted to 7% of the pups in the area. None of them suffered obvious injury or death.

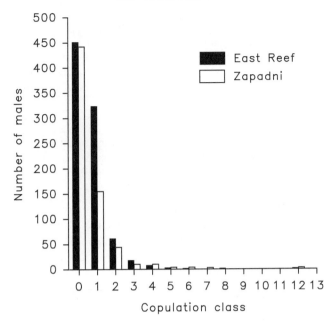

Fig. 4.4. Total lifetime copulations by adult males at East Reef (1974–88) and Zapadni (1975–84) study sites. Copulation classes are increments of ten except Class 0, which = 0 copulations (Class 12 = 111–120). N = 2,837 copulations for 672 adult males at East Reef; 2,317 copulations for 869 males at Zapadni (total of 1,541 males representing 2,274 male-years of territorial occupation). Average lifetime copulations: East Reef, 3.65 (SE = 0.28); Zapadni, 3.68 (SE = 0.39). Average copulations per year spent on territory (including males with zero copulations): East Reef, 2.48 (SE = 0.14); Zapadni, 2.49 (SE = 0.21).

Males tended to hesitate most during interactions that were least critical to territorial defense. Most cases of hesitation (45.2%) were during slow interactions with females, 24.7% occurred during male boundary displays, and 22.6% were during fast-moving interactions with females. Males were never scored hesitating or making a full stop during the most vigorous male interactions (fighting, stealing females, or nipping neighbors).

Lifetime Copulations

Males averaged only 2.5 observed copulations per year, which translated into 3.6 copulations for the average reproductive lifetime of 1.45 seasons (chapter 3) [38]. The range was quite broad (0–116 observed copulations; fig. 4.4). However, these values are not realistic measures of reproductive

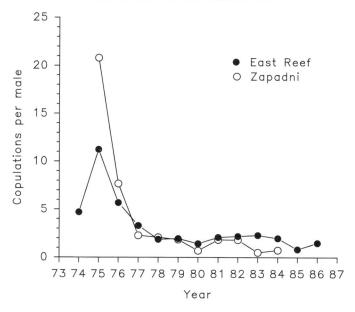

Fig. 4.5. Mean copulations per male at East Reef and Zapadni study sites by year. Sample sizes are given in the caption to figure 4.4. The male population averaged 95.2 animals per year per study site (SD = 40.7).

success because about two-thirds of females copulated unobserved at night (chapter 3).

Males that were seen for several seasons copulated about twice as frequently as males that were seen for only a single year [39]. In fact, many males that were present for only a single year were not seen copulating at all.

Copulation frequencies did not change with the relative age of males as reported for the Commander Islands population (Vladimirov, 1987). There males begin territorial occupation at age 7 years and peak in copulation frequency at age 9, an age that corresponds to the third year of tenure by our method. We found no such peak in the East Reef data, and a suggested peak in the Zapadni data was not significant [40].

As the number of females available per male in the population declined (fig. 2.4), the number of copulations per male also declined (fig. 4.5), as expected. Males breeding after 1977 experienced significantly lower average reproductive success than males before 1977 [41]. Both the declining female numbers and the increasing male numbers contributed to reduced numbers of copulations per male. The number of copulations per male was always smaller than the number of females per male that year because copulations were not recorded at night.

DISCUSSION

The male social system is marked by a high turnover rate. About 65% of all males identifiable on a site one year fail to return the next year. This proportion decreases to only 36% if the records of males that were identified for only a single year (mostly young) are removed. That is, about a third of the fully adult, experienced males change each year. Johnson (1968) found a 38% turnover rate using teeth from 405 males shot on territory and aged by tooth rings.

The turnover rate varied, but without trend, during this study. This lack of trend implies that the progressive changes in sex ratio and male/male competition that followed the end of the commercial kill were not intense enough to cause progressive shortening of the reproductive lives of adult males.

The territorial system is neither fragile nor susceptible to human disturbance, as was once believed. Males have strong proclivities to reside on specific parts of breeding areas and no other. Although they can be driven from these sites by humans (with varying difficulty depending on distance to the sea) they return to them faithfully, require few interactions with other males to do so, and will compete for no other site even when given the opportunity. This implies that the occupation of beaches by males is not a haphazard process but grows from an infrastructure in which specific males defend specific sites. Most of this infrastructure is repeated between years because two-thirds of the males return the next year and reoccupy most or all of their previously used sites. This specificity of males for sites gives the territorial system great resilience to human disturbance.

Male persistence on their sites over years is unexpectedly great. Most adults will not strategically shift to a "better" location if females vacate their territory. Failing to move seems maladaptive, but it is a common trait. They risk injury by moving and competing for a new site where they have no prior residence. Therefore, failing to switch suggests that the consequences of moving are worse to them than the consequences of staying. Their behavior is not as driven by the desire to increase copulations as we would expect (Miller, 1975).

Males are conservative in the amount of area they defend and the means of defending it. They usually occupy 3–6 subsectors (25 m² each) each year. They add a few adjoining sites in the second and third years of tenure, but thereafter defend only previously defended areas. This increasing conservatism may reflect declining reliance on fighting ability and increasing reliance on prior residence to defend territory. The advantage of prior residence may lie in learning to use terrain to supplement fighting ability. This switch may allow males to compensate for their declining physical prowess relative to younger, more capable opponents. Males arrive on their territories within an increasingly narrower (and earlier) window of time as they age. This allows them to precede other males and reoccupy their former site by prior residence rather than by fighting, and to precede the arrival of females. The frequency and intensity of the interactions needed to reclaim their areas, once driven off, differed between two study sites. Therefore, aggression in this species is not an innate quality but instead is shaped by aspects of the environment (number and closeness of neighbors, presence of females, time of season, and others).

Reproductive success is a more complex process for males than for females. Maximum female reproductive success is virtually assured by giving birth on a central breeding area in summer. However, males must survive longer than females before breeding, reach a certain size, and display a large number of behavioral and physiological factors to mate at all. Males that have larger territories, defend territory in the middle of the breeding area, have access to many females, weigh more, fast longer, stay on territory longer within the year, return for more than one year, and defend their territories more often tend to copulate more, on average, than males that score lower in each of these categories. However, high rank in any one category does not insure high copulation frequency, as shown by the lack of statistical correlation between any single category and observed copulations (except territorial defense at some sex ratios). Males must favor neither male/male nor male/female interactions to the exclusion of the other. They must not drive females away by persistent interactions, and they must detect and respond to estrus in order to mate. A low copulation frequency can result from an inadequacy in any one of the above categories, but a high frequency can only result from simultaneous success in all. All of these results are consistent with the conclusions from studies on other otariid species (Miller, 1974, 1975; McCann, 1980; Gisiner, 1985), and may be generally true for the family.

Few males ever achieve extremely high reproductive success. One male out of 1,542 was observed copulating 116 times in his lifetime, a minimal value given that most copulations occur unobserved at night. By contrast, we observed hundreds of males that had no copulations at all. The average for all males was 2.5 copulations per year, or 3.6 per lifetime. This average may be similar to that of all females, but the maximum attainable and the variability are greater. Male reproductive success is much more linked to the sex ratio than it is for females.

A question central to this research program was whether some change in behavior of animals could have caused the observed population decline (Anon., 1973). An analysis of eighteen components of male behavior, grouped into five broad classes, showed some correlation with population size. Territorial defense is the only category that could have affected reproductive processes, and it varied directly with male density. However, territorial defense is not likely to have caused a population decline because most of it occurred before reproduction began each year. In fact, presence of females reduced male/male encounters. In short, some aspects of male behavior changed with herd size but probably resulted from them rather than caused them.

Infanticide was not a reproductive strategy in this species as it is in some other mammals. Adult males do not intentionally harm pups, and make obvious efforts to avoid harming them accidentally. Perhaps males aid the survival of all young because, since they and the females they inseminate use the same site for several years, they cannot discriminate their own from the young of other males.

SUMMARY

Males defended exclusive, nonoverlapping territories with clearly defined boundaries. Individuals competed for only a single, very specific site, not for any site that was available. Because of this specificity, males that were experimentally driven off their territories quickly reoccupied the exact site, largely unopposed. The territorial system can be looked on as having an infrastructure in which the individual and the location are nearly synonymous. Site specificity reduced male competition for a given site and made the male breeding aggregation highly resilient to disturbance.

The turnover rate from one season to the next was about 34% for breeding males and 66% for all males. However, little turnover occurred from day to day within a season.

The amount of space a male defended varied with age, terrain, whether they had copulated, number of rival males, and previous years on territory.

Fasting ability affected reproductive success because fasting limited the time that males could defend territories among females. Fasts ranged from 1 to 87 days. Males that copulated fasted longer (39 days) than those that did not (30 days). But long fasts did not equate with a high copulation rate. The duration of fasting did not change with body size, age (after the first year), or changes in size of the male population. Males began fasting at an average body mass of 209 kg, and lost 32% of that mass while fasting (0.7% of initial mass per day).

Males adhered to established territories for years after females abandoned them. Only 10 of 1,542 males abandoned a territory and established a new one more than 10 meters from the first. That is, reproductive success did not seem a proximal motive in establishing or maintaining a territorial location.

Territorial defense passed through stages with age. Young, vigorous males defended as large an area as possible, given the limitations of terrain and other males. The location of females within their areas determined where they spent most of their time. In succeeding years they expanded onto areas adjacent to formerly defended sites, and then late in life stopped this expansion and used only the formerly defended sites. They arrived earlier in the year as they aged, which may have allowed them to precede other rivals on preferred sites, thereby using prior residence as an alternative to fighting for space. It may also have allowed them to precede females. Very old males defended small enclaves that were well defined by terrain but contained few females.

The behavior that males used to defend territory and the frequency of defending it depended on several factors. It was more frequent on flat than on rough terrain. It was greater when the male population was large, and it was less frequent in the presence of females. Therefore, male territorial defense responded to the environment and was not an inherent attribute of the species. Territorial defense did not diminish as males aged.

The most frequent means of defending the territory was a two-male display (82% of all male encounters). True fights were very infrequent (1.4% of encounters), and fighting ability (the ability to injure) did not assure fighting success. The role of fighting in territorial defense has been greatly exaggerated.

Females had a reducing effect on the frequency of male territorial defense. At both the population and individual level, male threat rates were lower when females were present than when they were not. Female presence or absence was more important than the absolute number of females present.

Male reproductive success was highly variable. Males averaged only 2.5 observed copulations per year and only 3.6 per lifetime (range 0–116 copulations per lifetime). True values are undoubtedly higher because we did not observe at night, when two-thirds of the copulations occurred. The copulation frequency did not change with male age, but it changed greatly with the sex ratio.

Males copulated more if they (1) had large body size, (2) had a large territory in a good location, (3) fasted longer, (4) had longer territorial tenure within and over years, (5) maintained both male/male and male/female interactions, (6) did not interact too often with females, and (7) had a high frequency of territorial defense (only at high sex ratios). However, possessing any one of these traits did not insure high copulation frequency. Males achieved high reproductive success only by exceling in all of these categories simultaneously. Inadequacy in only one area could reduce reproductive success substantially. Variability among males attested to the difficulty of balancing all of these factors.

Males sometimes inhibited interacting with other males if they ran the risk of injuring pups. They also walked in a way that reduced injury in the pups they contacted. The concern that male numbers would have a detrimental effect on pup survival in a normal herd (one lacking a kill for pelts) are probably groundless.

This study found no evidence that changes in male behavior caused the population decline from 1956 onward. Some aspects of male behavior changed with population size—but only following these changes, not preceding them.

CHAPTER 5

Male-Female Associations

THE DOCUMENT that established this study (Anon., 1973) questioned whether the population failed to recover from the herd reduction program because past kills had created a suboptimal sex ratio that affected the population through the ways males and females associate. Our study was not designed to answer this question directly because stopping the kill of males at St. George Island was expected to reduce the sex ratio, not increase it to its former level. However, the behavioral study could describe the determinants of male-female associations, and from that, judge whether a suboptimum sex ratio was likely to have developed in the past.

The data we collected took into account the range of interpretations of otariid mating systems that have been offered over time. The system was originally interpreted as a male-dominated "harem" system. More recently, it has been suggested that female mate choice may play a previously unsuspected role in otariid mating systems. We collected data that could address this wide range of interpretations.

The earliest writers held that northern fur seals had a male-dominated mating system (Steller, 1749; Veniaminov, 1839; Scammon, 1869, 1874; Bryant, 1870–71; Allen, 1880; Elliot, 1882; Nutting, 1891; Stejneger, 1896). The term "harem" was intended as an analogy to male-dominated human harems (Peterson, 1968). A harem was defined as an isolated group of females that was actively gathered by a male and kept separate from other groups by male herding behavior. Females were considered to be passive social agents, and males were considered to have exclusive reproductive access to all harem females. The slang terms "harem master" and "sultan" hinted at the social relations believed to exist in a harem.

The earliest modern writers on northern fur seals perpetuated this harem view of social organization (Bartholomew and Hoel, 1953), and laid heavy emphasis on the male in a model of the evolution of polygyny (Bartholomew, 1970). This emphasis persists in newer versions of the theory (Stirling, 1983). Although the harem concept was disavowed by Peterson (1968), the term still lingers (Jouventin and Cornet, 1980). This chapter addresses the harem issue through a search of daily census and distribution maps for evidence of "harem" formation.

The question of female mate choice in northern fur seals was first considered during the last century. Bryant (1870–71) stated that females did not seek particular males for mating. His work sparked a disagreement between

Darwin and Nutting (cited in Nutting, 1891) over this issue, a surprising development since neither of them ever observed the species firsthand.

Historically, female mate choice has been considered to be either direct or indirect. An indirect choice is considered to have been made when a female chooses a breeding area where males of high fitness reside but does not actively choose a specific partner there. Indirect choice is not a theory but a tautology (Peters, 1976); an indirect choice cannot be measured, only inferred, and therefore its existence can never be disproved. Whether indirect choice is a reasonable inference about the northern fur seal system can be judged by looking for factors other than males that affect whether a female uses a site (see below and chapter 7).

Direct mate choice is measurable in several ways. One way is to test whether females are sexually receptive to only certain individuals or classes of males (addressed in chapter 8). Another is to record whether over their lifetimes females copulate at many locations, indicating they are following the locations of "fit" males, or at the same site regardless of the male there (addressed in chapter 6). Still a third way is to look for evidence that females repeat mating with the same individual male in different years of their lives (present chapter). If a female actively chooses a male because of some phenotypic character that indexes fitness, she should choose that same male in succeeding years because that character and his fitness should remain relatively unchanged.

Boness (1991) asserts that some otariid systems that were formerly interpreted as territorial systems are in fact leks that are driven by female choice. We looked for evidence of leks in northern fur seals by comparing their characteristics against those of known lek-breeders, specifically: (1) males provide no parental care and control no resources vital to females, (2) males aggregate where females come to mate, (3) females have the opportunity to compare males, (4) males may evolve elaborate displays or coloration relevant to these comparisons, (5) male displays or coloration, and female choices for them, need not be adaptive, (6) a few males do most of the mating, and (7) females may compare several leks while making their choice (Bradbury, 1981; Queller, 1987).

Given all the above considerations, this chapter addresses a diverse group of questions: How are males and females disposed toward each other? How does each sex acquire mates? How do male-female associations result in the kinds of social groupings seen on breeding areas? Which sex affects those social groupings the most? How does male behavior change with sex ratio and period of exposure to females? Do males display for females? In this book, male-female relations are treated in empirical rather than evolutionary terms because this approach begins with fewer assumptions and may have more lasting value, as theories of mating systems come and go.

METHODS

If harems are the basis of the mating system they should predominate at the peak of mating in most years. We searched for appropriate groups (small female groups attended by a single male and separated such that no inter-group movement is likely) on the daily census and distribution maps discussed in chapter 2.

If leks exist, then females should somehow appraise males at the time of choosing a parturition site. We collected data on the ways females arrive and select parturition sites, and the ways males and females behave toward each other during those selections. We followed focal animals for brief periods, and recorded the rates of some interactions at East Reef and Zapadni in 1974, 1975, and 1977.

Prior to this study, the behavior of males toward females was reported in broad, nonspecific terms. We quantified it according to several specific classes [1]. We used a scan sampling technique to quantify the behavior of males toward resident (not arriving) females.

We used only selected focal males, about ten per site per year, in scoring these male-female interactions. All were large males defending territories among the females. During uninterrupted one-hour blocks, observers scored the categories that best fit the male's behavior every 5 minutes [2]. No data were collected before the male population reached its peak (1978), so the effect of high-population sex ratio on male-female interactions could not be tested. As a substitute, we analyzed the data according to the number of females in the male's territory during each sample (see note 2).

The question of how males acquire mates was partly addressed by recording female stealing (kleptogyny) throughout the study. In these events, a male would leave his territory, seize a female in his teeth, and carry, drag, or throw her back into the territory (fig. 18 in Peterson, 1968; fig. 6 in Francis, 1987). Females were almost always injured and sometimes killed this way (Peterson, 1965), and pups were sometimes trampled (chapter 4). Francis (1987) estimated that a female had about a 50% chance of being stolen at least once in her lifetime.

We recorded all instances of kleptogyny to determine whether it varied with population size [3], and compared the copulation frequency of males that stole females against that of the male population at large. We tested the hypothesis that males that steal females have few females in their territory. We also tested the hypothesis that males that steal females have low copulation frequencies (e.g., they are peripheral males with little access to females).

Female mate choice, as indicated by repeat copulations between the same individuals, was tested by identifying the male with which females associated at estrus in each year of their reproductive lives [4]. Most copu-

Fig. 5.1. Groups of females attended by single males on Zapadni Beach below the study site, St. George Island, in late June 1975. Females arrived singly at many points along the beach and formed into nuclei around resident males. Such nuclei probably gave rise to the "harem" concept of the mating system.

lations occurred unobserved at night, so the parturition site was used as a proxy for the location of copulation. This is a reasonable substitution because most females usually did not change locations between parturition and copulation. Repeat copulations were also found on our list of observed copulations for known females.

RESULTS

"Harems"

Female groups that could be called "harems" (but which are here referred to as single-male groups) occurred every year of the study (fig. 5.1). At East Reef (a shoreline breeding area) they formed early in the season (mid-June) and then the groups fused as new females arrived (fig. 5.2A). By 12 July, the median date of mating, virtually no single-male groups existed; female groups were large enough to span the territories of several males. By the first of August, females formed into a single, unbroken band along the beach that encompassed the territories of most adult males.

At Zapadni (inland breeding area), single-male groups developed only after the peak of mating. Early in the season arriving females formed a solid mass that had no divisions along the territorial boundaries between adjacent males (fig. 5.3). Females moved freely anywhere there were other females. After most mating had occurred and females were making foraging trips, the female mass progressively loosened and eventually separated into small, discrete-appearing single-male groups by the end of the season (fig. 5.2C).

The existence of single-male groups varied with population size. As the population declined from 1974 onward, the single-male groups at East Reef persisted longer into the breeding season than when the population was larger. In 1988, most single-male groups there were still traceable by 12 July, and many were still separate by 24 July (fig. 5.2B). The Zapadni population declined too fast to make a comparable comparison.

Leks

Males did not advertise their physical attributes as would be expected in a lek system. If male displays and calls served as advertisement (Boness, 1991), they should have increased after females arrived. Instead, the arrival of females suppressed the number of male displays (chapter 4). Most male displays and calls were directed either at neighboring males or females that had already selected a rest site on shore. In fact, males often lay prone, silent, and inconspicuous at the critical moment when newly arrived fe-

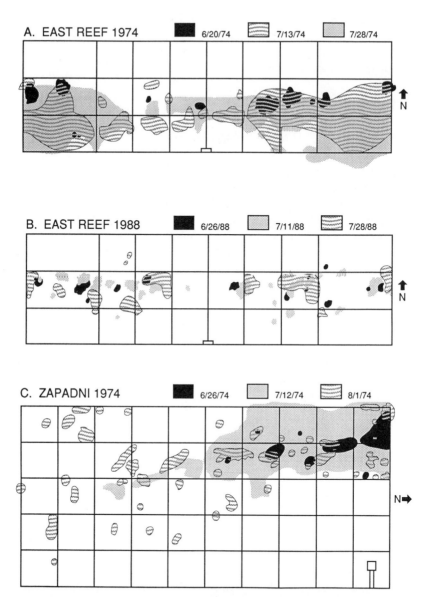

Fig. 5.2. The effects of time of season, population size, and terrain on groups of females that were attended by a single male (small irregular circles). Panel A shows that at large population sizes, single male groups occurred early in the season and coalesced into larger groups later in the season. Panel B shows that at small populations, single male groups persisted throughout the season. Panel C. shows that on flat terrain, single male groups formed only late in the season.

Fig. 5.3. Distribution of the female population at Zapadni study site on the same day shown in figure 5.1. Females arrived in masses, forced by terrain through a single access corridor that is out of view. Females formed into a single aggregation without regard to male territorial boundaries.

males were selecting their rest sites. Arriving females first scanned the shore as if searching for landmarks. They would land on a male's territory acting tentative and watchful, and would flee back to sea if any nearby male moved (not quantified). In this sense, they behaved as if male presence was aversive to their use of a land site. The male tendency to remain inconspicuous when females were arriving seemed appropriate to the female tendency to flee from males. Male colors (black, gray, and brown) did not offer a strong contrast to the substrate that would attract females. Males controlled access to a major resource needed by females—the parturition sites.

Female Selection of Parturition Site

By watching the eye movements of landing females we could tell that they were visually scanning the terrain, not the males holding territories there. Once on shore, they did not randomly search for a rest site but instead appeared fixed and intent on movement to a specific inland destination (most moved perpendicular to the shore line). If males blocked their progress, they paced while facing inland, somewhat like migrating birds showing *Zugunruhe* (Wolfson, 1958). They entered male territories by default since these covered all the traditional parturition sites, but they treated males as impediments to their movement.

Most males did little to actively bring arriving females into their territories. They would wait until a female entered their territory, then move to cut off her path of retreat, in effect trapping those that had already committed to entering. Later, after males had acquired a nucleus of females, they tended to watch new arrivals from the inland side of the group without interacting with them. The number of males that females contacted at arrival depended on the layout of the breeding area (1.6–4.4 males per female arrival at East Reef and Zapadni, respectively) [5]. But, females had on average less than one interaction with each male (0.6 at East Reef, 0.2 at Zapadni) (see note 5). Inexperienced peripheral males were an exception. They would sometimes seize passing females and force them to remain in their intertidal territories for a few hours, interacting with them constantly.

Most males actively suppressed the attempts of the first arriving females to leave again (usually uphill or inland). They used open-mouth threats, blocking behavior, striking, and biting to keep females from departing for more inland areas. They stopped threatening whenever the females adopted a prone position. Males spent about 14% of their first hour with a female actively blocking (n = 3 males). Blocking did not continue after the female gave birth (often overnight) because new mothers rarely moved.

Male blocking waned quickly when many new females arrived per hour. About 77% of male blocking attempts at Zapadni occurred in the first half hour to one hour of contact. While blocking, males tended to lose more females than they gained (table 5.1, under "Blocking," Lost > Gained for two of three males). After males stopped blocking they tended to gain more females than they lost (table 5.1, under "No Response," Gained > Lost for two of three males). Therefore, males seemed to successfully establish an initial nucleus of females by blocking them. But after the group reached a certain size, perhaps ten females, blocking behavior apparently became counterproductive and decreased.

Discreteness of Female Groups

After an initial nucleus of females formed, the subsequent growth, movement, and density (see chapter 6) of the group were determined by females. When female groups were new, arriving females found rest sites on the group's outer edges. About a week after the group formed, when some

Table 5.1

Effects of Three Kinds of Male Behavior on Changes
in the Number of Females in the Territory

	Blocking[*]	No Response	Fight
	[Male C3, 1.5 hours observation, 4 females at start]		
Retained	4/6	—	—
Gained	4/6	22	—
Lost	4/6	19	—
	[Male HH, 1.2 hours observation, 25 females at start]		
Retained	4/8	—	—
Gained	0/8	3	0
Lost	5/8	7	7
	[Male BC, 2.0 hours observation, 1 female at start]		
Retained	10/17	—	—
Gained	8/17	32	17
Lost	51/17	21	1

Notes: Blocking = male herding behavior; Fight = any kind of male/male interaction; Retained = residents remained; Gained = new arrivals joined the group; Lost = residents departed. Denominators are the number of blocking attempts.

[*] Of the thirty-one blocking attempts for these three males, twenty-four (77.4%) occurred in the first half of the observation period for each male.

group members had mated and left for sea, new arrivals found parturition sites in the group's center. Males had no effect on this pattern. They could temporarily force females to lie where densities were unusually high (Francis, 1987).

Whether groups remained separate from each other depended on the behavior and number of females. If enough females arrived and sought parturition sites, the corridors separating groups closed despite any male effort to keep females within territorial boundaries. If not, the corridors remained open because of female reluctance to cross open spaces, not because males successfully prevented it. In 1974, when males were sparse and female were numerous, single-male groups of nearly two hundred females formed, often covering 100 square meters. These groups often had discrete boundaries, a low rate of intergroup movement, and high female densities. A single male could not have imposed these traits on so many females at once. Instead, females remained together as a cohesive group by mutual inclination.

The tendencies of females not to cross open spaces, not reside in territories that lacked other females, and not to rest far from other females all decreased throughout the season. All these tendencies were most pronounced early in the year when females had not mated and when male

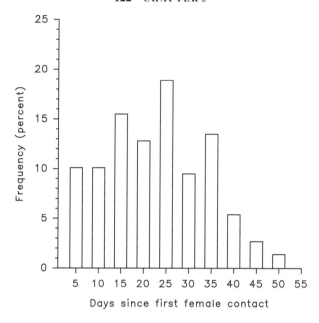

Fig. 5.4. Total days spent in contact with females during 148 male-years of territorial occupancy (122 different males).

attention to females was most assiduous. When females returned to shore from their first foraging trips, they spaced farther from other females and began to move more among groups. Simultaneously, males began to reduce their herding behavior (Francis, 1987). The cumulative effect of these changes was that groups gradually became less discrete and eventually merged entirely. This process took 6 weeks to spread across an entire central breeding area because females mated and made their first foraging trips in that period.

Male-Female Interactions

Almost all male-female interactions could be characterized as dominance-subordinate relations. Females sought mainly to evade males. Males sought primarily to assess female receptivity and secondarily to suppress female resistance and movement. Females acted subdued around males and moved submissively (chin lowered, neck arched) without making eye contact. When confronted, they faced the male with an open mouth and sidled away defensively. They bit only when they were being stolen or during copulation. They were more aggressive after they mated, sometimes making offensive open-mouth lunges at males. Females rarely approached

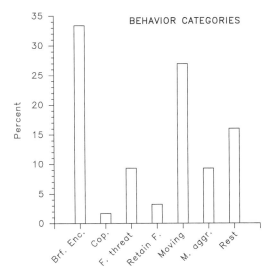

Fig. 5.5. Percent occurrence of seven classes of male behavior in 1,024 hours of focal male observation (n = 4,993 interactions for 122 different males). Brf. Enc. = brief encounters; Cop = copulation; F. threat = female threatened; Retain F. = male retains female; M. aggr. = male aggressive encounter. See note 1 for definitions.

males. But they allowed males to sleep beside them if the male initiated no interactions. Therefore, females appeared to evade male-female interactions and not male proximity.

Males appeared to assess receptivity by sniffing the nose or open mouth proffered by females (chapter 8) in most interactions. They also appeared to sniff the ano-genital area and threatened, bit, or struck any females that resisted these attempts. Male-female interactions were relatively simple and redundant. Females avoided males that interacted with them too frequently. Young males seemed more liable to lose females this way than older males.

The number of days that males had contact with females during a season was equivalent to about 80% of their total days of territorial tenure. Most males (89% of sample [6]) had contact with females for ≤ 35 days per year (fig. 5.4). The males spent the other 20% of their time ashore establishing territory by interacting with males before the females arrived.

About 48% of all male behavior, grouped into the broadest possible categories, was female related (fig. 5.5) [7]. The most frequent of these was the "Brief Encounter" (see note 1 for definitions) in which the male investigated the female's nose or open mouth but showed none of the aggression that typified other male-female interactions. Aggression was most common when males threatened or herded females ("♀ Threat" and "Retain ♀"), but it could occur during any interaction, including copulation. Males

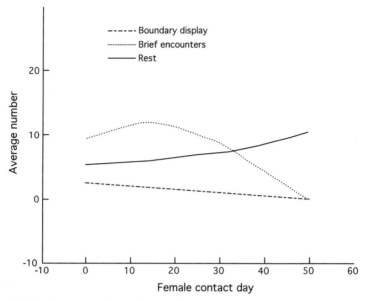

Fig. 5.6. Trends in three categories of behavior—Boundary Display, Brief Encounters, and Rest—by day of contact with females. Data for each category for all 122 males were plotted and lines were drawn using Lowess smoothing.

threatened females about as often as they threatened other males, a previously unreported finding.

Three classes of male behavior changed as a function of time spent in contact with females (fig. 5.6). The number of "Brief Encounters" declined (especially for males with more than 50 days of female contact); the number of "Boundary Displays" also declined (mean rate of 1.7 per hour); and time spent "Resting" increased [8]. That is, males stopped interacting with other animals and spent more time inactive.

Only one class of behavior ("Total Movement") changed as a function of time spent on territory, irrespective of contact time with females. Males moved more body lengths per hour when the population included more estrus females (fig. 5.7) and moved less late in the season after most females had passed estrus.

The declines in "Total Movement," "Brief Encounters," and "Boundary Display" and the increase in "Resting" were all signs that males were losing vigor. Other signs were that they lost body mass and developed loose folds of skin on the neck and chest. A few days to a week after males reached this state they abandoned their territory for the season.

None of the classes of male behavior showed trends when plotted against the number of females present. That is, the presence of females affected some male behavior, but the effect of females was not additive.

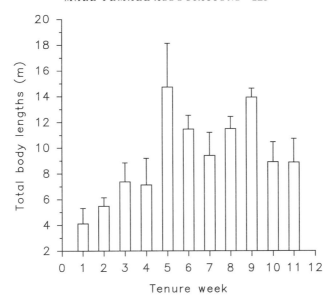

Fig. 5.7. Male movements in body lengths per hour as a function of time on shore. Length of movement was estimated for each of 122 males during 1,024 one-hour observation periods, and the results were collated by the day of tenure on which they occurred. A mean and SD (vertical bars) were calculated for sequential 7-day intervals throughout the stay on land.

Kleptogyny

Female "abduction" was a more accurate term than kleptogyny, because females were not always stolen from another male's territory. They were sometimes taken from the surf zone, or while in transit through male territories. The male usually seized the female by the skin of the back which always punctured and sometimes cut long gashes in the skin. Sometimes, early in the season when few females were present, two males would seize the same female simultaneously and tug in opposite directions, peeling back large flaps of skin and blubber. We saw three females killed in this way (see also Peterson, 1965; Francis, 1987), too few to correlate with population size.

Abductions of females were relatively rare (n = 403) [9]. About 15% of identifiable males took a female at least once in their lifetime. Abductions were normally distributed within the season (fig. 5.8), and peaked between 5 and 10 July, when most females were on shore. Most abductions occurred when males had few females in their territory (in 50% of cases 0 females were present; in 85% of cases ≤ 5 were present; in 2% of cases   20 females were present). Many abductions occurred during major disturbances, such

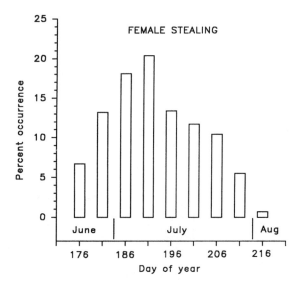

Fig. 5.8. The distribution of female stealing (kleptogyny) by males throughout the season. The figure combines 408 instances recorded during eight seasons at East Reef, two at Kitovi, and five at Zapadni between 1977 and 1988.

as fights or the forays of young males, when the territorial males were distracted from guarding their borders.

Males abducted the closest females, rather than those with immediate reproductive potential. Only 9% of abducted females were sexually receptive at the time (see note 9). This result suggests that males abducted females not for immediate copulation, but for some more general reason, such as to start a nucleus of females in their areas. The fact that some males abducted several females (mean = 1.6) was consistent with this interpretation (see note 9).

Abducting females was a moderately successful tactic for acquiring residents. Exactly 50% of 280 females that were checked at 1, 2, 4, 8, and 24 hour intervals after being abducted were still in the new position. We did not followed them for more than 24 hours because most females using a site that long remained to give birth or mate.

Males that abducted females were not always those that had low copulation frequencies, that is, marginal males with poor access to mates. In years that males abducted at least one female, they averaged 2.6 copulations (SE - 0.27, range 1–22), a rate that was close to the population mean (2.5 copulations per year for all males; chapter 4). That is, in terms of reproductive success, males that abducted females represented a typical cross section of all males in the population. The data do not indicate whether males that abducted females were younger than other males.

In summary, abduction did not happen often in the population, it was done by a small number of reproductively typical males multiple times within a season, and it may have served to start new female groups.

Female Choice through Repeat Matings

Relatively few females demonstrated mate choice by sharing a subsector (assumed mating) with the same male in different years. On average, females shared a subsector with 1.2 unique males per year of mating [10] (see also note 4). Only 29 females (of 286) shared a subsector with the same male in two different years. Because of the turnover rate among large, adult males (average rate of 37% for East Reef and Zapadni; chapter 4), only 106 of the 286 females in the sample were likely to have had the same partner available to them in 2 years. Therefore, the 29 females that paired with the same male twice represented 27.4% of the females that had the opportunity to do so. This is a conservative estimate [11].

The conclusion that mating with the same male in different years was not common was supported by copulation data for the population at large. Of 113 known females that were seen copulating over the years, only 8 were seen copulating in two successive years, and only one of these (12.5%) mated with the same male twice (result not corrected for the repeat availability of the same male). The reason that so few known females were seen mating was that most matings occurred at night.

These two lines of evidence suggest that actively choosing the same male in succeeding years, perhaps because of that male's fitness, was not widespread. At most it occurred in a quarter of the females that had the opportunity to do so.

DISCUSSION

Male-female associations were relatively simple. Females avoided males and complied with male attempts to investigate them sexually. Males were intent on assessing receptivity in large numbers of females and were aggressive toward those that moved or resisted investigation. The most frequent interaction between the sexes (brief nose-to-nose encounters) were usually not aggressive, but all others were (males threatened females). We did not measure male-female interactions over years as the sex ratio changed. But measures within a season showed no major changes in male behavior as the number of females in their territories increased.

The "harem" concept of the mating system was not supported by data from maps. If single-male groups dominated the mating system, they would exist at the peak of the mating season. But in fact, they occurred only before or after the peak, or any time females were sparse. Over historic time this condition has not occurred often. Therefore, most mating has probably occurred when single-male groups did not dominate the breeding aggregation.

Open corridors between female groups, formerly thought to signify harem boundaries, are in fact only artifacts; they exist only when the population of females is small, such as early in the season. These corridors contribute little to reproductive isolation of groups except early in the season. Corridors exist because females are reluctant to cross open spaces, not because males effectively stop female movements using herding behavior. Corridors sometimes exist around groups that are far too large for males to control, and they may persist longer into the year than male territoriality. Corridors do not infer reproductive exclusivity of groups.

If groups do not represent harems, then what is their significance? They may reflect female attempts to reduce the risk of injury from males. Attraction to other females, avoidance of open spaces, and avoidance of being the only female in a territory all tend to reduce the local sex ratio and thereby the risk of injury that attends each male-female interaction. Females actively seek the centers of groups to avoid males (Francis, 1987). Groups undergo a slow transformation throughout the season from small, dense, discrete knots to a large, loose, amalgamated mass. The immediate cause of this change is that individuals space farther from neighbors after estrus than before it. The ultimate cause may be that postestrous females no longer attract much attention from males and therefore need less of the protection that groups afford. The docile demeanor of females while complying with male sexual investigation may also protect them from the aggressive retaliation that resistance triggers.

Northern fur seals lack the kinds of behavioral displays, physical appearance, and social interactions that typify lek breeders. Males are cryptically colored and have no special displays for females that could have advertisement value. Male-male threats that could serve that function are suppressed when females arrive. Males remain inconspicuous at the very time when a display would best serve a lek breeder, namely when arriving females are choosing a mate. Finally, males control access to a female resource, namely parturition sites.

Females do not show attraction to or compare males as in leks. Instead, they show wariness and avoidance of males. If females were using interactions to compare males, they would be expected to venture into the open and have more than the 0.2–0.6 interactions per male that we observed.

We found poor evidence for direct female mate choice through repeat mating of the same pair; only 27.4% of females with the chance to do so mated with the same male in different years. Their failure to be more selective could be interpreted as another means of reducing the risk of injury from males. Evasion or resistant behavior by females often leads to injury from males. Perhaps the benefit females would gain by rejecting some males in favor of others would be offset by injuries they receive in the process. By accepting any available male, females could quickly dispense with the need to mate and simultaneously keep male-female interactions, and risk of injury, to a minimum.

For males, the frequency of investigating females was a balance between the risk of losing females through harassment and the risk of losing reproductive success through failure to detect estrus. Males may learn this balance through experience, as females avoid males that interact with them too often. This avoidance shows that female choice may exist in the species, but only in a negative sense. Also, the effect is small; few males per season are avoided by females, and usually only briefly because they quickly change the way they interact with females.

Females have access to males of high fitness without choosing individuals directly. Males compete for space among females wherever they land. This competition insures that females, through no mate selection process of their own, have available as mates only males that other males cannot exclude. Presumably such males are the most fit, although no measures exist. Male competition may restrict genetic variability among potential breeders more than any mate selection that females could make from among the survivors, and more than among females as a group (Miller, 1975).

The key to females having access to males of high fitness is to arrive at highly predictable times and places (chapters 3, 6) and not to deviate from that pattern over years. Persistence and predictability on the part of females provide the basis of male competition on traditional parturition sites. Thus, females assure contact with the winners of male competition merely by gathering predictably.

The mating system is neither male nor female dominated. It is an interaction of different tactics used by the two sexes. The females' tactic is to gather predictably and persist in a given pattern through most of their lives. The males' tactic is to compete for access to the sites that females predictably use, relying on large size, fighting, and fasting abilities that evolved relative to this competition. The winners of the male competition apportion the resource (parturition sites) on the basis of territory, and remain on their territory for a lifetime, even if females abandon it (chapter 4). From the female standpoint, dense female groups minimize contacts with males; from the male standpoint, they increase the number of potential mates in the territory.

SUMMARY

The "harem" concept does not fit northern fur seal mating aggregations because, demographically, single-male groups are the exception rather than the rule. They predominate only when females are sparse. Corridors between adjacent female groups are not harem borders but artifacts of the population size having little if any reproductive significance. The harem concept implies a male-dominated system, which northern fur seals do not have.

The "lek" concept of mating systems also does not fit northern fur seals. Female arrival suppresses most male displays that would be useful for self-advertisement. Also, males lack conspicuous colors and adopt inconspicuous behavior at times when lekking males would display. Most importantly, males control access to a female resource (parturition sites). The northern fur seal mating system is not as female dominated as the notion of the lek implies.

Female mate choice, as measured by the tendency of females to mate with the same male in different years, is weak. Only 27.4% of the females that have an opportunity to repeat mating with the same male in different years do so. Females do not actively choose males, although they occasionally avoid one that herds excessively, and they never approach males in a sexual context. Females do not choose mates but may choose parturition sites.

Most males gain females not by actively gathering them, but by blocking the departures of volunteer residents. Most males have few if any interactions with arriving females, as fits female antipathy toward these interactions. Male herding declines when a small nucleus of perhaps ten females is present (usually in the first hour). All mate-finding relies on males initiating frequent, brief investigations of evasive females.

Much of male behavior is specific to insuring contact with females. About 80% of a male's time on shore occurs after females arrive; about 48% of all male behavior is related to interacting with females. The frequency of male interactions with females decreases over time, as does the distance they move while interacting with females. Male behavior did not change with local sex ratio, although the ratios observed here were much less extreme than during the herd reduction program.

Avoidance of male-caused injury is a major factor in female behavior, with important implications for the mating system. Their docile demeanor during interactions with males, attraction to other females, formation of groups, avoidance of open spaces, avoidance of being alone with a male, and decreasing density late in the season may all be a result of females avoiding injury by males. Females do not avoid male proximity, only male-female interactions.

Female attraction to other females is here identified as a major factor in the northern fur seal mating system. Female attraction is responsible for the initial formation of most female groups, and for their subsequent growth, movement, density, and degree of separation from other groups.

The main factor that drives this mating system is repeated, predictable use by females of parturition sites that are selected in the context of the attraction that females have for other females. Despite males being aggressively dominant to females, and female behavior aimed at reducing the risk of injury from males, males do not dominate the mating system. The most important factor in the mating system is that females gather predictably, which provides a focus for male competition. Males superimpose their own social structure upon this gathering of females. But this has much less of an effect on the mating system than the predictable arrival of females has. The timing and location of mating, and the density, dynamics, and longevity of mating groups, are determined by females.

CHAPTER 6

Behavior of Adult Females

THE PREVIOUS chapter concluded that in the northern fur seal mating system, males are behaviorally dominant over females but have an ephemeral effect on the shore aggregation and virtually no effect on where, when, or how often females come ashore. The timing, location, and persistence of shore colonies depend on female, not male, preferences. Therefore, understanding this mating system largely depends on understanding the factors that shape the behavior of individual females. The present chapter examines these factors through long-term observation of individuals and experiments on captives.

Predictability in the time and place of female gathering, the key factors upon which the male territorial system is founded, depends on tendencies of individual females. The female population arrived between 15 June and 14 July each year for 14 years (fig. 3.2), but about 55% of individuals arrived in a 10-day span specific to them (table 3.1). That is, the population was composed of individuals with a more narrow and specific arrival tendency than shown by the aggregate. What factors affect the arrival date of individuals? Is age important? Does arrival vary with the female's reproductive status (parous or not)? How tightly linked are arrival and estrus, and does the date of estrus also vary with age?

The places where females gather may be more predictable over years than the dates when they gather. Some breeding areas have been used continually for at least two centuries. About 78% of females show philopatry (return to their natal site to mate; Anon., 1973; Baker et al., 1995). The remainder disperse to other central breeding areas. For these two groups of females, what factors determine how individuals use space? Do individuals have site preferences that are more narrow and specific than for the population at large, as they did for arrival date? How precisely do individuals return to the same parturition site? Does this precision differ for animals showing philopatry and those that do not? Is this precision affected by age? How many suckling sites does a female use, how far are these from the parturition site, and what proportion of total land available does this represent? Do individuals continually expand to new land sites with age? Do terrain or population size affect how females use parturition or suckling sites? Does the spatial pattern of parturition sites suggest whether individuals are making mate choices? Moving among different sites could indicate a search for the most "fit" males; residence on the same site for periods exceeding average male life expectancy could indicate that the site is more important than the male on it.

Females are largely solitary at sea (Kajimura, 1980; Kajimura and Loughlin, 1988; also see note 1 in chapter 3). They are at sea for all but about 38 days of the year (each makes an 8-day perinatal visit and about fifteen subsequent 2-day visits for suckling per year). Each day on shore they are likely surrounded by strangers because the foraging schedules of females are individualistic, and neighboring females make brief visits to sites other than those where they suckle their pups. Under these conditions, what kind of social behavior has evolved? Specifically, how do females act toward one another? To what aspect of the shore aggregation (pups, females, males) do they give the highest priority, as indicated by time spent interacting? How different is the behavior of females from that of males? Is female behavior affected by population size or terrain?

Females on shore must protect their pups from other females. Neonates are highly precocial and thermally independent (Blix et al., 1979), but are often bitten by neighboring females. Such bites are the leading cause of trauma deaths in neonates (trauma deaths = 17% of those that die on land; Keyes, 1965; Anon., 1969). Pups move out of female groups when their mothers are absent but rely on their mothers' protection when they reenter the group to suckle. How much time do females spend in aggressive conflicts? Do females leave the group at parturition as a means of protecting the pup?

Group formation is fundamental to females; few of them remain alone during the regular breeding season. Residence in groups may give them some protection from injury by males (chapter 5). But is avoidance of males the driving force behind group formation, or is it only a spinoff of a more basic attraction of females to other females? Female attraction has been reported as the basis of social organization in fallow deer (Clutton-Brock and McComb, 1993), kob, and lechwe (Deutsch and Nefdt, 1992).

We conducted experiments on captive females to address several fundamental questions about group formation. Specifically, do groups form or remain together if not forced by a male? Is the spacing within a group determined by the male or by the females? Does group density in captives change seasonally as it does on the central breeding area (chapter 2)? Will females remain separate from a captive group to protect their pup if no males are present?

Female social behavior is also influenced by the presence of males. Because of the size differential, many male activities may separate the mother from her young and injure or kill one or both of them. Female social behavior must simultaneously protect mother and young from males while keeping the mother accessible for mating. To what extent does the female be-havioral repertoire change with the presence or absence of males?

All behavior that females direct toward other females and pups not their own, and most behavior they direct toward males, is agonistic. Play, grooming, help with young, thigmotaxis, or any other form of communal behavior among adults is absent. Female agonistic behavior is an important regulator of social processes. The natural occurrence of female aggression on the central breeding area was studied by Francis (1987) as part of this project. He quantified aggression at East Reef, Zapadni, and Kitovi (fig. 1.4) using the facilities described previously. His conclusions will be re-viewed as an introduction to female social organization and to experiments we performed on groups of captive females.

REVIEW OF FEMALE AGGRESSION STUDIES

Francis (1987) studied female aggression [1] as well as the effects of male herding on female grouping. He found that female aggression was frequent [2] but not usually intense. About 80% of encounters involved two females threatening each other simultaneously. Most encounters (87%) involved mild forms of aggression (open mouth threat), while more intense forms, like biting and pushing, were rare.

In most cases (52%), aggressive encounters had no apparent cause. Fe-males threatened each other regularly as if the trigger were an internal time schedule. Of the causes that could be identified, the most frequent (16% of cases) was the opponent moving or changing position. Pups also were a trigger for female encounters (aggression was more frequent when pups were present than when they were absent). Pups were threatened by neigh-boring females about 0.3 times per 15 minutes, and this caused the mothers of these pups to threaten in return. Non-mothers were involved in as many aggressive acts as mothers but initiated fewer of them, which confirms that the presence of young was a trigger for aggression.

Most aggressive encounters had an ambiguous outcome. Only 14% of cases had a clear conclusion (one animal moved away). In all other cases, opponents did not move far enough to prevent further aggression. They moved away from intense forms of aggression (biting) more than from mild forms.

Female aggression followed a seasonal trend; it was most frequent in early July when female density was high and most pups were newborns. Aggression was ten times more frequent on the day of parturition than on the day before or after. For this reason, the aggression rate declined to one-half by early August when new births became rare.

Female aggression declined seasonally along with the seasonal decline in group density (chapter 2). Female aggression and group density were not linked within quarters of the breeding season, except at one site. That is, density was a less consistent contributor to aggression than might be expected. Size of female showed no clear effect on aggression rate or outcome.

Males did not often suppress female aggression. Males came between two aggressive females in only 5 of 1,423 encounters. Francis (1987) concluded that male herding behavior, which peaked from 9 to 14 July, increased female intragroup density. He also concluded that males could force females to adopt a rest site with a higher density (more neighbors) than the rest sites females found for themselves. He stated that females usually attempted to avoid contact with males, often by seeking the centers of female groups where higher densities decreased the likelihood of male/female contact. He attributed this avoidance to decreasing the risk of being stolen by a male. He calculated that females had a 50% chance of being stolen once in their lifetime.

Francis (1987) concluded that female aggression functioned to defend space and to defend pups from the attacks of neighboring females. He speculated that the selective advantage of female aggression was that it enhanced pup survival, but he acknowledged that other functions may exist because nonmothers are also frequently aggressive.

He concluded that male herding behavior was partly but not fully responsible for dense female groupings because female groups persisted after males left the breeding area in August. He reported that in the absence of a male, captive females formed groups (described below) out of fear of captivity. He eliminated both predator avoidance and mate selection as the possible ultimate causes of grouping.

METHODS

Time and Location of Mating

Important information about the affinity females have for specific land sites and about their reproductive histories comes from recording the time and location of mating. These and other data were recorded on history cards for individuals [3]. Card entries included the female's behavior and location using the same numbered 5×5 meter subsectors as for males (chapter 4). Smaller subsectors, although they better suited the females' movements, were impractical for field use. The first arrival of the season was especially important because it denoted the timing of reproductive events [4]. It was identified by a special coded entry on the history card. Most females arrived at night, so the first card entry usually came a few hours after arrival.

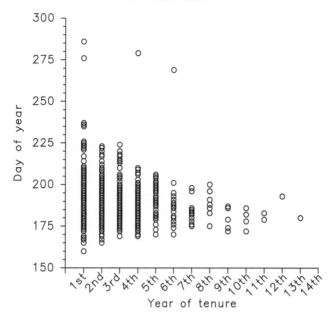

Fig. 6.1. Changes in the arrival date of individual females as they aged. Based on 814 female-years of data from East Reef.

To examine how the dates of estrus varied within a lifetime, we had to estimate these dates from other events because copulations often occurred unseen at night (table 3.2). From the records of females with good arrival, parturition, and copulation records, we determined that estrus usually occurred 6.6 days after arrival and 5.3 days after parturition for mothers, and 3 days after arrival for nonmothers [5]. We used these estimates, or actual observations, to assign a date of estrus for each marked female that was seen arriving or giving birth, and compiled this information into a frequency distribution [6].

The parturition sites that individual females used were important measures of site fidelity from year to year, and they were the center of the suckling area used by a female within a season. The parturition site was recorded as a subsector number [7].

Behavior

Behavior was recorded each time an entry was made on the history card, usually once daily with additional entries made whenever noteworthy events occurred [8]. These many brief observations were used to construct a rough time budget for females.

Captive Studies

Experiments were performed on some aspects of female behavior among small groups of captives. Captives were held in two enclosures; one was a permanent cage compound [9] constructed at our laboratory (see chapter 1). Experiments were performed there on the effects of separating females and their young from others, and on the effects of males on female behavior [10]. A second (temporary) enclosure was constructed on the tundra near the East Reef study site and was used to quantify aggression among females in the absence of a male [11].

Births

To determine whether perinatal aggression was a disruptive factor in the aggregation that might change with population size, we observed births carefully in 1974 [12]. Births appeared not to be disruptive, so these detailed data were not collected in subsequent years.

Sample Size

The sample size reported in various measures often departs from the total of 1,051 females observed because not all records were suitable for answering all questions [13]. In all cases, female records were excluded from analysis if the female's identity was uncertain (from faulty marking), or if she used a site that was blocked from view and was thus seen irregularly.

RESULTS OF OBSERVATIONAL STUDY
Timing of Arrival

Females made their first appearance for the season in an increasingly narrow window of dates as they aged [14]. Just as for males, most of the variability in the date of first arrival was for young females (fig. 6.1). The variability decreased until about tenure year 9, by which time most first arrivals were occurring in a span of about 15 days (21 June–5 July).

Just as for males, the loss of variability in arrival dates resulted from individuals changing their behavior with age. When the data set was restricted to females that were seen for three or more years (n = 257), the resultant plot was nearly identical to a plot for all females, as shown in figure 6.1. That is, no matter how long females ultimately mated, they

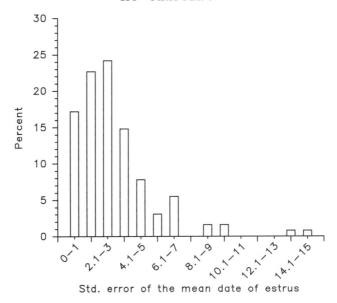

Fig. 6.2. Lifetime variation in the date of estrus for 128 individuals (119 at East Reef). Most dates of estrus were estimated from arrival or parturition dates.

tended to be variable in arrival date when they were young and to reduce this variability as they aged. Most of the scatter in the plotted arrival dates was therefore caused by variation among younger females.

The dates of arrival also depended on whether females were parous or nonparous; mothers arrived significantly earlier than non-mothers at both sites (by 6.1 and 4.6 days at East Reef and Zapadni, respectively) [15]. This fact was first reported by Bigg (1986) based on the examination of anonymous animals killed at sea.

Timing of Behavioral Estrus

Copulations occurred in the population for at least 6 weeks of the year. But most individuals mated in a period of 8 days that was specific to them (SE of 3.9 days around a mean date; fig. 6.2) [16]. That is, individuals were much more specific in arrival and mating date than was the population they formed. Most females that tended to mate in mid-June were not likely to mate in late July, and vice versa.

Females usually mated near their preferred date as long as the sequence of parturition events remained the same. But their dates changed by 7–8 days if this sequence changed (table 6.1). Females that were either parous (sequence 1, table 6.1) or nonparous (sequence 2) for 2 years running

Table 6.1

Differences in Arrival Dates Corresponding to the Sequence of Birth and Nonbirth
in Two Successive Years for Known Females

Sequence No.	Year 1	Year 2	n^a	Mean Diff. in Arrival Date[b]	Tukey Grouping
1	Parous	Parous	30	-0.31^c	b
2	Nonparous	Nonparous	29	-0.87	b
3	Parous	Nonparous	30	$+7.70$	a
4	Nonparous	Parous	30	-8.30	c

[a] Except for Sequence 2, these sample sizes are subsets of larger samples but are reduced here for the Tukey HSD test. See note 17 for details.

[b] All units are in days: $-$ values = later in year 2 than in year 1; $+$ values = earlier in year 2 than in year 1.

[c] Tukey HSD test: mean square error = 96.63; 3 df; F = 13.27; $p \leqslant .0001$ numbered locations.

showed no difference in their arrival dates between years 1 and 2 [17]. However, those that were parous in year 1 and nonparous in year 2 (sequence 3) arrived 7.8 days later the second year. When they were parous again in year 3 they arrived 8.3 days earlier than in year 2 (sequence 4). That is, the same individuals shifted later when being nonparous and about the same amount earlier when they resumed being parous. The mechanism by which arrival date is tied to reproductive status is not known. Bigg's data (1986) did not show this labile nature of arrival dates for individuals.

Use of Space

The precision of returning to the same site for parturition in different years differed at the two study sites, probably because the distance to the sea and the nature of the terrain differed. The mean distance between the centers of subsectors where parturition occurred for females at East Reef was 8.3 meters (fig. 6.3) and for Zapadni 17.6 meters, a significant difference [18]. Thirty-nine percent of East Reef females gave birth in the same subsector in 2 years (zero distance), and two of them (nos. 273 and 331) gave birth in the same subsectors for 5 and 6 years, respectively. Often they gave birth on the exact rock or hollow in different years. Other females were less precise; several females at Zapadni had parturition sites nearly 100 meters apart because the female population was withdrawing from that site (fig. 2.13).

Females at the two study sites differed significantly in the number of suckling sites they used throughout lactation and in the distances among these sites. Females at Zapadni used more suckling sites than females at East Reef (table 6.2, "Avg. No. Sectors Occupied"), apparently because

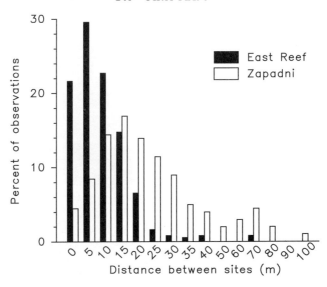

Fig. 6.3. Distances between the centers of 5 ∞ 5 meter subsectors in which parturition occurred for individuals in successive years based on 445 measures for 220 females (East Reef, 365 measures for 176 females).

Zapadni was unobstructed by rocks [19]. Suckling sites were farther from the parturition site at Zapadni (average 21 m) than at East Reef (average 11 m) [20].

Female site fidelity, measured by the ratio of new to old suckling sites used per year, increased as they aged [21]. From their third year of tenure onward, females used fewer new sites than in their first two years (table 6.2, "Avg. Ratio New/Old Sectors"). That is, as they aged they selectively suckled on sites they had used previously instead of expanding to new suckling sites. However, the total number of sites they used seemed not to change (no trend in "Avg. Sectors Occupied," table 6.2). In this regard females differed from adult males that used fewer sites as they aged, as well as returning to previously used sites with age (chapter 5).

Females sometimes moved to a new suckling site because their previous site was occupied, or covered by mud, but sometimes they moved for no apparent reason. Their suckling sites did not conform to male territories in any obvious way, either before or after estrus. Some females, especially large ones, would occasionally drive away any other female they found using their preferred suckling site. We did not quantify this competition for sites.

Females used about the same amount of space for suckling as adult males defended as territory. Multiyear females used more sectors in their lifetimes than multiyear males (compare tables 4.1 and 6.2), largely because they lived longer and had more years to accumulate suckling sites. Also, females and males were similarly affected by terrain differences.

Table 6.2

Mean Number of 5×5 m Grid Sectors Used and Proportion of
Total Sectors Available That Were Used by Adult Female Fur Seals at
Two Study Sites at St. George Island, 1974–88

Year of Tenure	Sample Size (n)	Avg. No. Sectors Occupied	SE Sectors Occupied	% of Grid Occupied	Avg. Ratio New/Old Sectors	SE New/Old Sectors
			East Reef			
First	443	4.6	0.2	3.9	1.00	0
Second	221	5.0	0.3	4.2	1.01	0.08
Third	152	4.4	0.3	3.7	0.35	0.05
Fourth	106	4.0	0.2	3.4	0.23	0.03
Fifth	64	4.0	0.3	3.4	0.12	0.02
Sixth	21	3.7	0.5	3.1	0.15	0.05
			Zapadni			
First	204	6.6	0.4	5.5	1.00	0
Second	71	5.3	0.4	4.4	1.04	0.13
Third	35	5.4	0.6	4.5	0.47	0.08
Fourth	16	5.3	1.0	4.4	0.23	0.03

Note: Results are shown only for those years having a sufficient sample size.

These were unexpected findings, given that males were territorial and fe-males were not.

Females visited some of their suckling sites more often than others (table 6.3) [22]. They spent most of their time alternating between two sites but spent a larger share of their time visiting their third through eleventh ranked sites than males (compare table 4.2). That is, males used two sites to the near exclusion of several others, whereas females spread their visits more evenly among all the sites they used. Females tended to visit their most frequently used sites early in the season when the pup was young and immobile, and to visit the less often used sites later in the season when the pup could easily follow.

Behavior

Females spent most of their time ashore being "Inactive" (scored in catego-ries that involved no or very little movement; table 6.4). Comparison of tables 4.5 and 6.4 suggests that females spent less time "Inactive" than males (65.5% for females vs. 92.9% for males). But many females that were inactive were scored as "Suckling." If all "Suckling" scores were changed to "Inactive," and "Departing/Returning" scores were deleted (fe-

Table 6.3

Frequency of Sighting Known Adult Female Fur Seals at Numbered Locations

Study Site		Location					
		1^*	2	3	4	5	6–11
East Reef	Mean	50.1	20.1	10.8	6.2	3.9	1.8
	SE	3.2	0.6	0.5	0.5	0.4	0.2
Zapadni	Mean	42.6	18.9	11.1	7.1	5.1	2.8
	SE	3.3	1.0	0.4	0.4	0.4	0.3

*Mean percent of all sightings at the first through eleventh most commonly used locations on the study site. Sample sizes: East Reef, 479 females (5,061 female-years); Zapadni, 581 females (7,987 female-years).

males could leave daily, males once per season), then females and males were much closer in their "Inactive" scores.

Mothers that were on shore spent more time with their young (55.4%) than they did alone (44.6%) [23]. However, this result represents the early season (June, July) only. As pups matured they spent increasing amounts of time with other pups and returned to suckle from their visiting mothers only at intervals.

Females differed from males mainly in their interactions with pups. Thirteen percent of female behavioral scores were related to pup care (table 6.4), but male contact with pups was incidental (table 4.5). Females also scored higher in Departing and Returning from sea because males did not make feeding trips. Females spent little time interacting with males or other females (1.1% and 1.4%, respectively; table 6.4). These scores do not suggest that complex social relations were being maintained.

Few important behavioral changes occurred as the female population declined over the years [24]. Fewer females were scored in the Departing/Returning class over time, and more females were scored in the pup care category. These changes correlate with a decrease in the length of female feeding trips from 1979 onward (to be discussed in chapter 10). As the length of sea trips decreased, females spent less time at sea and more time ashore with their young.

The behavior of females at the two study sites differed significantly in only two categories. Females at East Reef were more often scored departing/returning, and females at Zapadni were scored more often in reproductive activities [25]. These differences were relatively minor, and may have resulted from differences in our ability to sight females at the two sites [26].

The way females interact with each other (holding frequent, low-intensity threats with nebulous beginnings and inconclusive endings; Francis, 1987) may be the way females respond to one another under any circumstances. For example, the social interactions between the first two females

Table 6.4

Time Budget for Adult Female Fur Seals Taken from the History Cards of
547 Individuals from East Reef and Zapadni Study Sites from 1974–1988

Behavior	Percent [d]	Behavioral Class	Percent
Rest[a]	23.1		
Rest[b]	13.4		
Rest[c]	10.7		
Groom[a]	0.8	Inactive	65.5
Groom[b]	3.5		
Upright[a]	8.7		
Upright[b]	4.5		
Upright[c]	0.8		
Suckling	9.3		
Interact with pup	1.2	Pup care	13.0
Calling pup	2.4		
Resp to pup	0.1		
Interact with male[a]	0.9		
Interact with male[b]	0.2		
Interact with female[c]	0.3	Social interaction	2.6
Interact with female[b]	0.2		
Interact with other pup	0.1		
Interact with female[a]	0.9		
Moving[a]	5.0	Moving	5.0
Departing	3.5		
Returning	7.3	Depart/Return	12.1
First arrival	1.3		
Birth	0.9	Reproduction	1.8
Copulation	0.9		
Total	100.0		100.0

Note: Only females seen more than ten times were used.

[a] Pup absent

[b] Pup present

[c] Pup not visible

[d] Percent. For each female we calculated the proportion of observations in each behavioral category. Values in this column represent the mean percent for all females that scored more than zero in that category.

on the beach one year [27] showed the same principles as described for groups by Francis (1987); the mother initiated most threats that were of low intensity and had neither a clear cause nor clear resolution. Pairs of females caged in the absence of a male acted the same way. In all cases, females were attracted to the presence of other females but threatened them frequently whenever they were near.

Parturition

Birth was a relatively unobtrusive event that had little impact on the behavior or structure of the breeding aggregation. Females often delivered the young with no excessive movement, aggression, or change of posture that would affect other animals or attract the attention of observers. Furthermore, they usually vocalized only after delivery. For these reasons, births were not recorded after 1974.

Births for marked females occurred on average 1.3 days after arrival [28]. Most of them (56%, n = 39) were caudal presentations, but the chances of seeing a cephalic presentation were lower because they did not take as long to complete. The female began calling to the pup immediately, and the pup began to answer within 4 minutes (n = 24). These calls were highly stereotyped (Insley, 1992). In six births the pup began suckling 31 minutes after birth, although initial attempts were weak, poorly coordinated, and brief.

RESULTS OF CAGE EXPERIMENTS

Captive females quickly created a group and maintained close spacing even in the absence of a male. Groups that were formed in June had initial spacing of less than 1 meter between individuals, the group that was formed in late July (1983) had initial spacing of less than 2 meters, and the group that was kept from early to late July 1974 increased spacing from about 1 to about 2 meters over time. That is, captive females spaced similar to noncaptives at the same time of season in the absence of a male. Females often rested their heads on other females' bodies but avoided their heads. Spacing seemed unaffected by the amount of human disturbance in the vicinity (it was the same at the laboratory as at the undisturbed field compound at East Reef). Groups that were created in early July and in late July differed in spacing, but the females (newly caught) would have had equal initial fear of captivity (this appears to negate the hypothesis that captives group out of fear; Francis, 1987). Captive groups formed quickly even when the individuals came from different breeding areas (likely to be strangers). Males could not move captive female groups once they formed.

Both mothers and non-mothers had a strong preference not to be separated from their group. When individuals were moved two cages (6 m) away from a group, they performed stereotyped pacing movements along the intervening cage wall and tried to climb it. They did not pace when separated from others by only a wire fence, suggesting that proximity prevented pacing. Separated mothers paced even if their young were present, but if they were separated from their young as well they vocalized loudly while pacing. No matter how females were introduced into a common cage, either as a group of after being held in isolation for 48 hours beforehand, they gathered together at the first opportunity. Individuals wandered from the group more often than females wander on the central breeding area, especially estrous females and non-mothers. However, they rarely slept separated from the group unless forced. The group would move to join a female that was unable to move because of a weak neonate. Females with free access to water pools did not enter them until after they had passed estrus, in the same way that females on breeding areas do not go to sea until after estrus.

Female vocalizations and vigilant behavior (visual scanning) were greatly reduced when a male was present. Without a male, the female group looked watchful and vocalized at many foreign sounds (such as truck traffic). When a male was present, females were silent except during extreme disturbance, such as a human entering the cage. They also stopped visual scanning and quick head movements, which gave them the docile appearance described previously (chapter 5). All females, including those in estrus, stopped moving around the compound when a male was present. Females showed a wide range of individual behavior, from placid to highly aggressive, only when males were absent. Near a male they appeared slow, deliberate, and almost uniform in temperament. Estrous females showed homosexual mounting with pelvic thrusting in the absence of a male (chapter 8). These mountings were never seen in captivity when the male was present, and was seen only once in 15 years on the breeding area.

Captives interacted in the same ways as females on the breeding area. The frequency of interaction was comparable (3.8 threats per female per hour, compared to 2.8 to 3 per hour reported by Francis, 1987) [29]. Most threats among captives were caused by movement within the group, most threats were of low intensity, the outcomes were usually vague, and no winner or loser could usually be identified. Captive females were not equally aggressive despite having similarly aged young.

DISCUSSION

The timing, location, persistence, and character of breeding groups result from the tendencies of individuals. The timing of arrival is ultimately determined by a physiological response to the light cycle that establishes the date of implantation (Temte, 1985; Spotte and Adams, 1981). This response causes the population to arrive at breeding sites within a 6-week period. But individuals arrive within 4 days either way of a date that is specific to them. That is, figure 3.2 (population arrival) does not show the probability of individual arrival. It is essentially a histogram of the number of individuals that have a tendency to arrive on all the different days of the season. The same is true for the curve for copulations by date (fig. 3.6).

Arrival dates of individuals are narrower than for the population, and vary with age and reproductive history. As individuals age they restrict their arrival to an increasingly narrow time frame. Also, they arrive 7–8 days later in years when they do not bear young than in years when they do. This results in mothers preceding nonmothers in the northward migratory procession (Bigg, 1990), and arriving and mating earlier. Judging from the fact that the median arrival date (chapter 3) did not change during 14 years of this study, the population age structure and overall natality can be inferred not to have changed, either.

Analysis of parturition and suckling sites shows that females have a preference for a small portion of a given central breeding area to which they return yearly (site fidelity). The size of the area varies according to several factors (below), but is always much smaller than the dimension of the central breeding area (example, East Reef study site = 3,000 m², but mean distance among parturition sites of different seasons = 8.3 m). Females may pass through but will not reside on nonpreferred parts of the central breeding area, even though these areas may contain many other females. That is, "suitable" space for the species is very different from suitable space for each individual. For that reason, the population is properly viewed as a collection of individuals, each of which is on a site (18 m or less in diameter) for which it has a special affinity. A population is not an amorphous mass having fidelity to a general area of thousands of square meters. A testable hypothesis is that northern fur seals have an affinity for smaller sites than do other species of otariids (see review in chapter 7).

The distance among an individual's parturition sites over years varies with individual preference combined with environmental influences, such as the number of males present, time of season when females arrive, and whether the population is stable or declining. Males are important because they may block females from reaching sites they would otherwise reach (hence increase distances among parturition sites). Time of season is important because it affects the number of females present (when many are present, individuals can move from group to group and settle nearer the previous year's site than when females are sparse). Population trend is important because if the population declines (as at Zapadni; fig. 2.13), preferred sites may be farther inland than female groups extend and may not be reachable for an arriving female. Parturition sites used during a population decline show that females can be flexible in the sites they use if they are forced. But these measurements contain an artifact and should not be used as characteristic of the species under stable conditions.

If all females returned to preferred sites faithfully for years, inbreeding would undoubtedly occur. However, this may not occur because at least 22% of females move to a non-natal breeding area to bear young and mate (Anon., 1973; Baker et al., 1995; see also chapter 7). Each breeding area contains females that were born there and those that were not. The averages reported here for fidelity to parturition sites undoubtedly include both kinds of females.

There was no pattern in the female use of parturition sites that suggested they were in any way related to males. Females used the same part of the breeding area for longer (often more than 5 years; fig. 3.13) than individual males spent breeding (average 1.45 years, usually less than 5 years; fig. 3.11). This pattern suggests that the site is primary to females, not males. A precedence of site choice over mate choice was also reported for the New Zealand fur seal (Miller, 1975).

All female distributions are a balance among conflicting tendencies to use a preferred site, form groups, and avoid being alone with males. In large, stable populations at peak season, females can reach preferred sites while remaining in a group. But when few females are present (early season), residence in groups takes precedence over use of a preferred site if the female must be alone with a male to reach it. Arriving females form a series of small groups at intervals along the beach. When the population is small, these initial groups persist through most of the season. When it is large, groups coalesce and females move inland. When the population declines, female avoidance of males results in females abandoning a site (example, fig. 2.13) rather than remaining thinly scattered over it.

Group formation results from an intrinsic attraction of females to other females. Grouping is not imposed by male presence because groups continue to grow long after males stop herding (at a group size of about ten). Captive females form groups in the absence of males, and being separated from these groups is aversive to females. Therefore, herding only reinforces preexisting tendencies of females to form groups. Francis (1987) attributed grouping in captives to fear. This is reasonable caution, but it does not account for the finding that groups that were formed at different times of the season had different spacing, whereas if they grouped out of fear the spacing would have been similar. Such strong attraction seems an incongruous trait for animals that are mostly solitary for 330 days of the year.

Female attraction is more likely an evolutionary adaptation to avoidance of injury from males than a proximate reaction to male herding behavior. By gathering into dense groups, females can reduce the likelihood of contact with males (Francis, 1987). Other behavioral traits, such as acting subdued, usually remaining silent, giving birth unobtrusively, moving neither far nor fast, showing little vigilance, and using less than their full behavioral repertoire (such as homosexual mounting) also make females less noticeable. Together this suite of traits may reduce the frequency and intensity of contact with males and the risk of injury that all male/female interactions carry (chapter 5).

Grouping as an adaptation to reduce injury from males also explains why females form the densest groups at the time of season when other females are most aggressive (around parturition) and their own newborn are most vulnerable to injury. High densities give females the greatest protection from the males that join them in time for the postpartum estrus. Apparently the risk that their young will be injured by other females is smaller than the risk that they themselves will be injured by males if they separate from the group for parturition.

Females do not inhibit aggression toward some individuals in a way that suggests kinship ties. Social bonds among nonrelated animals (pairs resting or foraging together) are not apparent. Brief coalitions of several females against one seem not to exist. Social dominance, with clear winners and losers of aggressive conflicts, is not obvious. Females are gregarious (attracted to other females) but not social in the sense of forming social bonds.

Agonistic behavior ultimately functions to protect neonates. Since most threats serve a preemptive function, ensuring that all neighbors remain at least the length of an outstretched neck from the young, they are given frequently but are of low intensity. Personal space for the mother seems irrelevant because each allows others to rest in body contact as long as their heads are not near the young. Aggression may affect the rest sites of females of different age or reproductive status (Vladimirov and Nikulin, 1991), but aggression does not segregate females the way it does males.

SUMMARY

The variables that shape the behavior of individual females in any year include age, terrain, and whether they are parous or nonparous. The constants that shape their behavior are a brief stay ashore (38 days per female per year), attraction to other females, and avoidance of interactions with males.

The female aggregation is a collection of individuals, each expressing a narrow range of preferences for arrival and mating date, and for parturition and suckling site. Arrival time becomes less variable with age in females, as in males. At a particular age, as long as the female's reproductive status remains unchanged, average arrival date varies by less than a day over several years for most females. But arrival occurs 7–8 days later in years when females change to nonparous from parous, and the same amount earlier when they revert to parous the following year. Thus, changes at the individual level explain the earlier observation that, as a class, mothers arrive before non-mothers. About 80% of females mate in an 8-day range around a date that is particular to them. A curve of copulations by date depicts the number of females that tend to mate on the various days of the season.

Preferences for parturition and suckling sites are marked. Each female uses only a small fraction of the total space available on a central breeding area (< 20 m diameter). About 39% of females use exactly the same parturition site (to the meter) for two or more years; others are more variable. Females can change their preferred site if the population declines strongly. Individual habits, terrain differences, and population size and trend may affect the location and size of the area an individual female uses. The collective effect on the female population of individuals having narrow preferences in arrival dates and parturition sites is that breeding is highly predictable in time and place.

Females use their preferred site much longer than the average reproductive lifetime of males, suggesting that mate choice as not a determinant in the selection or use of a parturition site. Females seem willing to mate with any male that resides on the site they prefer.

Females are attracted to other females and join them to form groups. They are not forced to do so by male herding behavior because groups form in captivity when no males are present, groups continue to grow on central breeding areas after males stop herding, and they persist after males have departed for the season. Male herding behavior only reinforces a preexisting tendency of females to group. Separation from the group is aversive to females in any setting. Gregariousness to this extent seems incongruous for animals that are solitary most of the year.

Female attraction is a previously unrecognized phenomenon in this species, or in any otariid. It explains much of the social behavior and group formation that were formerly attributed to males.

Despite being gregarious, females are not social in the sense of forming adult social bonds, coalitions, pair bonds, dominance relations, or kinship ties. Social interaction is reduced to its essentials, as appropriate to the brief period of social contact each female has per year. Even care of the young is minimal; at peak season (July) mothers are with their young on only 55.4% of the sightings, and the proportion decreases as the pup ages.

Female interactions are always agonistic despite the underlying female attraction. Aggression regulates female social processes (Francis, 1987), although it does not appear to change female spacing or distribution much. Female aggression appears to protect the young by preventing neighbors from moving such that they can threaten the young. Female aggression is frequent (especially on the day of parturition), usually of low intensity, and lacks an obvious cause or outcome. These principles apply equally to groups or solitary pairs of females. Aggression is most frequent in July, when most females are giving birth, intragroup density is highest, and neonates require the most protection. Females do not avoid this aggression by leaving the group for parturition as other otariids do.

All female distributions can be explained as various balances among conflicting tendencies to use specific parturition sites, form a group, and avoid being alone with males. Groups of females attended by a single male, formerly called harems, are thus not social units that have a particular reproductive significance. They are only one of several possible ways females balance the above tendencies, depending on number of females present, distance from the sea, and other factors.

PART THREE

Processes Fundamental to the
Mating System

Part Two described the northern fur seal mating system from the standpoint of behavior that was directly observable on central breeding areas. Several other traits, adaptations, and processes are also essential parts of the mating system, although they are more difficult to measure. Part Three examines three of these subjects in detail. Chapter 7 discusses two key processes that keep populations breeding in the same location over centuries: site fidelity and philopatry in individuals. Chapter 8 discusses the core of reproduction; estrus and estrous behavior in females, female behavior toward males, and the male's role in female receptivity. Chapter 9 discusses the ontogenetic changes that underlie the male territorial system. These subjects can be considered the mainstays of the mating system. The mating system includes at least two other essential features that this book does not address: navigation and fasting.

CHAPTER 7

Site Fidelity and Philopatry

SITE FIDELITY (repeated return to a non-natal site over years) and philopatry (returning to the natal site) are related phenomena. Together they explain the tendency of northern fur seals to persist on the same central breeding and landing areas for centuries, or conversely, to avoid colonizing new islands (two in 200 years; chapter 1). They also explain the resilience these areas have to human disturbances. Both sexes show site fidelity and philopatry, but the latter has not been documented well in either sex because it requires following animals from birth to adulthood.

Persistence in land use can be seen on a small spatial scale. For 15 years of this study, female fur seals failed to use a 20 ∞ 20 meter section of beach immediately seaward of the knoll on which we built the observation blind at East Reef on St. George Island. A map of "harem" locations in 1914 (Osgood et al., 1915) showed an identical absence of females seaward of the same knoll. This small area, different in no apparent way from the surrounding beach, may have been avoided by six decades of fur seal cohorts.

Chapter 6 advanced the hypothesis that northern fur seals are exceptional among otariids in the size of area that individuals use, and the consistency of its use. Adult males defend an average maximum space of 110 square meters (chapter 4) but spend most of their time on half that much land. Few males (0.6% of 1,541) move their territory by more than 10 meters any time in their lives. Females on shoreline breeding areas use parturition sites that are on average 8.3 meters apart and suckling sites that are on average 11 meters from the parturition site (chapter 6). Antarctic fur seals, which have a similar mating system and maternal strategy, have parturition sites close together (6.4 m; Lunn and Boyd, 1991; Boyd, 1996), but suckling sites hundreds of meters apart (Doidge et al., 1986). Unlike northern fur seals, antarctic fur seals have colonized many new islands and beaches in the past 30 years (Budd and Downes, 1969; Payne, 1977; Shaughnessy and Goldsworthy, 1990; Bengtson et al., 1990). Within a season, Steller sea lions suckle mostly within 20 meters of the parturition site, but do not use suckling sites faithfully year to year (Gentry, 1970). Also, the parturition sites of at least a few females vary by hundreds of meters over several years (Gentry, pers. obs.).

The importance to the mating system of individuals returning annually to a small site, especially when they have closely timed arrivals (chapter 3), is that social gatherings and reproduction become highly predictable. Competition among males centers on the predictable arrival of females. The winners of this competition remain to do most of the mating and become, by definition, the most fit males. By this route, predictability may be important in sexual selection.

If predictability is important to the mating system, how does the species retain the flexibility to change its mating sites on the scale of millennia? The islands where northern fur seals bred, and the sea locations where they foraged, must have changed many times in the past 5 million years as islands arose and sea level changed with glacial cycles (chapter 1). The process by which individuals form attachments to land sites must make them conservative in the short term yet permit some flexibility in the long term.

To understand how site fidelity and philopatry affect the population, it is first necessary to quantify it in individuals. The most important question is: What is the rate of philopatry in males and females? It cannot be 100%, otherwise inbreeding would result. Therefore, a related question is: What is the rate of female transfer among non-natal breeding areas? Another related question is: What is the precision of philopatry (how close do animals mate to their own natal sites)? In philopatry, at what age does the lifetime attachment to the natal site begin? What factors are responsible for its expression? Do immigrant females show the same fidelity to a site as those that show philopatry? Do females that repeatedly rear young on the same site have the flexibility to move to nonpreferred sites? Do these nonpreferred sites include only traditional central breeding areas, or are unoccupied beaches included? Does the female's age affect her willingness to adopt a new breeding area?

This chapter attempts to answer all these questions. Some answers came from data collected by management since the 1940s. Others came from routine, long-term observation of undisturbed animals on breeding areas. Some answers required experimental manipulation. This chapter reports on four such experiments, including the development of site attachment in neonates and the factors that affect its expression in adult females.

REVIEW OF PAST RESEARCH

Males

Much effort has been expended quantifying how juvenile males use land because that use was central to the commercial take of pelts. Hundreds of thousands of tags were applied to one-month-old pups from 1941 to 1969 (Scheffer et al., 1984). Many of these tags were recovered 2–5 years later during commercial kills for pelts. Baker et al. (1995) analyzed these tag returns and simulated the kill with a mark/recapture study to estimate homing (degree of association with the natal site in juveniles). They concluded that homing increased at least to age 6, was greater on St. George than on St. Paul

Island, and increased late in the breeding season. None of these males were breeding, so homing in itself has no immediate reproductive significance.

Few good data exist on male philopatry because animals must be observed at birth and again 9–12 years later. Chelnokov (1982) reported finding twelve adult male northern fur seals defending breeding territories on the same sector of the breeding area where they were born and tagged. However, he did not report sector size, so the precision of philopatry is still unknown for males. This study needs to be repeated, noting exact sites of birth and subsequent breeding territory.

Females

Much of what is known about philopatry in females comes from animals that were tagged as pups and killed during the herd reduction program of 1956–68 (chapter 1) [1]. Of the approximately 315,000 adult females killed on land, 7,940 had been tagged at birth; 78% of them were killed on or near their natal central breeding area (Baker et al., 1995). However, brief visits of females to non-natal breeding areas likely confused the estimate of philopatry [2]. Until more data are available, a reasonable statement is that philopatry is approximately 78% in adult females. Little is known about the dispersal of the remaining 22%.

Philopatry in females varies somewhat with age. Baker et al. (1995) showed that philopatry increased between ages 2 and 5, and decreased after age 10. A testable hypothesis is that the tendency to make brief visits to other central breeding areas, not philopatry per se, varies with age. It may be only coincidental that the age groups with the greatest fidelity to natal site (8–9-years) also have the highest pregnancy rates (exceeding 85% for 8–13-year-olds; Lander, 1981–82).

The precision of philopatry in females is important because of the potential for inbreeding. Some central breeding areas are quite large (ca. 1 km long), such that philopatry measured by that scale would mean little in terms of inbreeding. Unfortunately, neither the exact location where tags were applied or collected were recorded during the herd reduction program. To measure what that precision might be, we recorded the sites on which some tagged pups were born and looked for them as breeding adults.

RESULTS OF OBSERVATIONAL RESEARCH

Female Dispersal

The size and shape of central breeding areas make it difficult to use tag resighting to measure individual dispersal. The dispersal of females was not studied during the commercial kill for pelts because the taking of females was prohibited [3]. Therefore, no systematic way of measuring female dispersal was available.

Some information on dispersal came from recording tags at our study sites that we had not applied. Of the 469 tagged females we recorded at the East Reef study site, 40 (8.5%) had been tagged as pups on some other breeding area (had not shown philopatry). Thirty-one of these were from the Russian islands, nine were from other breeding areas on St. George Island, and none were from St. Paul Island. These results reflected the tagging effort at the different islands; Russia tagged longer than the United States, so more animals of Russian origin were seen on our study sites.

The longevity of immigrant females was nearly identical to that of local females. The forty known immigrants averaged 2.4 years of tenure (range 1–6 years), compared to 2.3 years for females tagged at East Reef (table 3.3). This implies that when females disperse and mate at non-natal sites they may persist at these sites to about the same extent as females born there. That is, females that do not show philopatry are as site-faithful as females that do. Philopatry and site fidelity may be different phenomena; females may or may not show the former, but all of them show the latter.

Precision of Return

We attempted to measure the precision of philopatry for forty-five female and forty-nine male pups [4]. None of the males and only two of the females were seen mating as adults. One female (no. 453) gave birth as a 5-year-old in the same subsector (5 ∞5 meters)as her own birth. The other (no. 148) returned for 10 years and gave birth to eight pups an average of 9.8 meters from her own birth subsector, a distance that is comparable to the population average of 8.3 meters among parturition sites (see East Reef in fig. 6.3). With present data, 8.3 meters may be used as the precision of philopatry on this site because data from the kill of females (Baker et al., 1995) showed that 78% of females on a site were likely born there. However, the value is probably conservative because of the method used to measure it.

Brief Visits to Foreign Central Breeding Areas

We found that females on foraging trips landed briefly on areas other than those where their pup resided, a behavior that was formerly unknown. These brief visits are important because they mean that the time a female spends absent from a breeding area does not always equate with time spent foraging. Brief visits also mean that the number of resident females using a site cannot be closely estimated by counting those on shore there. Our evidence for brief visits does not show their frequency or duration, or the proportion of females that make them. But it does suggest the scope of the problem.

The tendency to make brief visits may have been widespread, as shown by the results of mass tagging females in August. In August of several years, we captured groups of 50–100 females at East Reef and Zapadni by surrounding them and permanently marking each member without first determining whether it was resident (suckling) there. Typically, up to 50% of females caught in these groups failed to return the following year (compared to losses of 15% or less for suckling females caught individually). Many of the females that were lost after mass tagging were later seen suckling a pup elsewhere on the island. Most likely, these females had been making a brief visit to our study sites on the days we tagged. It is not likely that they were our residents that changed breeding sites because such changes were rare. Even if natural mortality accounted for half of the losses (which would be high), the numbers still suggest that 25% of the females we caught in August may have been making a brief visit to our sites when captured. By extrapolation, a quarter of all females ashore in August may be resident at sites other than where they are seen. That fraction is high enough to warrant further research on the question.

The duration and frequency of brief visits were poorly documented. The data we have [5] suggest that brief visits may last 1–3 days and involve foreign breeding areas on the same island where the pup resides. Whether brief visits are made to other islands is not known, but it is possible. Resolving this question would probably require instrumenting females with radio transmitters and monitoring them from all available landing sites.

EXPERIMENT I: DETERMINANTS OF INITIAL
ATTACHMENT TO SITE

A serendipitous finding in 1975 suggested that pups may form an initial attachment to a site by 30 days of age. Of twenty-seven pups that we held captive in the laboratory for one month with their mothers (chapter 1), four of them voluntarily returned there a few days after they had been re-released on their natal sites [6]. They probably walked 2 km overland to get there because no pups were swimming then (July). We again returned them to the natal site, but within 3 days they reappeared at the holding facility. Apparently their mothers were at sea feeding. Several returned a third time, and one pup made nine separate returns to the village, some requiring less than 8 hours.

Clearly these pups had formed an attachment to the laboratory site. They could discriminate this site from others and navigate to it from at least 2 km overland. Most importantly, they were attached to a site on which they had not been born. That is, the parturition site was not important to the process of site attachment. Suckling or some other experience seemed to create the attachment.

We performed an experiment in 1977 to test whether suckling or contact with peers, the two main activities on land aside from sleeping, were involved in site attachment in pups. The experiment divided these two activities onto geographically separated sites and then forced pups to choose between them after a set interval.

Five pups were captured with their mothers after the perinatal visit and were moved to the laboratory. They were housed with their mothers in separate cages for two days, during which time they had no contact with peers. They were then moved to a separate cage 1 km away and kept for 2 days with peers but without their mothers. They were moved between the two sites every 2 days for the next 30 days. This alternation replicated the suckling and periodic fasting that pups normally undergo on the central breeding area, but it equalized the days spent in each activity to eliminate time bias.

After 30 days, pups were released at a point midway between the two holding facilities and their movements were followed. None of them approached or returned to the cage where peer contact occurred; all of them returned more or less directly to the laboratory where they had suckled. The test was repeated with two of the pups with the same result. All pups and mothers were then released near their capture sites. Mothers were prevented from departing for 24 hours to ensure that pups suckled on those sites before being left alone. This procedure was apparently successful because none of the pups returned to the laboratory.

This experiment suggested that the suckling experience may be more important in the pup's initial attachment to a site than being born or having peer contact there. If other factors are involved it is not clear what they are. The attachment seems to form by 30 days of age. Whether it forms earlier is difficult to test because pup mobility, upon which the test depends, is limited before that age. Whether the attachment persists until the first mating is also unknown.

EXPERIMENT II: STRENGTH AND FLEXIBILITY OF SITE ATTACHMENT

Undisturbed females returned repeatedly to the same small area to give birth and nourish their young, apparently for most of their lives, as documented above. Is this repetitious behavior also inflexible? How readily can females shift among sites when the population abandons an area (Zapadni in figs. 2.1 and 2.13), or the climate changes?

To measure the relative strength and flexibility of site attachment, we performed an experiment in 1977 that forced females to choose between the site they preferred (as indicated by parturition there) and the site to which their young had been moved. That is, this experiment opposed the two strongest motives females had for returning to land: attachment to site and attachment to the young.

We captured eighteen female pup pairs on the day of parturition on St. George Island, and 4 pairs on St. Paul Island, using long noose poles (Gentry and Holt, 1982). We transported fourteen pairs in hand-carried cages and by vehicle to East Reef, a distance by land of from 100 meters to 10 kilometers from the capture sites. There pairs were carefully released into existing female groups near enough to the observation blind that their movements could be monitored hourly. We transported four female/pup pairs to Kitovi (fig. 1.4), St. Paul Island, by air and released them near the observation blind there. Another four pairs were sent from Kitovi to East Reef. Of the twenty-two females translocated, four had been captured with the wrong pup, and they left the translocation site immediately upon release. They were excluded from the experimental results. (The four pups that were moved in error were returned to the original capture site and all were reunited with their mothers.)

The results showed that twelve of the eighteen females (67%) remained with their young on the translocation site, making multiple feeding excursions until mid-August, when observations ended. The portion that remained with their young was the same whether they had been moved between islands or elsewhere on the same island. The duration of feeding absences of translocated females were within the limits of resident females, suggesting that returning females readily found the translocation sites without extensive searching. Whether these females weaned at the usual time is not known. One of them (♀ 1066) returned to the translocation site voluntarily the next year, gave birth, and successfully reared her young. This was one of the few times that we saw females voluntarily change breeding sites (see chapter 2). Many of the other females were seen at their initial (preferred) site with newborn.

The six females that abandoned their young did so within one-half to four hours of being released at the translocation site and were back at their preferred sites within a few hours. The inter-island females were back at their preferred sites the next day. The pups of these six females were moved back to the original capture sites and were reunited with their mothers within a day. The females resumed their usual attendance patterns. The brief interval between the departure of these females from the translocation site and their arrival at the capture site suggested that they found their way without extensive searching.

These trials suggest that the attachment females have for specific sites is not as fixed and inflexible as the year-to-year return of individuals to breeding areas suggests. Instead, the majority of females appear flexible enough to change to another area in unusual circumstances. For most females, the location of the pup seems to have more influence on the site used than does their historical preference for a site. But the fact that a third of the translocated females abandoned their young suggests that individuals vary in the strength of site attachment, mother/young bond, response to capture trauma, or some combination of all three.

EXPERIMENT III: SITE FIDELITY AND COLONIZATION

Does the willingness of females to shift among sites to rear their young extend to unoccupied beaches? That is, will females disregard traditional central breeding areas if their young are moved elsewhere? This is an interesting question because if they will, the potential of establishing a new colony exists, which is a rare event in this species. If pups form a strong attachment to a site within the first 30 days of suckling there (Experiment I) and their mothers will rear them to weaning there, the attachment they form may induce them to return there for mating as adults. If a number of pups were so treated, a self-perpetuating colony could result. The conservation implications of learning the basic principles of colonization are significant.

In 1977 we moved females and their young to an unoccupied beach and compared the proportion that adopted this new site with the proportion reported in Experiment II. A similar experiment at Robben Island (Anon., 1971) showed that such females abandon their young. However, the translocation site was not carefully selected, and it was not determined whether translocated females died or returned to their preferred sites.

To ensure that the translocation site was adequate for rearing offspring, we used a former central breeding area, Little East on St. George Island, which was abandoned about 1914 (chapter 1). No animals were using the site in 1977. The site showed no evidence of having once been a traditional breeding area. Little East was a north-facing cobble beach backed by a 2-meter-high berm just to the east of 10–20-meter-high cliffs, and 1 km west of East Reef.

Eight mature females and their young were moved from various sites to Little East and observed [7]. All females remained with their young for 6 days, a normal perinatal interval, before going to sea. In that interval they were joined by a juvenile male, 5–6 years old, that may have mated with some of them. All females went to sea within one or two days of each other so the beach was unoccupied when the first of them would have returned 2–4 days later.

No females were seen on the beach after the first trip to sea. Some may have made brief visits to their young, but these were not detected by the infrequent observation schedule. All eight females were searched for and found at their original capture sites. Their young were returned to them, reunions were successful, and mothers resumed normal attendance patterns.

The last two experiments suggest that the location of the young is important enough to make most translocated females abandon a preferred site, but only when the alternative site is another breeding area. If it is not, females will abandon the site and their young. Obviously, some feature of established social groupings is more important to translocated females than the location of the young. That factor is clearly not the presence of males [8], but may be the presence of other females. Group formation is critical for females, and separation from groups is intolerable to captives (chapter 6). These translocation experiments should be repeated using enough females so that the site is never unoccupied. Physically, this would be a very difficult task.

EXPERIMENT IV: EFFECTS OF AGE

Possibly the females in Experiment III failed to adopt the unoccupied beach because they had had prior experience on central breeding areas. If so, naive females that were selecting a parturition site for the first time should be more accepting of a nontraditional site than females that had already raised young somewhere. To test this hypothesis, we did a final translocation experiment in 1982 using young females.

Eight young females and their young were moved to a new study site on the day of parturition [9]. They all stayed a normal amount of time before departing on the first foraging trip.

The results showed that six females returned to suckle their young at least once thereafter, and four of them made one or two other visits. After the first week, no two females were on shore at the same time. The two females that did not return were located at their original capture sites and, as before, their pups were returned to them. Juvenile males joined the females almost from the start.

The six females that visited their pups eventually stopped landing, but instead swam offshore and called to their young. Two of them landed briefly at night but spent days on the nearby East Reef breeding area. Pups soon left the translocation site and joined their mothers on East Reef where they remained at least until observations ended in mid-August.

In summary, six of eight young females appeared willing to land on a nontraditional site to suckle their young, whereas none of the eight mature females in the previous experiment did. Most of them landed without the company of other females. These results suggest that experience in rearing young may reduce the female's flexibility in shifting among sites, and that young females may be less dependent on the company of females than older ones. Whether a viable colony can be established using larger numbers of young females is unknown.

DISCUSSION

In their lifetimes, undisturbed individuals are faithful to small plots of land for parturition and mating. This tendency may stabilize breeding areas for long periods. Probably most individuals are flexible enough to change among breeding sites or other islands. This flexibility may allow the population to move when the climate changes, for example when sea level fluctuates or, rarely, to colonize new islands (San Miguel and Bogoslof Islands were colonized by females tagged on the Commander and Pribilof Islands; Peterson et al., 1968b; Loughlin and Miller, 1989). But for the majority of females in the population, flexibility is apparent only when they are forced by manipulative experiments to show it. Perhaps the environment is usually too constant to trigger the flexibility of which most individuals are capable.

The prevailing view has been that northern fur seals do not readily expand onto unused land areas because of extrinsic factors such as predators, unsuitable terrain, or competition with other species. This view assumes that animals are predisposed to expand and will do so unless prevented. The present work suggests that the opposite is more often true; most northern fur seals are predisposed to use the same site unless forced to change (rarely a few will colonize a new island). Because of intrinsic factors, most animals find or satisfy key conditions near their natal sites. Absence of these conditions makes all other sites suboptimal to them. The population distribution therefore should be expected to remain more or less the same unless many individuals are forced to change their preferred sites.

The findings that females persist on their sites unless forced off and prefer to remain among females (chapter 5) partly explain why the breeding groups changed as they did during this study (see figs. 2.12 and 2.13). The inland groups nearly disappeared while shoreline groups remained relatively unchanged. By extension of the experimental results, females that had preferred sites farther inland than the last female group would abandon that site to remain among females; to do otherwise would place them alone with males, which they avoid. This process would have left inland tracts of land occupied by males only. The effect would have been progressive as the female population declined further. Female groups always extended at least 10 meters from the sea during this study, so females with preferred sites within that band were not forced to abandon their sites (see fig. 5.2B). Thus, inland areas and shoreline areas may have changed differently because of the way individuals balanced their tendencies to pup on the same site, maintain contact with other females, and reduce contact with males.

This balance may also govern changes in distribution when populations expand. Females may find their preferred sites occupied by other females and may select the nearest available inland site as an alternative, causing the population to grow outward from established nuclei onto formerly unoccupied ground. In such expansions, northern fur seals will even occupy human-built substrates if they are adjacent to traditional areas (Kogan, 1970).

The attachment that individuals have for a site seems to develop in the first 30 days of life, most likely in response to the experience of suckling. Simply being born on a site is not relevant to the attachment process, except that under normal circumstances parturition occurs within the area used for suckling. Whether this attachment to a site occurs within a critical period, as in imprinting, is unknown.

This attachment appears not to continue into adulthood for about 22% of adult females, which move to non-natal central breeding areas. A reasonable surmise is that most pups wean and leave land having formed an attachment to at least part of the area in which they suckled. Pups become familiar with a wide area surrounding their suckling site through movements during their mothers' absences (Goebel, 1988). But the experiments here suggest that most pups will not form an attachment to sites on which they have not suckled.

Future research should try to measure the distance from the natal site to subsequent parturition sites for females, and to territorial sites for males. These distances, and the site attachment that forms in the first 30 days, may form the basis for breeding areas being perpetuated for generations. Circumstantial evidence suggests that the distance may be a small (a few meters) for females. But additional data are needed from broad, flat, inland areas when the population is not declining. Males should be less precise than females because they must aggressively compete with peers for space. Their territorial locations may reflect the size, age, and fighting ability of rivals for their preferred sites.

The female's preferred site is not essential to successful maternal behavior. Females appear to use a small land site for many years, but most of the translocated females (67%) would rear their young elsewhere. The surprise finding was that 33% of the translocated females would abandon their young to reside on their preferred sites. The island of origin was not a strong influence on females because the same proportion of translocated animals accepted a breeding area on another island as adopted one on the same island.

It is likely that the presence of other females is an important precondition for females to use a land site. Experienced females would attend their young on established breeding areas, but not on beaches that lacked other females. The presence of females may be important because group formation is fundamental to females and occurs even in the absence of males (chapter 6). Experiments suggested that the presence of female groups, the location of their young, and the location of their preferred sites were the factors, in that order, that determined whether a female used a given site. Females did not land except where these conditions were met. This implies that they did not avoid juvenile male landing areas because of the low fitness of males there (the marginal male effect; Bartholomew, 1970), but because those areas did not meet the three conditions required for landing. Presence of males appeared not to be a factor; males always joined the translocated female groups that later disbanded, and females avoided them as they did elsewhere.

Young, less-experienced females seemed more flexible than older ones in that they would at least start to rear young away from female groups. Lunn and Boyd (1991) found similar differences in flexibility between young and older female antarctic fur seals. Perhaps young females are most important in colonization situations because raising young in the absence of female groups is unacceptable to the older females that make up the bulk of the population.

Data from northern fur seals support the hypothesis of Clutton-Brock (1989b) that where females do not disperse from their natal sites for mating, their age at first conception exceeds the duration of mating by their fathers. Males mate for only 1.5 years whereas age at first conception of their daughters is 4–5 years. Females could still mate with their brothers (usually half-brothers given the turnover rate in breeding males), which begin breeding at age 9–10 years, or even with their sons. If sister/brother matings were common, extensive male transfer would be expected. The extent of male transfer is at present unknown.

This work on site fidelity suggests that the central breeding area is the fundamental population unit to the individual. However, management reports population trends by island or archipelago. This grouping implies that geographic proximity somehow links them into common trends. There is no empirical evidence for such links. Fur seals respond to their immediate shore environment or to changes offshore, but do not readily move among breeding areas and therefore do not experience changes that occur on them. Because each breeding area is largely isolated from others by the site fidelity of its members, each breeding area should develop its own unique history, geographic pattern of age structure (Vladimirov and Nikulin, 1991), and possibly fecundity and mortality trends. Grouping breeding areas by island tends to obscure these processes through averaging.

SUMMARY

Probably most females and at least some males exhibit philopatry, i.e., returning to their own natal site as adults. Neither the extent of philopatry nor its precision can be judged well for either sex from existing data. Existing conclusions about the differences in philopatry by sex, island, year, or animal age should be taken as provisional. Conclusions about philopatry based on location at birth and death only are flawed by the finding that some females make brief visits to nonpreferred sites. The number that do so is unknown but could be as great as 25%. Conclusions about homing in juvenile males do not suggest the extent or precision of philopatry in adult males, for which the key measure is the distance of the territorial site from the original natal site. Prolonged observations of known individuals is the preferred approach for measuring philopatry.

The expectation that populations will expand into unused areas unless checked by extrinsic factors is probably untrue. Individuals have preferences for certain small sites that determine where they will reside. Therefore, an intrinsic factor (site preference) largely determines the population's distribution. This finding suggests that population distribution will tend to remain the same as long as the balance of factors that determines individual site use remains unchanged.

The precision of philopatry measured for females with known natal sites was similar to the precision of site fidelity for females with unknown natal sites. Some 8.5% of tagged females on our study sites were known to be immigrants, but were as site faithful year to year as females born there. Thus, site fidelity is high, regardless of whether females also show philopatry. Northern fur seals may use a smaller area for suckling than some species of otariids, but are equaled by some (antarctic fur seals) in distance among parturition sites.

No evidence exists that all breeding areas on the same island are linked by a common philopatry rate. Females and juvenile males move among land sites as if the central breeding area mattered more than the island on which it was located. That is, the movement patterns of individuals offer no justification for combining data from different breeding areas based on proximity alone. However, management routinely reports population trends by island.

The transfer rate of females among breeding areas is 20% or less, a relatively low rate. As predicted by theory in this situation, fathers stop breeding before their daughters reach reproductive age. But longevity patterns do not show that it is impossible for females to mate with their brothers or sons.

Experiments suggest that (1) philopatry results from experience in early life, (2) by 30 days of age neonates have formed an attachment to a site,

(3) suckling on the site is essential to the formation of this attachment, and (4) being born on a site and having contact with peers there are coincidental to the process of site attachment. It is not known whether this attachment is formed within a critical period, as in imprinting. This attachment process may explain how the "tradition" of using a breeding site is perpetuated across generations.

A translocation experiment showed that two-thirds of the females would rear their young on a site they did not prefer if they and their young were moved to it. The other third of the females abandoned their young if moved. This finding suggests that although females show long-term stability in the sites they use, most of them have the flexibility to change in the short term. That is, repetition in use of sites does not equate with inflexibility. Clearly, the relative strengths of attachment to site and attachment to young vary among females.

Another experiment showed that females require the presence of other females to adopt, even temporarily, a site on which to rear young. Young females were more inclined to adopt a nontraditional site than older, presumably more experienced ones. The low colonization rate in this species may result from a high degree of individual philopatry acting in concert with female attraction (colonial sites do not have preexisting groups of females for individuals to join).

The mating system centers around female site preference expressed in the context of attraction to other females. Females use a site based on the location of female groups, the location of the young, and the location of the preferred site. When stressed, females relinquish their preferred site first in order to remain with their young among other females. This preference explains why some inland areas were abandoned in the population decline from 1974 to 1988.

Estrus and Estrous Behavior

WITH JOHN R. HOLT AND CAROLYN B. HEATH

THE postpartum estrus is central to the northern fur seal's mating system (Sadlier, 1969). Estrus directly affects the timing and synchrony of breeding and the efficiency of reproduction, and indirectly affects male reproductive strategies, sexual dimorphism, and the type and extent of polygyny that can exist. Estrus and embryonic diapause (Sandell, 1990) account for the annual timing of reproduction. The synchrony of estrus among females gives males a focus for mate competition (Bartholomew, 1970; Ralls, 1977; Stirling, 1983) and enables males to monopolize females, thereby contributing to the evolution of sexual dimorphism for size.

Understanding estrus was very important to the St. George Island Program goals (Anon., 1973) because estrus mediates all the effects that a changed sex ratio (fig. 2.5) and changed social behavior might have on the population size. Estrus, and the ability of males to detect and mate with estrous females, may set the limits to human alterations of the northern fur seal sex ratio. In this species, estrus occurs 5.3 days postpartum (Craig, 1964; Peterson, 1968) and is highly synchronized (most adults mate in a 30-day period). But pregnancy rates can exceed 90% in some age classes (Lander, 1981–82), even at ratios of forty or more females per male. This chapter discusses observations and experiments that were aimed at describing the stages and duration of estrus, and any compensatory mechanisms for missed copulations, that could account for this level of reproductive efficiency.

More is known about reproductive physiology in northern fur seals than in any other marine mammal. Good data exist on the cycle of physiological estrus (Craig, 1964; Pearson and Enders, 1951; Yoshida et al., 1978), corpus luteum (Yoshida et al., 1977), embryonic diapause (Enders et al., 1946; Daniel, 1974, 1975; Daniel and Krishnan, 1969), implantation (York and Scheffer, 1997), age at first reproduction and age-specific pregnancy rates (York and Hartley, 1981; York, 1983), spermiogenesis (Oliver, 1913; Murphy, 1970), and rates of population increase that result from individual reproduction (Lander and Kajimura, 1982). However, estrous behavior is so poorly known that it is not usually discussed in reviews of mammalian sexual behavior (Beach, 1976).

From 1975 through 1981 we observed copulations in an undisturbed population and conducted experiments on estrus in the laboratory. The main questions we asked in the field were as follows: Which sex initiates or terminates coitus? How many times do females mate per estrus? How long does copulation last? What are its stages? In the laboratory, our initial question was, Do females enter estrus a second time if they fail to mate in their initial postpartum estrus? The follow-up question was, Does estrus last longer, or do its stages change, if the female does not copulate immediately postpartum? What signals (sounds, postures, odors) does the female give at estrus? Does the male affect the duration of estrus? If so, what physical cues (size, odor, behavior) are involved? How fast does sexual receptivity begin? Finally, we asked whether females reject males of low fitness (the "marginal male" effect; Bartholomew, 1970). That is, does female mate choice explain the exclusion of small males from breeding? The research emphasis was always on processes at the level of the individual or pair because these are the basic elements that determine reproduction at the population level.

Unless stated otherwise, the term "estrus" here refers to behavioral receptivity to mounting. Behavioral estrus is usually brief (reportedly 24 hours in northern fur seals: Bartholomew and Hoel, 1953; about 5 days in the northern elephant seal: Cox and Le Boeuf, 1977). However, physiological estrus (hormonal and cytologic changes) may last 4–15 days in fur seals (Bigg, 1979), and may recur within the year as a "false rut" (Bigg, 1973).

SUMMARY OF PREVIOUS RESEARCH

About 95% of northern fur seal females come into estrus from 26 June to 26 July each year. The numbers ashore and the number coming into estrus peak in the same brief period, 7 to 14 July, every year [1]. Most females that enter estrus after 26 July are young (Bartholomew and Hoel, 1953), with mixed or all black vibrissae (Scheffer, 1962). They tend to mate with young, small, peripheral males that replace adult males on territory starting in late July (Vladimirov, 1987). Therefore, estrus in adult females occurs before the end of July and is attended by adult males, while estrus in young females occurs after July and is attended by younger males.

Although estrus is evident in the population for more than a month, individual females enter estrus on a particular date, having a variance of less than 5 days (fig. 6.2). A few have a modal date of estrus early or late in the season, but most females have it between 7 and 14 July. The modal date of estrus may shift over a lifetime because females arrive in an increasingly narrow time frame as they age (fig. 6.1). The date of estrus may be greatly affected by a change in reproductive status. Estrus shifts 7–8 days later in years when individuals do not give birth, and shifts an equal amount earlier in years when they do (table 6.1).

Estrus is closely timed with other events. It occurs 6.2 days after the female arrives, 5.3 days after parturition, and 1.2 days before departure for first foraging (chapters 3, 6) with little variance. Nonmothers enter estrus 3 days after arrival. No females copulate after the first foraging trip, as in sea lions (Gentry, 1970; Heath, 1990) or New Zealand fur seals (Goldsworthy, 1992).

Most females mate at night (Antonelis, 1976), which suggests that most sexual receptivity begins after dark. Copulations peak near dawn and dusk and occur at a low rate during the day (fig. 3.10). Estrus may be terminated by a single copulation (females average 1.3 copulations per year; Bartholomew and Hoel, 1953) of brief duration (5.9–7.1 min depending on male sexual vigor; Bartholomew and Hoel, 1953). Even though there are two peaks of follicular activity per year (Craig, 1964), only one period of behavioral estrus has been observed. Ovulation appears to be spontaneous (Craig, 1964; Bigg, 1979).

Males are efficient at detecting and copulating with estrous females. Precopulatory interactions are brief. After touching noses, the male investigates the female's perineum while the female assumes a lordotic posture (DeLong, 1982); mounting follows immediately (Peterson, 1968; Bartholomew and Hoel, 1953). The blastocyst enters diapause until November, when lactation ends and the female leaves land for the year (Craig, 1964).

PART 1: FIELD OBSERVATION OF MATING BEHAVIOR

At the start of this study the available description of copulation was not detailed enough to support our experiments. We needed a better description of normal male and female copulatory behavior. Therefore, we made detailed observations of copulations on the central breeding area in 1974.

Methods

Observers recorded copulations at the East Reef study site for 16 daylight hours per day throughout the 1974 breeding season. A special form was used that included many categories of data [2]. Every observed copulation was recorded, but not all of the data were collected for each one.

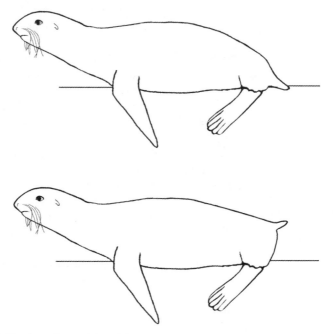

Fig. 8.1. Northern fur seal female's posture at rest and in lordosis.

Results

The following description of copulations comes from our sample of 411 copulations involving thirty-nine known males. Males initiated 97% of 148 copulations that were observed from the start. Estrous females did not approach or solicit males, seem aware of them, or avoid being approached by them as anestrous females did. Most females copulated with the male resident at their parturition site. A few females wandered aimlessly away from the parturition site at estrus onset and usually got no farther than the next territory before being mounted.

Males approached females while making a pulsed sound like the bark of a dog after laryngectomy (termed a "wicker" by Peterson, 1968). After the pair silently touched noses, the male made a woofing sound termed a "low roar." At that point, anestrous females would lunge at the male with an open mouth. Males would recoil and then try to press their noses against the female's neck or back. This touch elicited further lunging or even biting. Males often bowed their head to the ground in an apparent attempt to investigate the female's anogenital area, which caused anestrous females to turn, facing the male.

Estrous females acted very differently toward males. The initial touching of noses was often prolonged (20 sec or more) and the male clearly appeared to sniff (the nostrils flared rhythmically). Estrous females did not lunge but remained motionless with their vibrissae flexed forward, intertwining with those of the male. They showed no response when the male pressed his nose to their neck or back, and they assumed the lordotic posture when the male nosed their sacral area. Lordosis (fig. 8.1) consisted of tilting the pelvis forward until it was horizontal, and lifting it 10 cm or more above the ground.

Males with large numbers of females in their areas moved around briefly touching noses with random females at frequent intervals. Those with few females interacted less often but looked at their females frequently. Their eye movements revealed that males were selectively watching females that we knew (from their date of parturition) were approaching estrus. They rarely glanced at anestrous females.

Estrous females had a characteristic vocalization (estrus call) like the prolonged bleat of a lamb (DeLong, 1982), although this call was almost inaudible to us over the ambient noise. Some estrous females wickered like the male, simultaneously panting and flexing the vibrissae forward. Estrous females kept their eyes half closed and were rarely aggressive, even during anogenital investigation (AGI)when the male sniffed and licked the perineal area. Following AGI,the male hooked his chin across the female's back at the scapulae, which induced the female to lie prone. The male placed one front flipper across the female's back, rotated 90 , and mounted. Virtually all mounts were dorsal/ventral. Some males bit the nape of the female's neck during the mount.

Copulations had three distinct phases: precopulatory, intromission, and rhythmic pelvic thrusting (RPT).The precopulatory phase, described above, lasted on average 4.7 minutes. The intromission phase was highly variable, depending on the number of mount attempts made. The total interaction, from precopulatory contact to final dismount, averaged 7.2 minutes (SD = 0.05, n = 267).

The number of mount attempts made by the male during the intromission phase depended largely on the roughness of the terrain. Most copulations (67%) involved several mounts (mean = 2.5 mounts, SD = 2.5, n = 267). Those with a single mount usually occurred on flat ground, while those with several occurred on rough ground where the female's pelvis hung into depressions. Therefore, multiple mounts without intromission were not a prerequisite for insemination as in some other mammals (Diamond, 1970), but resulted from irregularities in the terrain. Between

Fig. 8.2. Northern fur seal male's posture during insertion attempts and during intromission.

mounts, males induced females to move by touching their front flipper or shoulder with the barred teeth but without biting.

After intromission, which was easy to recognize from the angle of the male's pelvis (fig. 8.2), females usually lay unmoving with eyes closed; 15% of them (n = 201) made the estrus call. Males behaved as follows. Initially, they made side-to-side pelvic movements with a few slow pelvic thrusts. They wickered, alternated between standing erect over the female and lying atop her, and made small stepping motions with the rear flippers. After a few minutes, males made 8–10 bursts of deep pelvic thrusts, with about 6–8 thrusts per burst. The interval between bursts became progressively shorter until pelvic thrusting was continuous and rhythmic [3]. Rhythmic thrusting lasted on average 1.3 minutes, at 164 thrusts per minute (n = 131, SD = 19). Older males usually did not perform rhythmic thrusting, only the bursts of thrusting that preceded it.

Ejaculation was not discernible [4]. At some time after intromission, the penis glans inflated to a spherical shape (fig. 8.3) that was still present at

Fig. 8.3. Penis of the adult male northern fur seal showing the inflated glans.

dismount. Males would not dismount between intromission and rhythmic thrusting except when a foreign male entered their territory.

Males terminated 79% of 341 copulations by dismounting while the females remained motionless, often with eyes still closed. In the remaining 21% of cases, females struggled and bit the male's chest and neck before the dismount. Most of the latter cases occurred at the beginning of the season when males copulated with greater vigor and attentiveness. Regardless of which partner terminated the copulation, the female was usually aggressive toward or evaded the male immediately after the male dismounted.

Few of our marked females copulated more than once. We saw 165 copulations for 144 marked females (1.2 copulations per female; 15% of the marked females copulated more than once). One female copulated three times, the remainder only twice. Some second copulations occurred 24 hours after the first, and some may have occurred at night unobserved.

Males could copulate twice in rapid succession. Several males copulated with a second female 10 minutes after dismounting from the first. The testis is compartmentalized, presumably to facilitate serial ejaculations (Murphy, 1970).

Summary

Field observations gave no evidence of female choice of mates in this species. Males initiated 97% and terminated 79% of all copulations, unlike some other otariids (Sandegren, 1976; Heath, 1990). A few females wandered away from the parturition site at the onset of estrus, but they did not move far and never sought out males. The onset of estrus was almost impossible to detect except by watching male reactions. About 15% of females gave a special vocalization (estrus call) during estrus. Pending discovery of the number of ejaculations during coitus, copulation was either type 11 or 12 in the taxonomy of Dewsbury (1972). Multiple mounts resulted from uneven terrain and were not part of a stereotyped copulatory sequence. No genital lock formed (Le Boeuf, 1967). Matings lasted on average 7.2 minutes, including a mean of 4.7 minutes of precopulatory interaction and 1.3 minutes of rhythmic pelvic thrusting at a rate of 164 per minute (total of about 200 deep pelvic thrusts per copulation). Females became aggressive soon after the male dismounted. Estrus was terminated by coitus in all but 15% of females (1.2 copulations per female).

PART 2: RECURRENCE OF ESTRUS

How do northern fur seals maintain pregnancy rates in excess of 90% for some age classes when the overall sex ratio exceeds forty or more females per male? One answer is that the overall ratio is never seen because most females are at sea on any given day [5]. But at inland areas, where females are forced through a narrow access corridor (fig. 5.3), some males gain one hundred or more females in a single night. Under such circumstances, do some females fail to copulate on day 5 or 6 postpartum? Is their estrus prolonged or does it recur later in the year? That is, does the species have some compensatory mechanism in estrus that keeps pregnancy rates high?

A second behavioral estrus during the year does not seem precluded on physiological grounds, although Bartholomew and Hoel (1953) did not observe one. Estrogen levels remain high throughout July and August (Daniel, 1974). Craig (1964) reported two peaks of follicular activity in northern fur seals, one in June–July and another in mid-August. A similar second peak has been termed a "false rut" (Rand, 1955; Bigg, 1973).

We conducted an experiment in 1975 to determine whether, for females that fail to copulate immediately postpartum, estrus can be extended beyond day 7 postpartum or recurs later in the year. The experimental design involved preventing captive females from copulating on day 5 or 6 postpartum, releasing them to a captive male at 5-day intervals after the estrus period, and observing for signs of renewed sexual receptivity.

Methods

Over a 4-week period, we captured twenty-seven females and their young on the day of parturition (indicated by pink, moist placentae) using a long noose pole (Gentry and Holt, 1982). The pairs were transported in boxes to the laboratory (chapter 6), tagged, and released with other pairs in outdoor cages. Seven days later females were allowed access to tanks of clean seawater inside the adjacent building where they were offered dead herring and squid. Those that rejected this diet were released, and those that accepted it were held for an additional period before being released to the test male. The test male was an adult animal of unknown copulatory experience that was captured from a territory at East Reef containing ten females. The male's 5 ∞6 meter cage was used as the test arena.

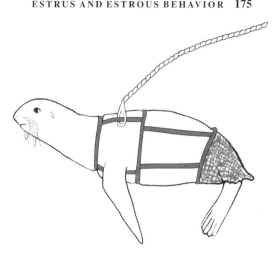

Fig. 8.4. A female northern fur seal wearing a harness and perineal shield to prevent intromission during test pairings with males.

We presented females to the male on days 1 (controls), 10, 15, 20, 25, or 30 postpartum. There were seven females in the 10-day group, and four in all others. Due to nonsynchrony in parturition dates, several females were with the male at one time. No females were tested beyond 30 days because the test male, which was fasting, became emaciated and ceased interacting with females late in July (just as wild males do).

Before release to the male, each female was removed from her cage with a noose pole, held by the neck on a yoke-type restraint board (Gentry and Holt, 1982), and fitted with a nylon harness. The harness included a perineal shield made of plastic-impregnated nylon (fig. 8.4) which prevented intromission but did not prevent normal movements and postures. Control females were released without shields as a test of the male's ability to copulate normally in this setting. Notes on behavioral interactions were spoken into a tape recorder and were later transcribed. Data were collected for 60 minutes after the female's release into the male's cage, and for another 60 minutes 24 hours later. Females were removed by pulling on a rope attached to the harness. Females and their young were then released as close to their parturition sites (recorded on a map on the day of capture) as feasible. No females were presented to the male more than once.

Results

We did not observe copulation or mounting for any of the experimental females. One female in the day 10 postpartum group allowed the male to approach closely, but did not permit AGIor any other precopulatory behavior. All other females lunged at the male or evaded interaction by moving away. By contrast, we observed copulation for three of the controls, and normal precopulatory behavior for the fourth. The latter session ended before the male tried to mount.

Summary

Twenty-three females that were prevented from copulating on days 5 and 6 postpartum showed no signs of sexual receptivity 10–30 days postpartum to a male which copulated normally with three of four control females. Therefore, it did not appear that estrus was prolonged or that it recurred if the females failed to copulate immediately after parturition.

This conclusion was confirmed by subsequent observations. None of the twenty-three experimental females that returned in the following year (1976, n = 9) bore pups despite being freely available to males on the breeding area after we released them. Between 1976 and 1980, we prevented eight other captive females from copulating, and released them onto breeding areas where they could resume normal contact with males 9 days after parturition. Only one of these females bore young in the year following treatment [6].

PART 3: DURATION AND BEHAVIORAL STAGES OF ESTRUS

Given that estrus does not seem to recur if the female fails to copulate 5 or 6 days postpartum, is the single estrus period prolonged as a means of maintaining high pregnancy rates? Do females undergo marked changes in behavior before and during estrus that increase their chances of being detected by males?

We tried to answer these questions through observations at Zapadni, where females arriving from sea were forced through a narrow access corridor that created extremely high sex ratios. Probably no better site existed in the Pribilof Islands to see copulations missed as a result of high sex ratios. On 30 June 1976 we marked thirty-five females in the vanguard of the rapidly arriving female population and observed them 16 hours per day for 7 days until all had departed for sea. We witnessed copulation for only twelve, few of which showed any behavioral changes preceding estrus. Those we missed may have copulated at night. It was apparent that we could not measure prolonged estrus by observing wild animals, so we conducted a second round of experiments in July 1976.

Methods

We captured eighteen mother/young pairs on the day of parturition and caged them as described in Part Two, except that we gave them unlimited access to seawater tanks. We observed females casually until day 4 postpartum when we removed them singly, fitted each with a harness and perineal shield, and released them into the test arena attached to a rope. The test male had no females in his territory at capture, although he was as large as other adults.

We used the first ten females mainly to define the behavioral categories to be used for scoring test sessions (table 8.1) [7]. We paired the others with the male for 15-minute test sessions, or until the pair performed the behavioral component ♂ *Mount*/ ♀ *Lordosis* (used as the operational definition of estrus [8]), whereupon we ended the test. After estrus started, we tested females a maximum of four times daily (3–4 hours apart) until they were no longer receptive. During each test, we kept score on a time-ruled check sheet at one-minute intervals [9], while remaining hidden from view. After all tests had ended, we decided how to calculate the duration of estrus, and how to estimate its onset and end [10].

After each test session, we subjectively judged the female to be "fully receptive," "nonreceptive," or "partly receptive," according to the behavior just seen, and scheduled future tests accordingly [11]. Later we calculated a receptivity quotient (RQ) from the behavioral scores as a numerical check on these judgments [12].

While handling restrained estrous females we noticed that we could elicit lordosis by lightly stroking the labia with a gloved finger or touching the skin over the pelvic bones. We used this response to measure three separate phenomena: (1) the duration of estrus (if sexual receptivity ends when lordosis can no longer be elicited by human touch, the duration of estrus could be measured without using a test male; we restrained two females four times daily until we could no longer cause lordosis by touch, and then released them into the male's cage without perineal shields); (2) extent of the sensitive area (we screened eleven females for the extent of the pelvic area over which palpation would elicit lordosis [Noble, 1979a]); (3) effect of the male (in some other mammals [Kow and Pfaff, 1973–74; Noble, 1979b], contact with a male strengthens the lordosis response in females; we tested the same eleven females for the strength of their lordosis response [rated as low, medium or high] before and after test sessions with the male).

Soon after females were found to be nonreceptive in tests with the male (always by day 9 postpartum), they and their offspring were released close to their parturition sites. No feeding in captivity was required.

Table 8.1

Definitions and Descriptions of Behavior Used in Scoring the Sexual Behavior of
Northern Fur Seals in Captive Experiments

1. ♂ *Approach*. Male moves to within 1 m of female.
2. ♂ or ♀ *Nose*. Either animal touches the nose of the other with its nose while flexing the vibrissae forward.
3. ♂ *Touch Neck*. Male presses his nose against any part of female's body anterior to the scapulae.
4. ♂ *Touch Back*. Male presses his nose against any part of female's body between scapulae and pelves.
5. ♂ *Anogenital Investigate (AGI)*. Male licks, noses, or sniffs the perineum, tail, or heels of female's rear flippers.
6. ♂ *Initiate Female Movement (IFM)*. Male mouths the dorsal surface of female's front flipper or bites the scapular area. See 14.
7. ♂ *Aggression*. Male threatens or lunges with an open mouth, strikes with the chin, bites, or bowls female over.
8. ♂ *Wicker*. A pulsed, nonvocal breath sound produced by male, resembling the bark of a dog after laryngectomy.
9. ♂ *Mount*. Male stands astride female, supporting his weight on his front flippers; usually *more canum*.
10. ♀ *Open Mouth Lunge (OML)*. Female lunges as if to bite but does not make contact.
11. ♀ *Evade*. Female moves away from male a short distance or pivots to remain facing him.
12. ♀ *Soft Bite*. Female gently bites and tugs male's skin without apparent attempt to injure.
13. ♀ *No Response*. Female remains immobile in response to male actions.
14. ♀ *Change Position*. Female moves forward or pivots pelvis slightly while being mounted. Response to no. 6 above.
15. ♀ *Lordosis*. Forward (downward) flexure of the lumbar spine, which raises the pelvis and perineum off the ground.
16. ♀ *Aggression*. Female rushes at male or bites hard in an apparent attempt to injure.
17. ♀ *Appease*. Female initiates soft bite without any preceding action by male.
18. ♀ *Estrus Call*. A vocalization given during estrus that is softer, higher pitched, and less directed than the pup attraction call which it resembles.
19. ♀ *Sleepy*. Female keeps eyes half closed, seems oblivious to surroundings.
20. ♀ *Pant/Flex*. Female pants with open mouth while flexing the vibrissae strongly forward.

Results

Estrus began 6.0 days postpartum in captives (n = 13, SD = 1.2), compared to 5.3 days on the breeding area. All females left their young and entered the water pools for the first time the day after estrus ended, just as wild females depart for their first feeding excursion.

Fig. 8.5. An estrous northern fur seal female flipper-waving during a rainstorm as if overheated, and showing the behavioral components ♀ *Pant/Flex*, and ♀ *Sleepy*.

The spontaneous signs of estrus were easier to see in captives than on the breeding area. For example, all captives produced the estrus call that was recorded for only 15% of copulating females in the field. Some of them also wickered loudly, and others produced a moaning/snoring sound (not previously reported) that was inaudible beyond a 5-meter distance. All estrous females had a characteristic look [13]. Some wandered slowly about the cage with neck held stiffly, eyes closed, vibrissae flexed forward, and swinging their head slowly from side to side. Many of them waved their rear flippers as if they were heat stressed, even during rainstorms (fig. 8.5). No females directed their movements toward the male in the next cage, but some of them froze and went into the lordotic posture when the male wickered. Females that had reacted strongly to loud background sound (truck noise) before estrus showed no response

Fig. 8.6. *Left*: An estrous northern fur seal female holding a low intensity lordosis posture in response to mounting by her 5-day old male offspring which mounted correctly and made coordinated pelvic thrusts. (Same female as in figure 8.5.) *Right*: An estrous northern fur seal female holding an exaggerated lordotic posture in response to anogenital investigation by another estrous female while being pulled from the cage with a noose-pole.

to it after estrus began. Not all females showed all of these spontaneous signs of estrus, but all females showed more than one from this list.

Lordosis appeared to be reflexive in that it was easy to elicit by touch even when females were forcibly restrained by the neck. Several females assumed the lordotic posture when mounted by their own young (fig. 8.6, *left*) or in response to AGI from other estrous females (fig. 8.6, *right*). Estrous females that charged us when we entered their cage would stop and present lordosis if we touched them on the sacral area with a pole.

The area within which light pressure or scratching caused lordosis included the dorsal side of the sacral and pelvic areas. Females touched on the flanks would tilt their pelvis but not present lordosis. Scratching over the lumbar and thoracic areas had no effect, but firm pressure there caused strong lateral flexure of the spine. Light stroking of the labia and heels of the rear flippers, along with rubbing the ventral surface of the coccyx, elicited the most intense lordosis. Light scratching on one side of the tail caused the coccyx to lift in an arch and deviate to the opposite side and the opposite side heel to rise off the ground. This might have been a compensatory sequence for helping the male achieve intromission.

The mucous membranes did not become discolored or engorged, nor did they produce unusual secretions during estrus. The labia were always black and gaped open loosely. Estrous females sometimes strained as if defecating, causing the vagina to evert slightly and expose the mucous membrane lining.

Females stopped assuming the lordotic posture to human touch before they stopped being receptive to the test male. The two females for which we tried to estimate estrus through touch alone copulated immediately when paired with a male.

Contact with the male increased the intensity of all female responses. Females that showed the component ♀ *Sleepy* became almost comatose when the male mounted. Even when no intromission occurred, the co-malike condition was so profound that these females had to be physically carried back to their home cage. The eleven females showed more intense lordosis to human touch after contact with the male than before (74% of thirty-one tests), and the same intensity in the other 26% of cases. Typically, they changed from merely lifting the coccyx to full lordosis after contact with the male. The same females tested during anestrus showed no such effects; all of them made lateral avoidance movements of the pelvis whenever they were touched (twenty-two tests). Females did not enter estrus 3 days sooner if influenced by the male than if they began spontaneously (contra Bigg, 1979).

By our method, the duration of estrus in the absence of copulation was 34.3 hours (n = 7, SD = 12.9). The end of estrus was estimated to within about 6 hours because the average interval between the last "receptive" session and the first "nonreceptive" session was 12.3 hours (see note 10). This number indicates that females usually went out of estrus overnight.

Virtually the entire behavioral repertoire changed during estrus in type and frequency (fig. 8.7). The best single indicator that females were either fully or partly receptive was that they showed no response when the male touched their neck with his nose (table 8.2) [14]. At any other time, females lunged at the male or bit if males nosed their neck (table 8.3) [15]. The best evidence that females were nonreceptive was that the male was aggressive toward them, which they rarely were just before mounting.

The RQ scores showed that our subjective assignment of females into the nonreceptive, fully receptive, and partly receptive classes was valid (the scores were significantly different from one another [16]). About 78% of all behavior that occurred when females were fully receptive was in categories that were not seen at all when they were nonreceptive, and that were seldom seen (12% occurrences) when they were partly receptive. Females switched their behavior so completely when receptive that receptive and nonreceptive behaviors were almost mutually exclusive [17].

All the categories of behavior reported in figure 8.7 occurred in the wild, but not always as often. For example, the male made unusually persistent efforts to investigate the female's perineal area, possibly because the shield

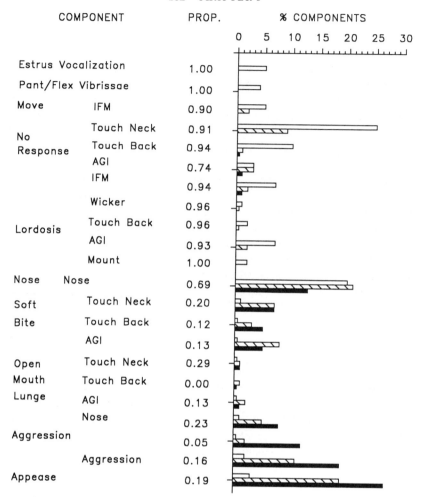

COMPONENT		PROP.
Estrus Vocalization		1.00
Pant/Flex Vibrissae		1.00
Move	IFM	0.90
No Response	Touch Neck	0.91
	Touch Back	0.94
	AGI	0.74
	IFM	0.94
	Wicker	0.96
Lordosis	Touch Back	0.96
	AGI	0.93
	Mount	1.00
Nose	Nose	0.69
Soft Bite	Touch Neck	0.20
	Touch Back	0.12
	AGI	0.13
Open Mouth Lunge	Touch Neck	0.29
	Touch Back	0.00
	AGI	0.13
	Nose	0.23
Aggression		0.05
	Aggression	0.16
Appease		0.19

Fig. 8.7. Single and dual ($\male$/$\female$) behavioral components scored during estrous tests showing the proportion of each response that occurred in the same test session as the component $\male$ *Mount*/$\female$ *Lordosis*, and the distribution of each component among three stages of receptivity. See table 8.1 for definitions.

prevented him from receiving normal olfactory or taste cues. Also, the male would only mount the female while facing in the observer's direction, and he appeared less able to move females around than males on breeding areas. Therefore, scores in the components $\male$ *AGI*, $\male$ *IFM*, and $\male$ *Touch Back* were higher than they would be in the wild, and female responses were inflated accordingly.

The onset of estrus (indicated by $\female$ *Estrus Call*, $\female$ *Sleepy*, and $\female$ *Pant/Flex*) most often began during a test session with a male (13/18 females),

Table 8.2

Classes of Female Responses to a Male, Their Occurrence in Different Stages of Receptivity, and the Contribution of Each Class to the Total Female Behavioral Repertoire

Female Response[a]	Fully Receptive[b]	Partly Receptive	Not Receptive	% of all Responses[c]
"Spontaneous"	100	0	0	8
Lordosis	95	5	0	11
No response	91	8	1	43
Move	90	10	0	5
Open mouth lunge	22	28	50	6
Soft bite	17	34	49	13
Aggression	5	10	85	4

[a] All responses defined in table 8.1. "Spontaneous" = ♀ *Sleepy*, ♀ *Estrus Call*, and ♀ *Pant/Flex* combined. This table sums all similar female responses without regard to the type of male behavior eliciting it.

[b] Stages of receptivity are defined in part 3.

[c] Does not equal 100% because of the exclusion of *Nose* responses (almost equal occurrence in all stages of receptivity), and of *Evade* (too few to analyze).

but it could also begin spontaneously (4/18 females began estrus with no prior male contact, 1/18 began estrus between test sessions). Our 15-minute test sessions were too brief for all females to make the full transition from aggressive to receptive behavior [18], but most of them started the process.

When estrus began, the behavioral changes were very sudden. Within a 5-minute block, female aggressive components (♀ *Aggression*, ♀ *OML*, ♀ *Soft Bite*) could be replaced by immobility (♀ *No Response*), which could in turn be replaced by receptivity (♀ *Lordosis*) (table 8.4). Sudden shifts of this type occurred both at the start (n = 11) and end (n = 2) of estrus [19]. All these sudden behavioral shifts occurred after interactions with the male, not just from proximity with him.

Summary

Behavioral estrus lasts on average 34.4 (± 6) hours if females do not copulate. Therefore, prolonged estrus does not account for high pregnancy rates at high sex ratios. Females undergo profound behavioral changes at the onset of estrus. Probably all females give a special estrus call that is audible to nearby males on breeding areas but inaudible to distant humans because of noise masking. Estrous females pant, make a snoring sound, look overheated and sleepy, and flex their vibrissae forward. They assume the lordotic posture in response to male sounds, or when specific parts of their

Table 8.3
Areas of Female's Body Touched by Male's Nose, and
Rate of Interactions by Stage of Receptivity

Body[a] region	Fully Receptive		Partly Receptive		Nonreceptive	
	%[b]	Rate[c] (n/317)	%	Rate (n/135)	%	Rate (n/315)
Nose	21	1.3	21	0.9	27	0.4
Neck	26	1.6	14	0.6	10	0.2
Back	24	1.5	8	0.3	7	0.1
Anogenital	11	0.7	12	0.5	9	0.1

[a] Body areas are defined in table 8.1.

[b] % = proportion of all touches directed to various body regions in this stage of receptivity.

[c] Rate = behavior scored per minute of observation.

[d] Stages of receptivity are defined in part 3.

bodies are touched, whether the touch is from a pup, other females, or a human. During estrus the vulva does not change color (Miller, 1974; Bartholomew and Hoel, 1953), but mucous membranes may show as a result of straining. The extent to which males perceive and respond to the above stimuli is not known.

Some estrous females wander but neither direct their movements toward males nor respond appropriately to their surroundings. Therefore, wandering does not represent a search for or a comparison of males.

Females are aggressive or evasive to males before estrus and rapidly shift to not responding to them when receptivity begins. A "no response" to the male is equivalent to full receptivity. Many behavioral components change during estrus. The extent to which a given component's presence signals receptivity indicates how much its absence signals nonreceptivity. Contact with the male increases the intensity of lordosis and other responses. Mounting induces in females a comalike state so profound that voluntary movements may be lost. Female behavior changes rapidly at the onset and termination of estrus. About 74% of females shift to or toward full receptivity within a 15-minute test, apparently as a result of contact with the male. This change recalls the Whitten effect, in which the male actively induces the onset of estrus (Schinckel, 1954; Whitten, 1956a, 1956b; Parkes, 1960; Shelton, 1960).

The parturition-estrus interval appeared different in the laboratory than in the field (6.0 vs. 5.3 days), probably because of our testing schedule. Captivity did not change the tendency to enter water after estrus, or the kinds of behavior used in estrus. Some male behavior, and female responses to it, occurs more frequently in captivity. Studies in captivity permit a better analysis of the spontaneous signs of estrus than is possible on breeding areas.

Table 8.4

Changes in Female Responses to Males at the Onset of Estrus,
Suggesting the "Whitten Effect"

Female Responses[a]	A. ♀ #13 Minutes			B. ♀ #16 Minutes		
	1–5	6–10	11–15	1–5	6–10	11–15
Aggression	2[b]	0	0	1	0	0
Open mouth lunge	7	2	0	8	4	1
Soft bite	6	3	13	6	2	0
No response	1	2	6	0	24	13
Lordosis	0	0	0	0	0	2

Female Responses[a]	C. ♀ #15 Minutes					
	1–5	6–10	11–15	16–20	21–25	26–30
Aggression	3	0	0	0	0	0
Open mouth lunge	0	0	0	0	0	0
Soft bite	0	0	1	1	0	0
No response	0	0	3	11	11	3
Lordosis	0	0	0	3	9	6

[a]Terms defined in table 8.1.

[b]Entries represent total occurrences of each response in 5 minutes.

PART 4: ABBREVIATING EFFECTS ON ESTRUS OF COPULATION WITH VARIOUS SIZE MALES

The above research showed two important aspects of copulations in this species. One is that the male may induce the onset of estrus (like the Whitten effect), which has not been reported for any marine mammal or for many mammals in the wild (Conaway, 1971). The other is that coitus has a marked abbreviating effect on the length of estrus (seen in other otariids but not in phocids).

These two processes can be used to test the extent to which direct female mate choice exists in this species [20]. Specifically, can males of low fitness induce and terminate estrus as successfully as adult males, and do estrous females reject males of low fitness if they are available? If the "marginal male effect" acts in this species, females should actively reject juvenile, peripheral, and senescent males because of their low fitness relative to territorial males (Bartholomew, 1970; Cox and Le Boeuf, 1977). An alternative to female mate choice is that females are receptive to all males but are denied access to males of low fitness by male/male competition.

In 1980 and 1981 we conducted tests to address the following specific questions: (1) How quickly does receptivity wane and aggressiveness return following coitus with an adult male? (2) What is the shortest duration of estrus when both its onset and termination are linked to male contact? (3) Are females receptive to males that are usually excluded from breeding? (4) If allowed to copulate, can these males terminate estrus as readily as adult, territorial males? and (5) Do such matings result in pregnancies?

Methods

We captured forty-two females and their pups and housed them as described previously. Starting on day 4 postpartum, females were fitted with a perineal shield and paired with a test male while their behavior was scored as before. When they became receptive (an RQ score of < 0.01 was used to define nonreceptivity) they were treated differently, depending on which of the five questions they were being used to answer [21].

To address questions 1 and 2, receptive females were removed from the test arena, their perineal shields were removed, and they were paired with a "stimulus" male in the adjacent cage. After copulating with this male, they were removed, the perineal shields were replaced, and they were returned to the test male an average of 3.4 minutes later (SD - 1.3, n = 8) to reassess receptivity. These females were held no more than 48 hours after receptivity waned before being released with their young at the parturition site (captivity was no longer than the usual perinatal visit to shore).

Four females were paired with males of different sizes to address questions 3–5. Females were paired with juvenile and peripheral males [22] in the laboratory. But those to be paired with senescent males were driven to a male landing area, released wearing a harness with a rope attached, and a senescent male was brought up using a bamboo pole. After the interaction was scored, the females were returned to the laboratory and reintroduced to the test male (10 min elapsed time). We did not test whether juvenile, peripheral, or senescent males could terminate estrus if allowed more than one copulation with a given female.

Results

Peripheral males proved best for the role of test male because their persistence and high interaction rate gave us multiple chances to assess the female's receptivity. Territorial males could apparently assess a female's condition from a single, brief nose contact and would stop interacting before we could assess her responses.

Aggressiveness returned very quickly following coitus. Most females (7/9) were fully aggressive in the time it took us to handle and return them to the test male (3.4 min after copulation). Actually, five of them resumed aggression during the last few minutes of the copulation. One female was still partly receptive during the postcoitus test (RQ = 0.20), and one was fully receptive after coitus and remained so for 32 hours.

The shortest duration of estrus for most females was less than an hour (15, 31, 41, and 50 min for four females that entered estrus during contact with a male and had their estrus terminated by coitus). These durations are conservative because females were not always allowed to copulate immediately after they were shown to be receptive (they were subjected to other tests; see part 5). A fifth female (the same one referred to in the previous paragraph) continued to be receptive for 32 hours after coitus.

Juvenile males could not usually induce a rapid onset of female receptivity as larger males could. Unlike adult males, juveniles seemed to be deterred by the aggression that females directed at them. (The importance of male aggression in the process leading to female receptivity is discussed in part 5).

Once females had entered estrus (had allowed a test male to mount them), they showed full receptivity to males of all sizes (peripheral males 30/30 pairings, juvenile males 15/16 pairings, senescent males 2/3 pairings). The RQ scores of females paired with juvenile, peripheral, and senescent males were not significantly different from those in thirty-four pairings with adult, territorial males (ANOVA, p > 0.93).

Juvenile, peripheral, and senescent males could not terminate receptivity by copulation as effectively as territorial males could. Thirteen of fifteen females that copulated with such males [23] remained fully receptive to a test male immediately after coitus. That is, these males were much less able to terminate estrus than were territorial males, which failed to terminate estrus in only 22% of captives and 15% of females on the central breeding area (see part 1).

Despite the fact that copulation with juvenile and peripheral males did not abruptly terminate estrus, these matings were nevertheless fertile. In 1981, we observed nine females that had copulated with juvenile (n = 8) or peripheral males (n = 1) the year before and with no other males. Seven of these females produced young in 1981. The other two probably did not, although they were seen infrequently.

Summary

Captive females entered estrus by a process resembling the Whitten effect in 39% of forty-nine cases for which we saw the onset of estrus in 1976, 1980, and 1981. The proportion entering estrus this way may be greater on breeding areas where females have more frequent contact with males. Estrus may be as brief as 15 minutes if it is terminated by coitus. Receptivity wanes and aggression returns fully by at least 3.4 minutes postcoitus (a conservative estimate). Aggression resumes during the latter minutes of copulation for many females.

Estrous females are as sexually receptive to juvenile, peripheral, and senescent males as they are to prime territorial males. Contrary to theoretical prediction, females do not reject these males of supposedly lower fitness. They usually mate with adult, territorial males because other possible mates have been excluded by male/male competition. Females show no evidence of an ability to discriminate among males once they are receptive. They present sexually to pups, other females, and humans. Once they are receptive they will mate with any male present.

Juvenile males are unable to trigger the fast onset of receptivity in females, and juvenile, peripheral, and senescent males are generally not able to terminate it with a single copulation, as territorial males routinely do. These findings could be interpreted as evidence for mate choice if rapid entry into or rapid termination of estrus were mandatory for impregnation. However, this was not the case. Most females that copulated with 45 kg juvenile males gave birth the following year despite not entering estrus abruptly or having their estrus terminated by copulation. That is, ovulation did not depend on a rapid onset or termination of estrus. It appeared to be spontaneous, just as it is among females that do not copulate at all (Craig, 1964; Bigg, 1979). Juvenile males (45 kg, 3–4 years old) were fertile (contra Murphy, 1970). Females were kept in captivity until their receptivity had ended, so none of them would have copulated with an older male after release.

PART 5: PHYSICAL STIMULI INVOLVED IN ONSET AND TERMINATION OF ESTRUS

Given that males and females respond to each other very quickly at the onset and end of receptivity, what are the sensory cues that mediate these responses (Caldwell et al., 1989)? In 1980 and 1981 we tried to identify some of these cues as an aid to understanding the mechanisms that ensure high pregnancy rates at high sex ratios.

Female odor may be important in male detection of estrus as it is in other mammals (Huck et al., 1989). Northern fur seal males make brief nose-to-nose contact with females during their interactions. Can they discriminate estrous and anestrous females using only olfactory cues from the female's facial area? We tested males for this ability. We also recorded the type and frequency of male behavior during interactions to determine whether the behavioral sequence itself may have been important to the rapid onset of receptivity.

Some male characteristics may act as stimuli which trigger a rapid onset or termination of receptivity in females (as in land mammals; Diamond, 1970; Goldfoot and Goy, 1970; Carter, 1973; Lodder and Zeilmaker, 1976; Slimp, 1977; Irving et al., 1980; Carter et al., 1989; Erskine et al., 1989). For example, male odor is important in mediating the Whitten effect in land mammals (Parkes, 1960; Carter et al., 1989). All male otariids have a strong odor (Hamilton, 1956). Fasting northern fur seal males have a musky, fruity breath odor, somewhat resembling the ketone breath of fasting humans (Cahill et al., 1970) [24]. Juvenile males lack this odor unless forced to fast for more than 2 weeks.

The male's pelvic thrusting pattern may act as a physical stimulus that terminates female receptivity. Mice have a "vaginal code" in which male thrusting pattern is critical to the induction of pseudopregnancy (Diamond, 1970) [25]. We looked for evidence of a vaginal code in northern fur seals by testing whether vaginal stimulation alone could terminate estrus in the absence of other male stimuli, as it can in some other mammals. We also compared the pelvic thrusting rates of males that succeeded and failed to terminate estrus with a single copulation.

Other adult male characteristics, such as body weight, size of the penis glans, and hormones in semen, may play a role in terminating estrus. We tested whether any of these correlated with rapid termination of estrus.

Methods

To determine whether males could discriminate estrus just from the female's face, we placed the noses of females into a $4 \infty 8$ cm hole cut in the otherwise solid wall of a male's cage and called him over by knocking on the wall. For the next 5 minutes we noted (on a stopwatch) the number of seconds that the male's nose was within 25 cm (estimated) of the female's. Females were put to the fence during pre-estrus (day 4 postpartum), estrus (day 6), or postestrus (day 8). The stage of estrus was determined before each test by pairing the female with a test male, as described before. Some females were presented at all three stages, none was presented twice in the same stage, and all had their noses exposed to a different male than the one we used to test their receptivity.

To test for the existence of a vaginal code, we artificially stimulated the vaginas of six females known (by pairing with a test male) to be in estrus. The females were held on a restraint board while a lubricated PVC rod, 20 cm long, and having a smooth $8 \infty 3$ cm knob at the tip (slightly smaller than the spherical glans of an adult male) was inserted into the vagina. The rod was moved laterally for 3 minutes and then thrust longitudinally for 2 minutes at 160 thrusts per minute in imitation of adult male penile move-

Table 8.5

The Ability of Adult Male Fur Seals to Detect Estrus from Odor Alone

Male[a]	Anestrus			Estrus[d]		
	Mean	SD	N	Mean	SD	N
Jr.	8.7	7.5	7[b]	26.0	16.9	8
Bea	69.3	57.0	15[c]	126.3	71.4	7

Notes: Detection ability is shown as the number of seconds within a 5-minute test session that males remained within 25 cm of the female's nose when all other cues to receptivity were precluded. Stage of receptivity was determined through tests with a male different from those listed.

[a] Differences between the males in time spent near females was significant (ANOVA, anestrus, p = .016; estrus, p = .004).

[b] All tests made before estrus onset.

[c] Includes seven tests before estrus onset and eight after estrus waned.

[d] Differences between stages of estrus were significant for male Jr. (ANOVA, p = .028), and barely significant for male Bea. (ANOVA, p = .057) due to large variability.

ments. Immediately thereafter, females were paired with a test male to assess their receptivity. Pairings were repeated four times per day until all signs of receptivity had ceased.

If weight were an important factor in a male's ability to terminate estrus, then males that could normally terminate estrus should fail to do so if prevented from using their weight. We prevented males from bearing their weight on females during coitus by presenting estrous females under a support (part of a steel drum) that conformed to body contours while leaving the female's pelvis exposed.

These combined tests used all forty-two females mentioned in part 4. Some females were used in several tests when treatment in one would not contaminate the results of others. The number of animals in some tests was smaller than ideal because of the physical effort these manipulations required.

Results

Males spent more time near the noses of females that were in estrus than those that were not (table 8.5). For one male, the difference (significant) was nearly threefold, and for the other (almost significant) it was nearly double. The males differed from one another in the amount of time they spent near female noses at either stage of receptivity (table 8.5). We could not detect any odor from the female's nose, even at a distance of 2 cm (Block et al., 1981).

Active male suppression of female aggression was important to the rapid onset of female receptivity. Tables 8.2 and 8.4 show that female aggression waned before receptivity began, but they do not show the male behavior that preceded it. One male [26] sequentially suppressed female aggression by striking or threatening until all resistance had ended. Males used slightly different techniques, depending on their size and experience. Juvenile males withdrew when females lunged at them, which may largely explain why they could not induce a rapid onset of receptivity like larger males (part 4).

Artificial stimulation of the vagina did not terminate receptivity. Most of the females accepted vaginal stimulation with the same lordosis and comatose immobility as we observed during normal copulations with male fur seals. However, all six females were fully receptive to the test male after vaginal stimulation, and remained so for an average of 33.3 hours (SD = 13.8) afterward.

The physical stimulus responsible for the rapid end of estrus was a combination of the species-typical thrusting pattern and large body size (likely related to size of penis glans). Males with a normal copulatory pattern (a pelvic thrusting rate of 160/min, and an RPT phase lasting 1.5 min) that were of small body size (four juveniles and five peripheral males, reported in previous section) usually failed to render females unreceptive after a single copulation (8/9 cases). Conversely, large males that had an atypical pelvic thrusting pattern were also generally unable (7/8 copulations) to terminate estrus. These large males were in two classes; senescent males (100 thrusts per min and no RPT phase, 2/3 failed to terminate estrus), and adult males copulating with females in the steel drum apparatus (either 100 thrusts per min, or no RPT phase, 5/5 failed to terminate estrus). These tests suggest that large size combined with normal thrusting pattern are the physical stimuli that lead to termination of receptivity with a single copulation.

Tests with the steel drum device were inconclusive about the role of male weight in terminating estrus because of the atypical thrusting pattern they caused. However, senescent males applied approximately the same weight to the females as did territorial males, but receptivity did not end in two of three copulations.

Semen and its hormones were apparently irrelevant to the rapid termination of receptivity. Sperm cells were found in vaginal swabs from eight females after nonterminating copulating with juvenile (6) and senescent (2) males. Microscopic measurements showed that the largest cells in each sample were of mature size (about 54μ; Oliver, 1913).

The strong, musky odor of adult males was apparently irrelevant to the abrupt ending of estrus. The odor was noted during four copulations with juvenile males, seven with peripheral males, and two with senescent males. Estrus ended abruptly after only one of these thirteen copulations (with a peripheral male).

Summary

Males may use cues (possibly olfactory) from the female's nose or mouth to identify estrus. Brief nose-to-nose contacts are the most frequent type of male/female interaction. Captive males spent two or three times longer near the noses of estrous compared to anestrous females when all other interaction was precluded. We could not detect an odor.

Male aggression seems integral to male-induced estrus onset. Serial, aggressive suppression by adult males of female aggressive and evasive responses was associated with the rapid onset of receptivity (within 15 min), whereas withdrawing from female aggression (juvenile males) was not. Male aggression may force or facilitate the female's entry into full receptivity.

Northern fur seal females may have a vaginal code (Diamond, 1970). Copulations involving both adult body size (possibly penis glans size) and species-typical pelvic thrusting pattern usually ended in terminated estrus but copulations lacking one or both factors usually did not. Juvenile males (small size) with a normal thrusting pattern, and senescent males (adult size) with an atypical thrusting pattern were unable to terminate receptivity in their partners. Artificial stimulation of the vagina alone was insufficient to terminate receptivity in females. However, these trials may not have been conclusive due to the small size of the test rod.

Male weight, presence of semen, and male odor were irrelevant to the rapid termination of estrus.

DISCUSSION

Reproduction in northern fur seals includes no mechanisms, such as prolonged or redundant estrus, that compensate for missed pregnancies. The population maintains high pregnancy rates even at high sex ratios because both sexes have traits that contribute to maximizing the chance of impregnation while minimizing the time spent mating. These traits apparently function across a wide range of sex ratios. The efficiency with which males detect estrous is probably density dependent and could limit pregnancy rates when females are very numerous. However, we did not detect such a drop in efficiency because all our observations were made at very low sex ratios.

The traits that foster reproductive efficiency are as follows. Only a third of all females are ashore on a given day, which reduces the sex ratio within which males must detect estrus. Females form dense groups in male territories, which increases the chances of males finding estrous females. Males are efficient at detecting estrus in females, they can copulate multiple times quickly, induce the onset of estrus (the Whitten effect; see Schinckel, 1954; Whitten, 1956a,b; Shelton, 1960) and can terminate it by coitus. Most females mate once per estrus, they remain receptive for 34 hours or until copulation occurs, and estrous females are receptive to all males. Finally, individuals enter estrus at highly predictable times and places, thus spreading matings and permitting females to mate in an orderly succession rather than forcing them to compete in a chaotic scramble.

The extreme brevity of estrus and efficiency of reproduction have advantages for both sexes. Sexual presentation during estrus is reflexive, not conditioned on male fitness (estrous females will present sexually to pups, other females, and humans). When estrus onset is male induced and its termination is coitus induced, receptivity may last only 15 minutes. This brevity benefits males by decreasing the time spent with each mate and increasing time available for finding other mates, that is, it increases male reproductive success. In females, universal receptivity and brief, coitus-terminated estrus decrease the number of males contacted for reproduction, thereby reducing the risk of injury during male/female interactions (chapter 5). Brief estrus also reduces the female's risk of being stolen by solitary males (Francis, 1987), and it hastens her resumption of feeding.

The male's role in the onset of estrus is facultative rather than obligatory because females will spontaneously enter estrus in the absence of a male. Males can only trigger the onset of estrus in females between days 4 and 8 postpartum. The male's influence seems to be exerted through aggressive suppression of female resistance rather than through odor (Parkes, 1960; Bigg, 1979; Carter et al., 1989). Juvenile and senescent males are less aggressive to females than other males, which may explain their inability to trigger estrus onset. In the face of this suppression, females change from being aggressive to being fully receptive within 15 minutes. Experienced males will not interact extensively with females that are not near estrus. They seem capable of detecting estrus from odor as other mammals do (Huck et al., 1989), and perhaps use sounds and visual signs that are specific to estrus.

The male's role in the termination of estrus is also facultative. In about 85% of cases, receptivity ends after a single copulation, perhaps by some mechanism like the vaginal code (Diamond, 1970). If females do not copulate, estrus wanes after about 34 hours. Juvenile and senescent males usually are less able than adult males to trigger the onset of estrus, and usually cannot terminate it with a single copulation, probably because of their atypical thrusting pattern and/or penis glans size. (Possibly the diameter of the glans is the key feature; the length of the *os penis* is linear with age [Scheffer, 1950], and therefore cannot account for the different estrus-terminating abilities of territorial and senescent males.) Females that are receptive do not reject juvenile or senescent males and are impregnated by them.

If male-induced onset and termination of estrus were required for pregnancy, then the fact that only fully adult males can trigger these responses would be evidence for female mate choice. However, copulations that begin spontaneously and those that are not terminated by coitus nevertheless are fertile, implying that ovulation is not induced but spontaneous (Craig, 1964; Bigg, 1979). That is, stimuli from males that trigger the onset and end of estrus are separate from ovulation (Terkel et al., 1990). Female behavior may have co-evolved with the physical and behavioral characteristics of adult males, but spontaneous ovulation may have evolved in response to a short breeding season and relative scarcity of males (Conaway, 1971).

Active female mate choice appears not to exist in the northern fur seal. Given the opportunity, estrous females will freely mate with any male in the laboratory, on male landing areas, and, by extension, on breeding areas. Therefore the reproductive exclusion of young, old, and other possibly less fit males results from intrasexual selection (Wilson, 1975; Short, 1984). Females do not contribute to this selection process by actively rejecting any male. In the laboratory they appear unable to appraise male fitness and they have no behavioral mechanism for rejecting low-ranking males. This is not an artifact of captivity; copulations in the new San Miguel Island colony in 1968 were mostly performed by males of a size that is reproductively excluded at the Pribilof Islands (DeLong, 1982; see photographs in Peterson et al., 1968; Peterson and Le Boeuf, 1969; Le Boeuf, 1972). The female's equal acceptance of all males probably arose because available mates are usually those which have been "selected" by male competition.

The tendency of estrous females to wander merits discussion because it has been cited as evidence for female mate choice and lek formation (Boness, 1991). Most northern fur seal females do not wander at estrus. Those that do cover little ground, usually reaching only the next territory before being mounted. Wandering is the start of the comatose condition that females enter fully once they are mounted. Such females resemble sleepwalkers; their eyes are closed, they lack appropriate responses to their surroundings, and they appear unable to direct their movements. They make no discriminations but only react automatically. Females at this stage will present sexually to humans and will accept simulated copulation with a plastic rod. If they do not differentiate as to species, then wandering cannot be a mechanism for discerning fitness among male fur seals. Female Steller (Gentry, 1970) and Hooker's sea lions (Gentry and Roberts, in prep.) wander similarly. In all three species, wandering represents the onset of receptivity to all males, not an active search for the most fit individual.

Northern fur seals retain many traits of estrus seen in terrestrial mammals, but have developed some differences along with their aquatic adaptations. For example, stimulation of the back, flanks, pelvic area, and perineum elicit lordosis in fur seals as in rats and hamsters (Kow and Pfaff, 1973–74; Noble, 1979b). But the heels of the fur seal's feet also elicit lordosis because these moved adjacent to the vulva when the legs became included within the torso during the evolution of flippers. As in rats, the sensitive area increases in size and responsiveness at the peak of receptivity (Kow and Pfaff, 1973–74), especially after contact with a male. Female fur seals, like dogs (Le Boeuf, 1967) may deviate their tail bone on being mounted. The fact that their tail is too short to interfere with intromission suggests that tail deviation is a vestigial reflex in fur seals.

Coitus has a more rapid and profound abbreviating effect on estrus in northern fur seals than in most other mammals (Goldfoot and Goy, 1970; Slimp, 1977). A single intromission of 4-minute duration abruptly terminates receptivity within 3 minutes of dismount in fur seals. Most land mammals (except the chinchilla; Bignami and Beach, 1968) require longer or more frequent intromission for abbreviation to occur, and the effect is more decremental, often taking 10 hours to complete (compare rats: Blandau et al., 1941; cats: Whalen, 1963; sheep: Parsons and Hunter, 1967; hamsters: Carter, 1973; and guinea pigs: Slimp, 1977). Fur seals have no refractory period after estrus onset in which the abbreviating effect of coitus is absent. Differences in the production of sex hormones between fur seals and terrestrial mammals (Carter et al., 1989; Caldwell et al., 1989) are worth investigating.

SUMMARY

Northern fur seals have no mechanism, such as prolonged or repeated estrus, that compensates for missed copulation and keeps pregnancy rates high at high sex rations. High pregnancy rates depend on efficient male detection of estrus. Behavioral estrus (receptivity to the male) is as brief as 15 minutes when its onset and termination are affected by a male, and as long as 34 hours when males are absent.

Males initiate 97% and terminate 79% of copulations. Females do not seek mates. Females undergo extensive behavioral changes at estrus which males presumably identify. Estrous females have an unusual gait and facial expression, have a special vocalization, and probably produce an olfactory cue from their nose or mouth. A few females make unusual, undirected movement away from the birth site at estrus. Males detect estrus by watching females closely, and by making brief nose to nose contact with them.

Estrus onset is induced by contact with an adult male in at least 39% of cases. This resembles the Whitten effect in terrestrial mammals except that males trigger estrus through active suppression of female aggression, rather than through odor (males that are not aggressive toward females cannot induce rapid estrus onset). However, estrus begins spontaneously in females that have no contact with males.

Females undergo a more profound and rapid behavioral change at estrus onset than many other species of mammals. About 74% of females change from being aggressive to being fully receptive and accept a mount within 15 minutes. They first suspend aggression, then fail to respond to being touched, then accept perineal investigation, assume a lordotic posture, and finally become so torpid that many are unable to make voluntary movements. Even brief contact with a male greatly increases the intensity of lordosis and the extent of body area which elicits it when touched.

Some evidence for a vaginal code (Diamond, 1970) exists in this species. Copulations of normal duration (7.2 min) that have a terminal period of rhythmic pelvic thrusting (160/min for 1.5 min, or about 200 thrusts) will terminate receptivity abruptly in about 85% of females. Artificial stimulation of the vagina using this thrusting pattern is not sufficient to end a female's receptivity to males.

The physical stimuli responsible for terminating receptivity abruptly appear to be the size of the inflated glans combined with the species-typical male thrusting pattern (male weight, odor, and semen play no role). Males lacking in one or both of these traits (juvenile, peripheral, or senescent males) are less able to terminate estrus through copulation than robust, adult males. However, this difference is reproductively irrelevant because pregnancy results regardless of how estrus begins or ends, suggesting that ovulation is spontaneous.

The 15% of females that do not terminate estrus after a single copulation rarely mate more than twice. Copulations are type 11 or 12 (taxonomy of Dewsbury, 1972), depending on the as yet unknown number of ejaculations during coitus.

In controlled tests, females were as receptive to males of low social status (juvenile, peripheral, and senescent males) as to adult, territorial males. Once in estrus they appeared unable to discriminate, and had no mechanism for rejecting males of different fitness, or even males of their own species. Estrous responses seem to be reflexive because even humans can elicit them.

The characteristics of estrus are appropriate for maintaining high pregnancy rates at a wide range of sex ratios. Males need attend most females just once. Because estrus is brief, females rapidly leave the breeding area to forage after mating. This removal reduces the female's chance of injury during male/female interactions, hastens the feeding of young toward weaning at 120 days, and improves male chances of detecting and mating other estrous females.

The timing of estrus relative to parturition and departure, the kinds of behavior used, and the mother-young bond are affected little if at all by taking females into captivity. Captive females show receptive responses (including lordosis and acceptance of artificial vaginal stimulation) to humans, even when physically immobilized.

Ontogeny of Male Territorial Behavior

YOUNG MALE northern fur seals do not possess the behavioral or physio-logical traits that are required for prolonged territorial defense. Since male territoriality brings the sexes together for mating, the ontogenetic changes leading to territoriality carry a high selective premium. As young males grow and age, they change the ways they relate to space, their location, food, and each other. Learning these behavioral changes requires social contact and frequent practice. The ontogenetic changes in females are less profound than in males. Females join the breeding group at age 2 years, mate at 3–5 years, and thereafter maintain the same social role for life. The ontogeny of male behavior is as important to the mating system as the postpartum estrus in females.

Landing areas are the arenas in which male ontogenetic changes occur. Nonbreeding males of all ages gather there, except those of about 8–12 years, which reside at breeding areas. At the start of this study, the kill for pelts had removed so many males from the population that landing areas were depauperate. Several years passed before landing areas developed the size and structure that probably typified the species before contact with humans began.

The landing areas at St. George Island were affected faster and more profoundly by the cessation of the male kill in 1972 than were any other aspect of the population. They began increasing in 1973 by an annual incre-ment of at least 4,000 animals (the approximate number taken in the 1972 kill; National Marine Fisheries Service, 1973). By 1975, the change in numbers of juveniles was obvious. By 1977, a new landing area had formed at East Reef (chapter 2), and animals on all landing areas had begun to segregate by size. By 1978, landing areas were as large and complex as they would be at any time in the next 10 years.

Growth of the landing areas also began to have a secondary effect on central breeding areas. In 1975, a noticeable increase occurred in the num-ber of juvenile males that made brief, tentative runs (called "forays") across the central breeding areas. Adult males chased these juveniles, with the result that female distributions changed, pups were killed, and more ag-gression occurred among adult males. By 1977, juvenile numbers had in-creased such that a foray might involve twenty or so animals simulta-neously. Similar disturbances probably occur in all otariids. They have been reported for South American sea lions under the term "group raids" (Campagña et al., 1988).

Because of the importance of ontogenetic changes in male behavior, the changes on landing areas, and the potential harm forays could inflict on breeding areas, we began a three-part study of nonbreeding males in 1977. The first focused on the movements of marked juveniles. The primary questions were: To what extent do individuals move among landing areas and between islands? What is their attendance behavior (on and off shore for feeding)? Do juveniles fast while on shore? What proportion of the juvenile population is on shore daily? Do marked animals maintain social bonds or affiliations with others?

The second part of this study involved direct observations of groups and social behavior of anonymous animals on landing areas. The study dealt with the ways behavior changes with size. We began it by asking: What is the typical male size structure on landing areas? Do males of different size move within their group, or do they keep different distances among them? How do males of various sizes relate to the terrain? How does size dominance translate into territorial behavior with age? What behavioral changes accompany this shift? What is the role of play in male ontogeny? How do males acquire blocking behavior that is related to herding females?

The final part of the study focused on forays occurring at the East Reef and Zapadni study sites. The questions included: What triggers a foray? When do forays occur throughout the day and season? What size males make most of the forays? Are forays coordinated group actions? If not, why do they appear simultaneous? Are forays a mating strategy? If not, why do juveniles risk injury to make them? Where do forays occur relative to where the same males defend territory later in life? Finally, why did forays decline as the population of adult males increased?

The present chapter attempts to answer all the above questions. The questions posed in the first of the three parts were addressed in a previous publication (Gentry, 1981b). They will be reviewed here briefly as background.

REVIEW OF PAST STUDIES

Populations on Landing Areas

The first experimental work on juvenile males began in 1896 (Scheffer, 1950), focusing on the growth and homing behavior of the males being killed. The social behavior of animals on landing areas, their size structure, and their movements to sea were not studied even though all of these affected the number available for killing. The most remarkable feature of juvenile male behavior, that they used the same areas despite 200 years of continual killing for pelts, was never studied.

These populations begin to grow in mid-May when early-arriving juveniles are driven off breeding areas by adult males. Their numbers increase until early July, level off or decline through July, then increase again in early August when 2-year-old males arrive. Accurate counts of juveniles are impossible after mid-August because some females, which resemble juvenile males in size, move onto landing areas then. The females usually form a group. Occasionally a female gives birth and mates there. Landing areas are abandoned in September when juvenile males move back onto central breeding areas to defend quasi-territories among females. Juveniles start their southward migration in mid-October.

Populations on landing areas also include some peripheral males (see definition, chapter 2), especially in June and July. In August, they move onto breeding areas where they defend territories and mate with nulliparous females (Vladimirov, 1987).

Adult males use landing areas briefly before they begin regular territorial occupancy on breeding areas in May and June. Some adult males visit these areas briefly if they have been driven from their territories by fights. Badly wounded and senescent males reside on them for long periods without leaving.

Dynamics of Landing Areas

Landing areas are populated in reverse order of age (Bigg, 1986), with 2-year-old animals arriving in August. Few yearlings return to land. The youngest cohorts are the most numerous due to natural mortality with age.

The number of seals on landing areas fluctuates daily as animals go to sea in the evening and arrive from sea in the morning. Weather affects this daily pattern. Rain or intense sun drives them to sea, and overcast skies with no wind favor their being on shore. Most animals are on shore in midday when, coincidentally, animals are least active (Gentry, 1981b; also see fig. 3.9).

Individuals spend about 75% of their days ashore at one or two sites (Griben, 1979; Gentry, 1981b), one of which is usually near their natal area (Baker et al., 1995). They move among different landing areas in an unpredictable sequence, landing on an average 4.2 of the 22 landing areas available on both islands within a season and staying for brief periods. At least 15–25% of juvenile males use landing areas on both islands (Griben, 1979). They did not emigrate to St. George Island to avoid the commercial pelt kills at St. Paul Island (Griben, 1979). Visits to landing areas other than near the natal site decline with age up to 6 years at least (Baker et al., 1995).

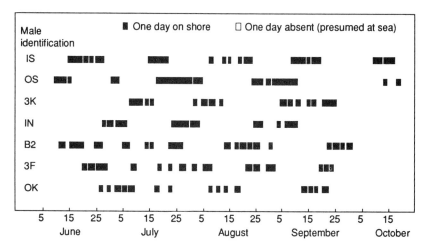

Fig. 9.1. Attendance behavior of seven marked juvenile males on landing areas of St. George Island in 1978 showing the clustering of days on shore into "visits" separated by relatively long periods at sea. To combine males with distinct (example ♂ 1S) and less distinct patterns (example ♂ 3F) into a single population, visits were defined as being separated by a minimum of 10 consecutive days absent. (Reprinted from Gentry, 1981b.)

The individuals on a landing area change daily, although the area may be occupied for several months. Juvenile males have an attendance behavior that is less predictable than that of adult females. They "visit" the island for a block of days (average 11.6, SD - 8.7), during which time they may be seen on several different landing areas. Then they are absent for variable periods, when they are not seen at all (fig. 9.1). Individuals make on average 3.1 such "visits" to land over an 80-day season [1]. They probably fast on these visits, because they lose 20–30% of body mass over a number of days (Baker et al., 1994). Their absences average 8.4 days overall, but this includes many 1–2-day gaps that are probably not long enough for foraging. Foraging probably occurs on less numerous, long absences (75% of 7,562 days of absence occurred during absences of ≥ 10 days; fig. 9.2). Weighing studies confirm that animals gain mass on long absences (Baker et al., 1994).

There is no evidence that social bonds or affiliations develop among those on shore, despite the fact that most animals have an affinity for a given site. Animals move to sea and back and among landing areas as individuals, not in groups, and pairs do not consistently rest or play together. Only about 19% of marked animals are seen on shore daily, with substantial daily and seasonal variation (fig. 9.3). Thus, most juveniles are at sea most of the time, and they form groups with dynamic membership when ashore.

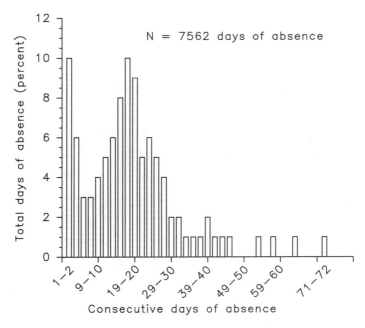

Fig. 9.2. The percentage of days absent that occurred during trips to sea of various duration in 1978. Most absences (52% of 1,106 recorded) were 1–2 days long. But about 75% of the days of absence (n = 128 males) occurred on very long (10 days) trips. (Reprinted from Gentry, 1981b.)

Juveniles are highly tenacious in using landing areas, however loose their individual attachments may seem. Frequent (weekly) kills were held on some of these sites for over two centuries without causing juvenile males to abandon them for safer sites.

METHODS

Size-Related Behavior

We studied size-related male behavior in 1977 and 1978 from an observation blind at Zapadni that afforded a clear view of males of all sizes. We defined five size classes, from very small (Class 1) to adult (Class 5), based on phenotypic characters [2].

We picked focal males when they arrived from sea, and scored their behavior in 5-minute blocks for one hour or until they established a rest site. We tried to follow all five size classes in equal numbers. In 1977 we recorded the sizes of all males with which these focal animals interacted [3], and the means focal males used to get a rest site. In 1978 we recorded

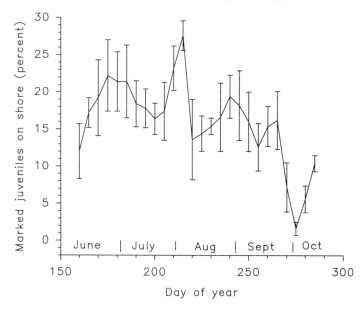

Fig. 9.3. The mean proportion of the marked juvenile male population (n = 128) that was seen ashore each 5 days throughout the 1978 season. On any day a few may have been ashore but unseen. Vertical bars are SD, calculated for each increment. The average of all 5-day blocks was 19%. (Reprinted from Gentry, 1981b.)

the occurrence of adultlike movements, regardless of the number or size of opponents.

We did not score "play" as a category of behavior, even though most of what we observed was playful. The difficulty of using play as a category is that the term implies intent [4], which often shifted within a bout. As an alternative, we scored the movement patterns (components) that comprised play, especially those that resembled movements used by adult males in fighting, threatening, interacting with females, and feeding. I avoid using the terms "play bout" and "play partner" because they also imply intent.

Forays

We recorded forays at the Kitovi and Zapadni breeding areas in 1978 to compare their frequency in populations where there was a kill and where there was none [5]. We also recorded them from 1981 to 1986 at East Reef, Zapadni, and Kitovi breeding areas, although never at all sites simultaneously, and not always over comparable dates. After 1978, we only recorded the forays of size class 4 and 5 males because forays of smaller males had become too numerous to record completely. Our intent was to record the general trend in forays over time.

We used the foray data to test the hypothesis that, as adults, males defended the same parts of the breeding area across which they made forays as juveniles. All subsectors visited by each identifiable male on a foray were listed with all subsectors the same males later defended as parts of adult territories, and the number of corresponding subsectors was counted. The probability that these correspondences occurred by chance alone was calculated for each animal using the formula

$$E = \frac{j!}{m!(j-m)!} \cdot \frac{(t-j)!}{(a-m)![(t-j)-(a-m)]!} \cdot \frac{a!(t-a)!}{t!},$$

where E = the probability of juvenile and adult using exactly the observed number of sites in common; j = number of sites used as a juvenile; a = number of sites used as an adult; t = total number of sites available on the study area; and m = number of sites that correspond between juvenile and adult years. E was calculated as a percentage for each animal. High values imply that correspondence of sites occurred through chance; low values imply that correspondence resulted from active selection by the adult for sites it used as a juvenile.

RESULTS

Size Segregation and Behavior

Before 1977, each landing area held only a single male group that included all but the largest males. In 1977, the size of groups on landing areas increased, males began to segregate by size, and the number of landing areas occupied increased (chapter 2). The smallest males (Classes 1 and 2) formed a group 100 meters or more inland from the sea, often in tall vegetation. Males of size Class 3 formed a group closer to the sea, usually on trampled vegetation. Males of Class 4 either lay alone widely scattered, or joined with Class 5 males on the beach, along the berm behind it, or adjacent to a central breeding area if one was nearby. These groupings were evident until mid-August.

The groups differed in density because individual spacing increased with body size. Small males often lay in contact with one another, large males spaced several meters apart, and medium-size males kept intermediate spacing.

Group dynamics also differed by size. Class 1 and 2 males did not maintain fixed positions within their groups; each animal had the potential to contact every other member. Class 3 males were similar, but much of their

Table 9.1

Size Preference during Interactions of Nonbreeding Males
at Zapadni Landing Area in 1977

Size Class[a]	Overall[b] χ^2	"Preferred" Size Class	Cell χ^2 of Preferred[c]
1	246.4	2	144.2
2	204.0	1	127.0
3	55.6	3	22.4
4	26.2	2	16.5
5	101.8	4	79.8

[a] Size class 1 = smallest, 5 = largest; see text.
[b] Vs. tabled $\chi^2_{(.001,4)} = 18.5$.
[c] Indicates the portion of the overall χ^2 contributed by the preferred size class.

behavior was associated with topographic features (logs, rocks). Class 4 males tended to keep fixed positions when resting with similar-sized males but moved around when resting with smaller males. Class 5 males kept fixed positions within their groups so that each had contact only with his immediate neighbors. The smaller males remained in a group except when they were moving to sea or back, and they never rested among large males. Large males, on the other hand, could rest alone or join groups of small males.

Size-related dominance accounted for many of the differences in these groups. Size decided the outcome of virtually every interaction on the landing area except those involving two Class 5 males. Size dominance was especially obvious in the way animals selected rest sites; 40.7% of males (n = 17) obtained a rest site by forcibly displacing another animal [6]. This is an unexpectedly high proportion given the large amount of space available. None of the seventeen displaced a larger animal, but three of them displaced an animal of equal size.

Size segregation in groups resulted partly from the size preferences animals had for opponents (preference is shown by relatively large cell χ^2 values in table 9.1) [7]. Males of three size classes preferred to interact with animals that were one size class different than themselves. The males of only one size class (Class 3) preferred to interact with same-size animals. The null hypothesis that animals had an equal probability of interactions with opponents of all sizes was rejected for all classes ($p < .001$; table 9.1).

The frequency of interactions (see note 3) declined with male size. Larger males spaced farther apart and their interactions decreased in frequency. Class 1 and 2 males had 478 and 501 interactions, respectively (adding focal and opponent roles together), while Class 5 males had only 153. We observed the size classes with nearly equal frequencies, so the difference reflects the reluctance of smaller males to select Class 5 males as

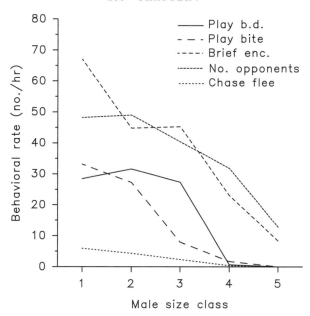

Fig. 9.4. The number of opponents faced by and the behavior of five size classes of male fur seals in 1978. The graph shows the total score for all observed members of a size class (23, 27, 25, 32, and 25 individuals observed in classes 1–5, respectively) per observation hour (range 9.1 to 11.8 observation hour per class). Class 5 males were adults (see note 2). The data were plotted as continuous to emphasize the continuum in male sizes.

opponents. Large males not only had fewer interactions, they had fewer opponents per hour when they did interact (fig. 9.4).

The type of behavior males used also changed with size. Neutral nose-to-nose greeting responses declined with size. Chase/flee, in which opponents chased and fled from each other in turns, also decreased with size (fig. 9.4). Larger animals seemed disinclined to adopt the subordinate role required by the "fleeing" portion of the chase/flee interaction. Play b.d. (boundary display) also declined with age, as did play biting (fig. 9.4). Interestingly, the decline in biting occurred at about the size when the canine teeth had become long enough to inflict serious injury (class 3). Class 1 and 2 males could puncture the opponent's skin, but not lacerate skin, blubber, and muscle as Class 3 and larger males could.

The most interesting change in behavior was the shift from size-dominance to site-specific behavior. Class 1 and 2 males moved widely, never residing on a piece of terrain for more than a few minutes. Class 1 males held chest-to-chest pushing contests interspersed with biting and lunging at opponents' foreflippers. Class 2 males were more likely to make fast head-

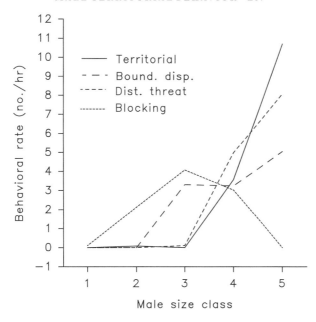

Fig. 9.5. Differences in the rates of some behavioral components as a function of male size in 1978. Class 5 males were adults. Hours of observation were the same as for figure 9.4.

feinting movements from the stereotypical "boundary display posture." Class 3 males behaved like Class 2 males except that they located their interactions in space using rocks or logs. The behavior of all three classes basically depended on the size of opponents.

The shift toward more site-specific behavior was apparent in Class 4 males. They spent long periods on a site and threatened others that approached it using the boundary display posture, much like territorial males. They rarely gave the metacommunication signal associated with play (see note 4). Instead, their behavior was more serious and adultlike than that of smaller males. However, they tended to remain on station for only a few days at a time.

Site-specific behavior was fully developed in Class 5 males (fig. 9.5). Territorial behavior (denying opponents access to a specific piece of terrain) was never seen in Class 1 males but was very frequent among Class 5 males. "Distance Threat," a vocalization bellowed at approaching animals (termed a "trumpeted roar" by Peterson, 1965, 1968), was diagnostic of site-specific behavior. It was not used until males began defending a fixed piece of terrain.

Blocking behavior, the movement used by adult males to herd females, was seen on the landing area only among medium-sized (Class 2 and 3) juveniles, and only when a size difference existed (fig. 9.5). Class 1 males usually lacked any smaller ones to block, and Class 4 and 5 males excluded from their areas the very males that might have triggered this behavior. Class 1 males were capable of herding, as evidenced by their herding of pups on the breeding area. The reason why blocking behavior was not seen in males of all ages on landing areas was simply lack of context (presence of a smaller animal).

Forays

Forays usually began with a trigger event, some disturbance that distracted adult males from their usual vigilance over territorial borders. Small males would lie prone around the periphery of the central breeding area until the territorial male nearest them was momentarily distracted, then dash onto the breeding area. Fights, air traffic, human movements, males abducting females, the arrival of a new male, and similar events all triggered forays. Also, the foray of one male could act as a trigger for others, with the result that a number of males often made a foray at nearly the same time.

The distance threat was responsible for a local disturbance spreading across the breeding area and allowing a foray of many males. Disturbed males made this call, and others in a wide area would look in the direction of origin and roar as well, thereby ignoring their own borders and allowing small males near them to start a foray.

Females sometimes used the disturbance caused by a foray to change their locations on the central breeding area. During several forays at Zapadni, twenty or more females moved uphill en masse when the male blocking them was momentarily distracted by a foray elsewhere.

Most males on a foray made tentative, aimless movements that in no way resembled the directed, resolute actions of serious territorial contenders. They spent most of their time looking in all directions as if disoriented or fearful. They never approached territorial males, and their contacts with females were brief and incidental to their path of movement.

Forays involving large numbers of males were disruptive but infrequent. In one year, 85% of forays involved either one or two males (Zapadni, 1978, n = 311 recorded forays), and only 1.7% were considered large (involving 7–20 males). Forays occurred during all daylight hours but tended to be least frequent near dawn and dusk, when most juvenile males were at sea.

Most males involved in forays were small, wandering animals. The proportions in size Class 1–5 were 4.7%, 19.5%, 39.8%, 27%, and 8.9%, respectively (n = 984 males, Zapadni, 1978). Therefore, almost two-thirds of the males making forays were too small (Classes 1–3) to keep permanent

Table 9.2

Forays per Day by Size Class 4 (Nearly Adult) and 5 (Adult) Males across
Study Sites at the Pribilof Islands (by year)

	Kitovi			East Reef			Zapadni		
Year	Forays	Days	Rate	Forays	Days	Rate	Forays	Date	Rate
1978	15	27	0.6				354	34	10.4
1981				14	35	0.4	34	37	0.9
1982				16	40	0.4	49	21	2.3
1983	4	16	0.3	18	31	0.6			
1984				12	39	0.3	9	24	0.4
1985	3	22	0.1						

Note: The greater rate in 1978 resulted from an increased population of juveniles
following cessation of the male kill, and the subsequent decline resulted from growth
of the peripheral male population.

attachments to sites. Most forays made by Class 4–5 males involved wandering animals, not peripheral males that were resident immediately
around breeding areas.

Forays happened more often when the population included many juvenile males (the size that made most forays) than when it did not. For example, forays were much more frequent at Zapadni in 1978 (following 6 years
of no kill for pelts) than at Kitovi (where the kill was ongoing) in the same
year (table 9.2).

Forays happened less often when the population included many peripheral males than when it did not. The near doubling of peripheral males at
Zapadni between 1978 and the 1980s (reported as males without females in
fig. 2.3) was accompanied by a marked decline in number of forays (table
9.2). Peripheral males actively suppressed forays by intercepting all approaching males. They drove away males their own size and intimidated
smaller males, thus preventing the accumulation of males that usually
made forays. In short, as the juvenile population increased due to the cessation of killing, small males made forays until some of their number attained
peripheral male status and thereafter prevented forays by others. The male
population was self-regulating in terms of forays.

The disturbance a foray caused depended on how vigorously the territorial males reacted to the intruder, which in turn depended on their size.
Territorial males only vocalized or feigned charges at small intruders but
chased and bit males that were large enough to be potential territorial chal-
lengers. In 72% of forays in which chases were recorded, the intruders were
size Class 4 or 5 (Zapadni, 1978, n = 40 forays involving sixty-nine chasing
males). Also, 72% of the males that were bitten while making a foray were
Class 4 or 5. A single large intruder could be chased by as many as five
territorial males. The disturbance to females and pups from these chases

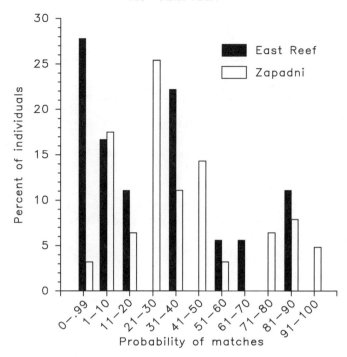

Fig. 9.6. The probability that the degree of correspondence between sites visited by juvenile males during forays and sites defended by the same males as adults occurred by chance. Figure combines eighteen males from East Reef and sixty-three from Zapadni.

could be substantial because adult males were heedless of other animals during a chase.

Forays were not a mating strategy for most males. Males on a foray established a new territory in only 2.8% of 1,251 of cases, and no males in these forays copulated. Territorial establishment and mating were reported at only a slightly greater rate in the South American sea lion (Campagña et al., 1988).

Most territorial males defended as territories parts of central breeding areas across which they had formerly made forays as juveniles. Eighteen marked males from East Reef and sixty-three from Zapadni were observed both as juveniles on forays and as adults on territory. The distribution of E for these males (probability that the degree of correspondence between juvenile and adult sites occurred by chance) was nonrandom and skewed to the left (fig. 9.6). About 78% of East Reef males and 64% of Zapadni males had E values less than 40%, indicating that correspondence was more likely to have resulted from active selection of sites than from chance alone.

Transition to Territory

Most males acquired breeding territories by first holding a territory on the periphery for one or more seasons. By 1982, 10 years after the last kill, the number of males following this strategy was so great that peripheral males had formed a band several ranks wide around most of the breeding areas. The innermost ranks abutted the territories of fully adult males that lacked females, and the outer ranks sometimes bordered landing areas. This band is probably a normal feature of all populations that are not subjected to a kill.

Peripheral males defended rest sites but not specific, identifiable boundaries around them. They never fought, they made occasional brief trips to sea, could easily be driven from their sites by a human, but rarely made forays like males of equal size that resided on landing areas. Peripheral males were a transition between juvenile and fully adult males in terms of size, temperament, and physical position. They usually acquired their first territory among females by replacing larger males when they abandoned their sites in late July.

DISCUSSION

The most remarkable feature of nonbreeding males (mostly juveniles) is that they did not abandon traditional landing areas even after two centuries of killing. The basis of this persistence may be that juveniles show no evidence of an ability to learn based on the experience (capture and death) of others. The individual is not tied to others on land through social bonds.

Some juveniles use surprisingly few of the land areas available, considering that they are free to wander anywhere. They spend most of their time at one or two sites, one of which is usually near their natal breeding area, and make brief sorties out to a few other sites. Others visit distant islands (40% of tagged seals taken in Russia in the 1970s were from the Pribilof Islands), and some remain to breed there. Therefore the male population includes individuals with conservative tendencies that foster mating in the same locations over time, and others with tendencies to wander that ensure the intermixture of seals from different islands. This balance may stabilize breeding locations while preventing inbreeding effects.

The ontogeny of male behavior involves a progressive shift from size-related dominance, in which site is irrelevant, to a territorial system in which site is paramount. Size-related dominance allows males to enter and leave aggregations without loss of social status. Territoriality allows males to remain fixed for long periods while mating. Territorial males respond to small males using dominance, not territorial, behavior.

Movements used in the above shift can be summarized as follows. Small males bite and push during fluid interactions that are unrelated to terrain. Medium-sized males inhibit biting in favor of lunging and feinting matches that are centered around topographic features that change often. Nearly grown males space far apart, defend an unchanging rest site, stop using the metacommunication signals of play, and interact mainly in ways that avoid body contact. Adult males defend identifiable boundaries, have mostly ceremonial interactions with fixed neighbors, and rarely contact males of smaller size. That is, males stop using potentially injurious behavior and increasingly isolate themselves with age. Mass and the ability to inflict serious injury also increase with age. The extent to which the first sequence is caused by the second is not known.

The shift from dominance to territoriality is mirrored in the size-related segregation seen on landing areas. The preference of Class 4 and 5 males for beaches and berms relegates smaller males to more inland areas. The smallest males (Class 1 and 2) move farther and more unpredictably than the medium-sized (Class 3) males, possibly because their behavior is not at all site-specific.

The final stage in the onset of territoriality is that males begin defending definable borders instead of rest sites, and the borders they establish are among or near females. This change is easily measurable by the firmness with which males defend their area against human encroachment.

The ontogeny of male behavior consists mainly of reorganizing adult-like, complex movement patterns that are present soon after birth. For example, by 5 days of age male pups can mount estrous females correctly and make coordinated pelvic thrusts (see fig. 8.6, *left*) without learning from adults. The form of a given movement pattern does not change much through maturation, but its frequency of use may change as behavior becomes site-specific; the metacommunication signals of play cease, and neutral nose-to-nose greetings wane at that point.

Forays involving multiple males are not a result of group action but are nearly simultaneous movements of unaffiliated individuals. They appear simultaneous because the factor that usually prevents them from occurring—the attention to boundaries of territorial males—can be suspended in all parts of the breeding area at once by a sudden disturbance. The simultaneous movement of many males onto the breeding area can give the false impression of concerted group action.

Coordinated group action is not likely in species having a life history like that of northern fur seals. Males act on their own behalf against similar-acting males at all ages, first in loose assemblages and later in territories. The skills needed in individual size dominance and defense of individual territories do not prepare the individual for the kinds of complex social bonds and conditional alliances that coordinated group action requires.

Most forays do not represent a reproductive strategy (Campagña et al., 1988). Fewer than 3% of forays ended in the establishment of a new territory, a low success rate if it is assumed that males making forays intend to mate. However, this is not a reasonable assumption given that most males that make forays are sexually immature, show only secondary interest in females, or are too small to defend a territory against adult males. Interpreting forays as a reproductive strategy forces the question of why sexually immature animals should risk injury for an opportunity to mate.

A testable hypothesis is that juveniles making forays and adults defending territories are both attempting to hold residence on the sites of their own birth. This hypothesis, if supported, would explain why adult males defend sites they visited on forays as juveniles and the low success rate of territorial establishment during forays. Forays may be part of a prior residence effect. By visiting their preferred sites yearly as they mature, males may establish priority over them with reference to their age mates and thus gain advantage in competing for those sites as adults.

The most parsimonious interpretation of multimale forays is that they are nearly simultaneous but unrelated visits to the central breeding area of males having dissimilar incentives. Forays may be linked to philopatry for small males; they may be a serious attempt to establish a territory only for the largest males. Size composition during a group foray is unimportant; neither small nor large males require the presence of the other to make a foray.

Forays of large groups are not likely in a stable or pristine population with a normal age structure. Such populations naturally have a large population of peripheral males that suppress forays by smaller males. Forays would only be common in populations in which these peripheral males were missing and young males were numerous, such as a few years after the kill ended at St. George Island in 1972 and at St. Paul Island in 1985. Peripheral males effectively create a buffer zone around breeding areas that increases their stability. The names applied to them by sealers, "idle" or "excess males," were highly inaccurate.

SUMMARY

Landing areas are important to the breeding aggregation because they provide the setting in which males acquire fighting and territorial maintenance skills and practice some of the movements used to interact with females.

The populations on landing areas are more dynamic than on breeding areas. Landing areas are used for a shorter season than breeding areas, much of the population there moves to sea and back on a daily basis, and weather causes hourly fluctuations. Also, small males make feeding trips. They visit landing areas about three times over an 80-day season separated by periods of 10 days or more when they are absent for foraging at sea. On average, about 19% of marked juveniles are seen on shore daily and the others are at sea. Data confirm that small males feed while absent and fast while ashore.

Animals on landing areas segregate by size because of size-related dominance. Small (young) males reside far inland, large males reside closest to the sea, and medium-sized males occur in between. There is no evidence of social bonds, coalitions, or kinship ties in these groups.

The transition from small to large body size is accompanied by (1) a change from high to low density (spacing increases with size), (2) a shift from fluid to highly fixed positions within the group, (3) a decrease in the frequency of social interactions, (4) a decrease in the number of animals interacted with, (5) waning of the metacommunication signal of play, and (6) a shift from using behavioral components that can injure to components that delineate space. This conversion of a size-related dominance system to a site-specific system prepares males to defend breeding territories as adults.

Juveniles leave landing areas to make brief dashes, here called "forays," across breeding areas during the breeding season. Because of chasing by adult males, forays may be highly disruptive and cause neonate mortality. Forays that involve multiple males may give the false impression of being a coordinated group effort. However, juvenile males are not prepared by early experience to carry out any coordinated group actions. Forays appear coordinated because many unaffiliated juveniles sometimes receive a chance to dash onto territories simultaneously when territory holders have their attention to boundaries distracted by a disturbance.

Forays are not a mating strategy. Only 2.3% of males making a foray ever establish a territory that way, two-thirds of the males making a foray are too small to defend a territory, most males on a foray do not try to contact females, and rarely if ever do any mate during forays. Data show that adult males defend as territories some of the very sites across which they made forays when they were juveniles. The hypothesis is advanced that forays represent young males trying to visit their natal sites.

Forays increase with the number of juvenile males and decrease with the number of peripheral males in the population. Peripheral males repulse all small males that approach the central breeding area from outside, and thus effectively buffer the breeding population from the disturbance their forays cause. An increase in forays following the 1972 cessation of killing at St.

George Island, and a subsequent decrease when peripheral males increased, was documented during this study. Numerous forays of large groups would not be expected in a stable population that was not subjected to a kill because it would include multiple ranks of peripheral males. The Pribilof population is now closer to a pristine condition than it has been for 200 years; it is unlikely that forays will increase there in the near future.

PART FOUR

The Maternal Strategy

——————————

The chapters in this section address four different aspects of the northern fur seal's maternal strategy. A maternal strategy is a system for extracting energy from the environment and delivering it to the young as the basis of their growth and survival to nutritional independence. Maternal strategies were a concern in this project because of the possibility that past human disturbance had altered a vital element in the system that produces new recruits. We looked at maternal strategies from several viewpoints.

Otariid maternal strategies are unusual among mammals in that they involve frequent shuttling between a highly productive marine area and a suitable shore area (Gentry et al., 1986a). The northern fur seal maternal strategy is an extreme type among the otariids in that the marine and shore areas used are at high latitudes requiring an annual migration. The short season at high latitudes truncates lactation, so the age at weaning is not flexible in this maternal strategy, as it is in some other otariids (Trillmich, 1990).

The first chapter in this section reviews the northern fur seal's maternal strategy (see also Gentry et al., 1986a; Trillmich, 1990) and presents new evidence that the pattern of milk delivery to the young (maternal attendance behavior) may respond to decade-scale changes in climate regime. Chapter 11 shows how slight differences in maternal attendance behavior may affect pup growth rates. Chapter 12 looks at individual variation in maternal diving behavior in different seasons of the year, and the stability of diving in different years. Chapter 13 compares maternal strategies at islands where the foraging environment differs, specifically, on islands that differ in the width of the continental shelf.

Female Attendance Behavior

FEMALE attendance behavior, alternating feeding trips to sea with suckling visits to shore, is one of the more flexible portions of otariid maternal strategy [1]. The particular challenge of northern fur seal females is to wean pups by a given date, perhaps at a given mass, by varying the rate or amount of energy delivered consistent with local foraging opportunities. Attendance behavior is their main means of regulating energy delivery, as it is for otariids that are not time limited. Attendance is under the direct control of the foraging mother, whereas maintenance metabolism, energetic content of food, and other factors in the energy budget are not.

Originally, it was proposed that attendance behavior varied with latitude because of environmental unpredictability (longest trips in predictable, subpolar areas; Gentry et al., 1986a; see also chapter 1). However, additional data made it clear that attendance behavior varied with local foraging conditions regardless of latitude (Goldsworthy, 1992; Figueroa Carranza, 1994; Francis et al., 1997; Guinet, pers. comm., 1996). The current search is for local variations in attendance behavior rather than for species-specific attendance patterns.

Attendance behavior is closely interwoven with the other aspects of maternal strategies, being a cause of some aspects and an effect of others. The central questions about maternal strategies reveal this complexity. For example, how does attendance behavior differ across the species' range, and by season of the year? What are the effects of maternal age and reproductive status on attendance behavior? How does the dive pattern at sea (i.e., energy acquisition rate) affect the attendance pattern? Does attendance behavior reflect the location of feeding (distance from shore)? Does attendance behavior vary with diet, and does diet change with the distance of feeding from shore? How is attendance behavior affected by El Niño/Southern Oscillation (ENSO) events that devastate other otariid populations (Trillmich et al., 1991)? How variable are visits to shore, and what factors affect their duration? How much does the age at weaning vary, compared to other otariids? And finally, if attendance behavior differs by island, does the pup growth that results from it also differ? The present chapter addresses all of these questions except the last one, which is the subject of chapter 11.

Another major question remains, namely: Do sea trips and shore visits vary over decades given that both are determined by the local foraging environment? Two papers have addressed this question. One found no significant change in attendance data over the period 1951–74 (Gentry and Holt, 1986), and the other found that the first trip to sea was shorter in 1984 than in the period 1951–84 (Loughlin et al., 1987). The studies used different methods, which begs the following question: Did the methods lead to these dissimilar conclusions, or did the attendance pattern in fact change between 1977 and 1984? To answer the latter question, we analyzed attendance data collected by the same method at three sites from 1974 to 1988.

This chapter begins with a review section that details the maternal strategy, shows the role of attendance behavior in it, presents a new comparison of attendance at different islands, reviews the effects of the 1982–83 ENSO event on the Pribilof population, and summarizes field experiments on the regularity of shore visits. The chapter then addresses the question of a possible change in attendance behavior between 1974 and 1984, relates this to the North Pacific climate regime shift in 1978, and discusses age-related changes in the attendance behavior of individuals.

REVIEW OF PAST STUDIES

Attendance patterns differ across the northern fur seal's range; there is no one pattern that can be called species-specific. One part of the attendance pattern, the length of the perinatal period, is the same everywhere that it has been measured (Peterson, 1965, 1968; DeLong, 1982; Yoshida and Baba, 1982; Vladimirov, 1983; Gentry and Holt, 1986), and 142 days at San Miguel Island (DeLong and Antonelis, 1985). But from the first foraging trip onward, differences are apparent. At Medny Island, females make average postmating trips of 3.5 days, while those at St. Paul Island are 9.8 days (table 10.1; see also Antonelis et al., 1990a; Loughlin et al., 1987). The latitude at these two islands is very similar and cannot account for this almost threefold difference. Visits to shore are everywhere shorter and less variable than trips to sea (table 10.1).

Sea trips of 3.5–9.8 days are in the moderate range for otariids worldwide. The Galapagos fur seal makes very short trips (a few hours; Trillmich, 1986), and temperate fur seals make long ones (up to 30 days) [2]. Ultimately, the ability of pups to fast while their mothers forage must affect whether a given colony will thrive. But few data exist on the fasting abilities of pups.

Attendance behavior changes with the season at every northern fur seal colony where it has been investigated (St. George, St. Paul, and San Miguel islands). As the season progresses, trips to sea become longer, but shore

Table 10.1

Comparative Attendance Behavior of Female Fur Seals at Various Breeding Islands

Island	Distance to Shelf Break (km)	Trip to Sea (days)	Visit to Land (days)	Source of Data[d]
Medny	17	3.5	1.4	Vladimirov, 1983
San Miguel [a]	2	3.8(1.1)	1.8(0.4)	DeLong, 1982
Robben	48	4.7	2.1	Yoshida & Baba, 1982
St. George [b]	31	5.8(1.8)	2.0(1.1)	Gentry & Holt, 1986
St. Paul [c]	115	7.1(2.1)	2.3(1.0)	Gentry & Holt, 1986
St. Paul [d]		9.8(2.8)	2.3(1.2)	Peterson, 1968

Note: Entries are population means for mothers only with SD reported (in parentheses) where available.

[a] Averaged over several years.

[b] Data from 1975 only, selected for its large sample size.

[c] Data from 1977 only, selected for its large sample size.

[d] Data from 1962, longest feeding trips reported for the species.

[e] Data from Antonelis et al. (1990a) not included due to its small sample size (n = 24 trips). Data from Loughlin et al. (1987) not included due to its focus on trips early in the season (shorter than later trips).

visits remain the same length or nearly so. These generalizations hold when environmental conditions are stable but may change if the environment changes (El Niño events or climate regime shifts; see below). A weak seasonal increase was also found in antarctic (McCann and Croxall, 1986; Goldsworthy, 1992), New Zealand, and subantarctic fur seals (Goldsworthy, 1992). Seasonal changes may accommodate the growing pup's needs and, in northern and antarctic fur seals, preparations for the mother's migration. As will be seen in chapter 12, diving behavior changes seasonally along with the attendance behavior.

Northern fur seals wean at about 4 months without regard to the local environment. Suckling lasts about 120 days at the Pribilof Islands (Gentry and Holt, 1986). At San Miguel Island, northern fur seals overlap with California sea lions, Guadalupe fur seals, and (formerly) Steller sea lions. All these species wean between 4 and 12 months of age. That is, northern fur seals apparently cannot extend weaning to one year, even in environments where other otariids do so. They also cannot extend weaning during bad years when El Niño conditions reduce the amount of food available (DeLong and Antonelis, 1991). Like antarctic fur seals, northern fur seals seem genetically constrained to wean at about four months (Trillmich, 1986).

Female northern fur seal pups are weaned at about 32% and males at about 38% of maternal mass at St. George Island. This proportion may vary little among islands as it is characteristic of fur seals in general (Gentry et al., 1986a). The extent to which pup growth rates reflect differences in maternal attendance behavior will be addressed in chapter 11.

The female's reproductive status has a profound effect on her attendance behavior. Females may be suckling a pup or not, and independently may be pregnant or not. (Pregnancy in the summer months consists of carrying an unimplanted blastocyst.) Pregnancy affects neither the duration of trips to sea nor visits to shore, but suckling affects both parameters as well as the predictability of movements between land and sea (Gentry and Holt, 1986). Suckling females become central place foragers, moving to sea and back on a predictable schedule. Nonsuckling females do not become central place forgers, and only occasionally use land for resting. They make a few long visits to shore interspersed with sea trips of highly variable duration. Lactation apparently leads to central place foraging because energetically it is the most costly phase of reproduction (Millar, 1977).

For mothers, the duration of shore visits are remarkably uniform, averaging 2–2.1 days with low variability (± 0.9–1.2 days; Gentry and Holt, 1986), which is characteristic of otariids [3]. We conducted a field experiment to test the hypothesis that the duration of the shore visit is determined by the female's ability to fast between feedings. We prevented pups from reuniting with their returned mothers for predetermined intervals and compared the duration of these visits against control visits of the same mothers when the young were not withheld. We found that females fasted three or four times longer when reunion with the pup was delayed than when their reunion was immediate (Gentry and Holt, 1986), thus negating the hypothesis.

To test the hypothesis that pup demand for milk determined shore visit duration, we altered pup demand by force-feeding pups with milk prior to the mothers' visits. As before, experimental visits (pup was pre-fed) were compared to control visits (pup not pre-fed) of the same mothers that were made before and after the experimental visits. The duration of experimental visits exceeded controls by an average of 67% (Gentry and Holt, 1986), thus supporting the hypothesis that pup demand for milk determines the duration of the mother's shore visit.

If pup demand affects female attendance behavior, and if male and female pups have different demands because they differ in size and growth rate, then the sex of the pup should affect the mother's attendance behavior [4]. However, it does not appear to do so. Macy (1982) found no effects of pup sex on trips to sea, number or length of mother's visits to shore, and number or length of suckling bouts. That is, mothers appeared to raise male and female pups with the same behavioral pattern. Isotope dilution studies show that male pups actually receive 61% more milk energy than female pups (Costa and Gentry, 1986). But since the duration of suckling bouts does not differ, perhaps male pups suckle more efficiently than female pups.

Differences in the length of trips to sea result in part from the way females dive while foraging. Three diving patterns have been described to date, termed "deep," "shallow," and "mixed" [5]. They differ mainly in the way depth changes with time, not just in depth. Deep dives occur on the shelf, and most shallow dives occur beyond the shelf where prey migrate to the surface at night (Goebel et al., 1991). Partly because of the distances involved, deep divers make shorter trips to sea than shallow or mixed divers (Gentry et al., 1986b) [6]. Foraging trip length is highly predictable for a given mature female (Gentry and Holt, 1986). But the population includes animals that use three different dive patterns, which explains why foraging trip length is so variable among mature females (Gentry and Holt, 1986). To date, the northern fur seal is the only otariid that has shown this diversity in foraging pattern.

Individual females show some site fidelity to foraging locations and may return to them on different trips to sea. Loughlin et al. (1987) followed VHF-tagged females by ship and aircraft, and Loughlin et al. (1993) tracked foraging females by satellite-linked Time-Depth Recorders (TDRs). They found that individuals fed in similar locations over six to eight trips to sea, moved to and from these locations over nearly the same routes, and animals from one central breeding area fed in very different locations. These animals, all adults, showed no random searching for foraging areas. It is not known whether young females show random searching or use the same foraging sites over time.

The dive records for individuals show some consistency from trip to trip, perhaps because the females forage in the same site repeatedly. On different foraging trips, females tend to leave shore at the same time of day, have a similar transit period before diving begins, begin diving at the same time of day, and make first dives of similar depth.

Attendance behavior in females differs by age; young females make longer trips to sea than older ones [7]. More importantly, they mix short and long trips such that their trip lengths are much less predictable than for mature females. The behavior of young females suggests that they lack foraging experience (Goebel, 1988). A working hypothesis is that young females lack a favored foraging location and diving pattern, and that once these are established both attendance and foraging behavior become predictable.

It may be no coincidence that the success of rearing pups also differs by age. Younger females that make long foraging trips are less successful at rearing young to weaning than are older females that make shorter trips (34.6% compared to 7.4% lost pups before weaning; Goebel, 1988).

Viewed at the population level, the distance to the edge of the continental shelf is generally an index to the duration of trips to sea there. Brief trips generally occur where the shelf is narrow (Medny Island; table 10.1), long trips generally occur where it is wide (St. Paul Island), and intermediate trips occur between these extremes (Robben and St. George islands). San Miguel Island apparently fits this generalization in some years (DeLong, 1982), but not others (Antonelis et al., 1990a). Yearly changes in the location of food concentrations may cause trip length to vary somewhat. The shelf edge may be biologically important because prey concentrate there (Smith, 1981), or because it is the nearest place where females can reach vertically moving prey in deep water. Most females at St.George Island (shallow plus mixed divers = 70% of females) spend most of their days diving for vertically moving prey. Clearly, deep water is required by most females. Near the Pribilof Islands they tend to concentrate at the shelf break and slope where deep water begins (Kajimura, 1980). For those females that do not feed near the shelf break, shelf width only represents the minimum transit distance to feeding.

The diet also differs with the distance to the shelf edge and the extent of the shelf. Where the shelf is small (Medny Island) the diet is mostly small pelagic squid found beyond the shelf break (Antonelis et al., 1997). At islands near the edge of a broad shelf (St. George Island) the diet is about half pelagic squid and half fish. Where the island is farther from a shelf edge (St. Paul Island) the diet is mostly fish. Therefore, broad shelves may support a large biomass of prey that fur seals more or less exploit, depending on the distance to the shelf edge. Narrow shelves apparently support a smaller biomass of prey but give fur seals ready access to deep water.

Attendance behavior may be affected by ENSO warm events in which warm water drives prey deep or displaces it horizontally. These changes have cascading effects on maternal foraging trip length, pup growth, and survival (Trillmich et al., 1991). The 1982–83 El Niño event, the strongest of the century, caused major changes in attendance behavior, decreased natality rates, and decreased pup growth rates in northern fur seals at San Miguel Island (DeLong and Antonelis, 1991). However, the same event had no measurable effect on attendance behavior, natality, or diving at St. George Island (Gentry, 1991). Apparently most northern fur seal breeding locations were north of the area in which this event affected the prey base. However, the survival of juveniles may be enhanced by ENSO events (York, 1991b). Over long periods, the survival of juveniles correlates with air and water temperatures, especially when warm temperatures occur in the autumn before their birth (York, 1995). The abundance of certain fish also correlates with these temperatures and may mediate the effect on fur seal survival.

METHODS

The duration of feeding trips was inferred from absences noted on the history cards of individuals. Absence from the study site was not always synonymous with feeding trip duration because females sometimes made brief visits (1–3 days) to foreign central breeding areas instead of returning to their young immediately after feeding (chapter 7). It was not possible to correct the database for brief visits. Therefore, absences reported here are the maximum possible feeding trip durations.

The durations of shore visits were also reported on history cards. All first visits of the season were deleted from the analysis because they included parturition and mating, whereas subsequent trips reflected mostly energy transfer to the young. No correction was made for days when females were on shore but unobserved. As a result, shore visit durations reported here are minimum values.

Attendance data were restricted in several ways to provide the best possible measure of interannual changes. Only the records of females with stable attendance patterns were used in the analysis so that the records of non-mothers (above) would be excluded. All absences longer than 15 days were deleted on the grounds that they represented our failure to see females during a shore visit (Peterson, 1968). Females with worn or broken tags were eliminated because their visits were easy to miss. Only trips to sea recorded between 19 June and 28 August were used because these dates let us compare data for the greatest number of years. Finally, foraging trips in November were included in the analysis by Gentry and Holt (1986) but are not included here.

RESULTS

The sample size was adequate for making long-term comparisons of attendance behavior [8]. One obvious result was that, over a 5-year period, females at different study sites had foraging trips of different length. Females from Kitovi made the longest trips, females from East Reef made the shortest, and females from Zapadni were intermediate (table 10.1). These trip lengths are shorter than reported previously (Gentry and Holt, 1986) because the present analysis included no data collected after 19 August when the longest trips of the year occurred. No interannual difference in trip length was found in a 5-year comparison of the three study sites, except that trips were shorter in 1983.

Over a 12-year period, feeding trip durations seemed suddenly to de-

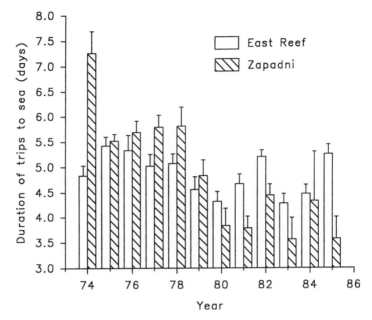

Fig. 10.1. Annual mean duration of 4,471 trips to sea made by 737 different females at two study sites between 19 June and 28 August 1974 through 1985. The period 1974–78 was significantly different from the period 1978–85 at both sites, but the sites did not differ from each other (see note 9). The 1982 results were probably not related to the 1982–83 ENSO event, which began in winter after the 1982 data were collected.

crease starting in 1979 (fig. 10.1). This change was statistically significant; trips were longer from 1974 to 1978 than from 1979 to 1985 [9]. The average decline after 1979 was 1.2 days (Zapadni, −1.5 days, East Reef, −0.3 days; both declines were significant). Over all 12 years, the sites were not significantly different from each other. A separate test showed that they differed before but not after 1979 (see note 9).

The decrease in trip length after 1979 was not tested in the Kitovi data because too few annual means (5 years total) were available. No major change was apparent from inspection. This may mean little, however, because no change was obvious by inspection when the East Reef sample was restricted to the same 5 years.

The duration of mothers' visits to shore became shorter after 1979 at East Reef, declining from 2.3 to 1.8 days [10]. However, shore visits at Zapadni did not change similarly (1.7 days before and after 1979). It seems that after 1979, females at East Reef cycled more rapidly (made both shorter trips to sea and visits to shore), while females at Zapadni made shorter trips to sea only.

Our records showed a weak tendency for young females to have longer trips to sea than older ones (confirming Goebel, 1988). We used relative age (years of appearing in the records) because absolute ages were not known for most females. Trips to sea were longer in the first two tenure years at Zapadni, but not at East Reef [11]. The age effect was distinct enough to be detected even though our subjects were not specifically chosen for that purpose as they were by Goebel (1988).

DISCUSSION

The duration of summer feeding trips by mothers decreased from 5.6 days in the period 1974–78 to 4.4 days in 1979–85, averaged over many females and trips to sea. The decrease was large and precipitous at Zapadni, smaller but detectable at East Reef, yet simultaneous at both. At East Reef, the duration of visits to shore also decreased, resulting in faster cycling between sea and land from 1979 onward. Nursing lines in the teeth of pups from St. Paul Island suggest that females made shorter trips to sea in 1978 than in any year from 1970 to 1981 (Baker, 1991). These lines of evidence suggest that a sudden change in foraging behavior occurred at both islands in the late 1970s.

The change in attendance behavior coincided with, but was not caused by, a change in the ratio of females to males at St. George Island. This ratio declined and reached a new plateau in 1978 or 1979 (fig. 2.4) because of an increase in adult males (killing ended in 1972 allowing more males to survive) and a decrease in females (for unknown reasons). The increase in male numbers is not a likely cause of shorter female feeding trips because adult males do not feed in summer, and any competition from males would increase, not decrease, the length of female feeding trips. The decline in the female population may have caused the change by reducing intragender competition for food. However, this would not explain the nursing lines in the teeth of pups at St. Paul Island where females were not declining in 1981 (Baker, 1991).

The change in female attendance behavior coincided with unusually warm weather. In 1979, Pribilof winds were unusually light, sea surface temperature was unusually warm, Pribilof air temperatures were warmer than in any other year from 1954 to 1986 (anomaly = + 6 C), and Bering Sea ice cover was the smallest of any of the same years (Niebauer and Day, 1989). However, the change in female feeding trips cannot have been a response to a single, unusually warm spring because the weather gradually moderated whereas female feeding trips still had not returned to their pre-1979 level by the time data collection ended in 1985.

Female feeding trips appeared to become shorter in years with large abundances of one-year-old walleye pollock. The 1978 pollock year class was one of the largest on record (Bakkala, 1993; Bakkala et al., 1987). Feeding trip length at both East Reef and Zapadni decreased markedly in 1979 when this huge pollock year class was one year old. The large 1982 pollock year class was followed by shortened fur seal feeding trips in 1983 (fig. 10.1). The 1984 pollock year class was large, and fur seal feeding trips in 1985 were shorter at Zapadni but not at East Reef (fig. 10.1). This coincidence is strong enough that in future research a causal connection should be sought between one-year-old pollock and fur seal feeding trip length. York (1995) suggested that abundance of age 0–1 pollock may explain the correlation between sea surface temperatures and survival rate of fur seal pups 2 years later.

Decreasing trip length coincided with increased abundance of pollock older than 0–1 years (Bakkala, 1993; Fritz et al., 1995). A good pollock year class may affect fur seal foraging for several years because fur seals, although preferring 3–20 cm fish (age 0–1 years), will switch to larger pollock when new year classes fail (Sinclair et al., 1994). For example, 1978 pollock were still dominant in fur seal stomachs in 1981 as 3-year-old fish (Sinclair et al., 1994). The 3 years of good pollock year classes were probably important to fur seals for more than a single year each.

Shorter trips may have resulted from prey moving closer to the islands. Much of the large 1978 pollock year class was reportedly centered around the Pribilof Islands (Traynor and Smith, in press). It is possible that other good year classes were similarly distributed.

Shorter trips coincided with a general increase in abundance of commercially valuable groundfish in the Bering Sea in the late 1970s. Between 1975 and the early 1980s, the total biomass of commercial groundfish in the eastern Bering Sea increased by about 5 million tons (Bakkala, 1993). Fur seals do not prey heavily on most of these species. However, groundfish trends may mirror trends in fur seal prey since both are affected by the same oceanographic conditions. But major increases in gonatid squid, deep sea smelt, sand lance, and other fur seal prey (Sinclair et al., 1994) could go undetected because fisheries surveys include only commercial species. If these surveys expanded to include marine biodiversity they could help determine whether simultaneous declines in fur seals, sea lions, harbor seals, and sea birds in the Bering Sea are linked to one another through the prey base.

On a wider scale, the increase in fish biomass and the decrease in fur seal feeding trip length correlated with a decade-long climate regime change in the North Pacific Ocean and Bering Sea. From 1977 to 1988 the Aleutian Low deepened and moved eastward, causing decreasing winds and warming temperatures in the eastern Bering Sea (see chapter 1); Trenberth, 1990; Niebauer and Day, 1989). The good pollock year classes from 1978 to

1985 may have resulted from this warming and reduced ice cover since both factors correlate with pollock recruitment (Quinn and Niebauer, 1995). Fur seal feeding trips changed in 1979, the same year that the regime shift had its greatest effect in the Bering Sea (Niebauer and Day, 1989). A similar 2-year lag is believed to have occurred in growth of fish resources in the Bering Sea as well. Such changes in the Bering Sea may affect the foraging of mothers more than the survival rates of their young (York, 1995) because weaned pups spend only 10.5 days in the Bering Sea before entering the North Pacific Ocean (Ragen et al., 1995).

A climate-related change in prey stocks also occurred in the central North Pacific Ocean where adult and juvenile fur seals feed from November to June. Sea surface temperatures decreased and chlorophyll a increased there starting in the late 1970s (Trenberth, 1990; Venrick et al., 1987). Landings and/or abundance of salmon and nonsalmon stocks (Beamish and Bouillon, 1993, 1995), as well as sardines in California, Japan, and Chile (Lluch-Belda et al., 1989), all followed trends in the Aleutian Low Pressure index. These stocks were high in the 1940s, low in the 1960s, and high again starting in the late 1970s, unrelated to fishing effort (Lluch-Belda et al., 1989). Six years after the 1977 regime shift occurred in the North Pacific, the long-term declining trend in northern fur seals at the Pribilof Islands stopped (1983). It is possible that the regime shift improved juvenile survival and ended the fur seal decline; about 5 years are required for newborn female fur seals to reach breeding age (York, 1983).

The hypothesis that fur seal attendance behavior is affected by climate change acting through spawning success and survival of fur seal prey species is testable. If fur seal feeding trips are short during periods of strong Aleutian Lows, then feeding trip length should increase whenever the Aleutian Low weakens and shifts westward for extended periods. We missed an opportunity to measure such a change by not collecting data after 1988, the year in which cold La Niña conditions weakened the Aleutian Low and ended the climate positive state that began in 1977 (Trenberth, 1990).

Changes in female attendance do not necessarily imply changes in fur seal population trends. Attendance behavior of females affects only the weaning weight of pups, which is weakly associated with postweaning survival. Male pups survive better if they are larger than average at weaning, but females do not (Baker and Fowler, 1992). Weaning weight may have a marginal effect on survival because the fat that pups acquire from their mothers cannot sustain them long after weaning, given their high metabolic rates in cold water. Instead, survival may depend more on their success at learning to forage and finding adequate food. Since survival of juveniles presently has the largest influence on population level in this species (York, 1991b, 1995), the study of female attendance behavior and its relationship to commercial fisheries is not likely to explain fur seal population fluctuations.

Extreme care should be exercised in inferring changes in the foraging environment from changes in fur seal attendance behavior (Gentry and Holt, 1986). The most reliable measures will come from a stable, marked population of females in which age, reproductive condition, foraging preference, and other variables are held relatively constant from year to year.

SUMMARY

Attendance behavior is an expression of central place foraging. It exists only for females that are meeting the energy demands of lactation. Nonlactating females either do not visit shore or visit so sporadically that they cannot be considered true central place foragers. Carrying an unimplanted blastocyst has no measurable effect on attendance behavior.

Attendance behavior lasts about 120–142 days in this species. Weaning at about 4 months is a genetically fixed trait that accommodates migration; it cannot occur earlier in the year even at more equable latitudes. Pup growth rate, the immediate result of maternal attendance behavior, is not a very flexible part of the maternal strategy. Pups are weaned at 32% (females) or 38% (males) of maternal mass.

Attendance behavior varies with the local foraging environment. No species-specific attendance pattern exists, and latitude has no effect on attendance behavior. The first visit to shore is uniform, but thereafter average foraging trip durations range from 3.5 (Medny Island) to 9.8 days (St. Paul Island). Periodic El Niño events may temporarily increase foraging trip length in the southern end of the range.

Attendance behavior varies with age. Trips to sea are unusually long and variable in the first 2 years of motherhood but thereafter show no further changes with increased age. Young females are less successful at rearing pups than older ones. The changes over the first 2 years may represent learning where and how to forage while lactating.

The duration of trips to sea usually increases over the season, perhaps driven by the increased energy needs of mother and young.

Diet changes along with differences in attendance. At islands with a narrow shelf, females have short foraging trips and take mostly squid. At islands near the edge of broad shelves, females take a mixture of fish and squid and make foraging trips of intermediate length. At islands far from the shelf edge, females make long foraging trips and take mostly fish.

The effects of the local foraging environment, female age, time of season, diet, and type of diving pattern used may be modified by the absolute abundance or distance to prey on a yearly basis.

Variability in the duration of female feeding trips is high in the population but low for individuals (except young ones). Variability in the population comes from females using three different diving patterns, which implies different prey types (this variability is not seen in other otariids), and seasonal changes. But individuals use the same diving pattern over time and may feed in similar locations, both of which contribute to low variability from trip to trip. Females from a given breeding area differ in dive pattern and feeding location; individuals may change their feeding locations on a seasonal basis. The flexibility females have in prey type and feeding location makes the calculation of an average feeding trip almost meaningless. Random searching for food probably plays a small part in foraging for mature females.

Shore visits are the least variable part of the attendance pattern (2.0–2.1 days) unless the environment changes. The duration of shore visits is probably determined by pup demand for milk.

Attendance behavior is not measurably affected by the sex of offspring, contrary to the prediction of sexual selection theory. Male pups receive 61% more milk energy per day than females, but do not suckle differently nor do their mothers visit differently. Apparently a single feeding pattern meets the needs of both sexes. Perhaps by weaning young well before the growth rates of the two sexes diverge substantially, mothers shift the energetic burden of dimorphic growth to the young.

ENSO warm events have negative impacts on attendance behavior and survival of fur seals in California. But the 1982–83 event, the largest of this century, had no effect on attendance and survival at the Pribilof Islands. It is likely that most of the breeding islands for the species are spared these effects by virtue of high latitude. Survival of weanlings may be enhanced during ENSO events.

Attendance behavior changed coincident with the 1979 onset of a climate regime shift in the Bering Sea. Feeding trips shortened from 5.6 to 4.4 days at two sites, and shore visits shortened from 2.3 to 1.8 days at one. That is, females cycled faster after 1979. These changes did not coincide with a changed sex ratio or warmer weather. They did coincide with an increase in abundance of age 1 and older pollock and other groundfish in the Bering Sea, the movement of pollock closer to the Pribilof Islands, and increases in salmonids and other species in the North Pacific. The climate change began in 1977 in the Pacific but the main effect in the Bering Sea was in 1979.

Changes in fur seal attendance behavior do not necessarily result in changes in fur seal population trends. Foraging changes may or may not be reflected in attendance behavior. Attendance may have a limited influence on the postweaning survival of young. Therefore, studies that link adult foraging behavior with commercial fisheries are not likely to explain fur seal population trends.

CHAPTER 11

Neonatal Growth and Behavior

WITH MICHAEL E. GOEBEL AND JOHN CALAMBOKIDIS

ONE MEASURABLE result of the maternal strategy is the rate at which neonates grow. Until pups begin taking solid food, which occurs only after weaning in this species, growth is based entirely on the energy in the mother's milk. Growth is the end product of the total energy gained by the mother from prey (including its abundance, size, and energy content), minus the metabolic costs of acquiring (diving effort) and delivering it (attendance behavior—transit to and from foraging areas), minus maternal and pup maintenance metabolism, minus the pup's metabolic cost of fasting in its mother's absence.

Relative to birth weights, the growth rates of all fur seals in the first 4 months of life are roughly similar. All double their birth weight within 40–93 days (median 66 days), have daily growth increments of 1.3–1.7% of birth weight, and are weaned at about 40% of adult female mass (Gentry et al., 1986a). In all species, growth changes at 4 months; the young of northern and antarctic fur seals are weaned, and those of temperate and tropical species show a decrease in growth rate. These relative growth rates are obviously achieved by different routes because the terms in the maternal energy budget (above) are probably not identical at any two islands. The growth rate of northern fur seal pups sharply increases after they molt in September (Boltnev et al., 1997).

The fundamental question in this chapter is whether size at birth and pup growth rates reflect relatively small inter-island differences in maternal attendance behavior and diet. Females at St. Paul Island make fewer, longer trips to sea than females at St. George Island (40 km away), resulting in longer pup fasts. They extend suckling by about a week, possibly to compensate for the fewer total days on shore these longer absences would otherwise cause (chapter 10; also Gentry and Holt, 1986). Females at St. Paul Island have primarily a fish diet while those at St. George Island take a mixture of fish and pelagic squid (Antonelis et al., 1997). At the time of this study, males were being killed for pelts at St. Paul but not at St. George Island. It is possible that food competition between females and juvenile males differed between the islands.

This chapter addresses the following questions: Does growth rate in the first 50 days of life (that is, before the molt; Boltnev et al., 1997) differ by sex at these two islands, on either a relative or absolute basis? (Growth in antarctic fur seal pups was found not to differ as a function of the mother's attendance patterns [Doidge and Croxall, 1989], but in that case diet did not differ.) Do birth weights differ by island, reflecting the different foraging patterns of the pups' mothers? Are heavier, male pups born first as an expression of maternal investment (Trivers and Willard, 1973)? What are the sex ratios at birth? Are the sexes born on different parts of the breeding areas? Is growth strictly linear with time? To address these questions, we selected two sites, Staraya Artil on St. George and Reef on St. Paul (fig. 1.4), where raised wooden catwalks above the breeding seals allowed us to mark and frequently recapture neonates for weighing. The main data were collected in 1980, although preliminary data were collected in 1979.

METHODS

Our method of measuring growth rates was to weigh known pups repeatedly from birth [1] to about 8 weeks of age. We caught pups using a 4-meter-long noose pole, lifted them onto the catwalk, determined their sex, weighed them on a spring scale to 0.25 kg, marked them [2], recorded the data [3], and released them to their mothers within a few minutes. Crews worked simultaneously at the two study sites. We spent mornings catching pups born the previous night, and afternoons reweighing previously marked pups. When new births decreased in August, we spent increasing time reweighing previously marked pups. Data collection began on 15 June and ended on 20 August 1980 because most pups had begun going to sea by then, and because their increased wariness of humans made captures difficult. Therefore, our data collection ended well before pups experienced the increase in growth rate at the molt (Boltnev et al., 1997).

Inter-island comparisons were based on pups for which birth weight and at least one subsequent weight were available (range 1–16 measures per pup). Most of the pups from which we obtained a single weight were those that died. Such pups usually weigh less at birth than those that survive (Calambokidis and Gentry, 1985). We excluded these pups from the records to focus the analysis on typical pups that were likely to survive. We compared "excluded" and "included" pups on the basis of sex ratio, mean date of birth, and birth weight.

The methods included two possible sources of error. Most pups were born at night and had taken small amounts of colostrum before the first weight was taken in the morning. The number of pups affected and extent of the error caused by colostrum are unknown. Also, the tags we used to mark pups probably reduced their growth rates, but so slightly that their effects were hard to measure against the background of individual variation (Calambokidis, unpub. data).

To determine whether male and female pups were born at different times of the season, we expanded the sample size by including a few pups captured at the same sites in 1979 (Staraya Artil, n = 55; Reef, n = 48). These pups were not weighed or reweighed.

We analyzed and compared growth rates by several methods. To show the range of absolute growth rates in the population, we regressed weight on age for each pup and compiled individual slopes into a frequency distribution for each sex. We also plotted daily mean relative weight gain as a percentage value from birth to a maximum of 50 days to look for fluctuations in weight over time [4]. Since the plot was generally linear but mean pup weights seemed to oscillate in a sine wave fashion, we added a sine variable to a linear regression to see if it would significantly improve the linear fit of the data [5]. These analyses were based on the assumption that we captured and weighed a random sample of the marked population each day. This seems a valid assumption because workers had no control over which pups moved within capture range, and they captured all pups that were reachable. Only those days for which the average was based on weights for three or more neonates were included in the plot.

Both absolute [6] and relative [7] weights were used to test for differences in pup weights between sexes and islands. We also performed a stepwise multiple linear regression (using only pups with three or more weights) to determine whether age, sex, island, and the interaction components contributed significantly to predicting pup weight.

RESULTS

Excluded Pups

A moderate number of pups were excluded because they were weighed only once [8]. Excluded and included pups did not differ in sex ratio at Staraya Artil, but at Reef more females than males were excluded [9]. Excluded and included pups did not differ in the mean date of birth at Staraya Artil, but at Reef excluded pups were born later in the season [10]. Finally, excluded and included pups did not differ in birth weight at Staraya Artil, but at Reef excluded pups of both sex were smaller at birth than included pups [11].

Table 11.1

Mean Weights in Kilograms at Birth for Northern Fur Seal Pups
at St. George and St. Paul Islands, in 1980[a]

Island and Sex	Number	Mean	SD
St. George Island[b]			
Females	180	5.19	0.611
Males	170	5.64	0.632
St. Paul Island[c]			
Females	163	5.32	0.657
Males	199	5.87	0.720

Results of Two-Way ANOVA

Source of Variation	Sum of Squares	Degrees of Freedom	Mean Square	F Stat.	p Value
Total	359.90	711	0.51		
Main effects	52.61	2	26.31	60.69	0
Sex	44.53	1	44.53	102.73	0
Island	5.81	1	5.81	13.40	0
Two-way interactions					
Sex/island	0.39	1	0.39	0.91	0.340
Residual	306.89	708	0.43		

[a] Includes only pups for which a birth weight and at least one subsequent weight were obtained. See text.

[b] Staraya Artil breeding area.

[c] Reef breeding area.

Birth Weights

Males averaged 5.6 to 5.9 kg at birth, and females averaged 5.2 to 5.3 kg (average of 0.6 kg difference), a significant difference; table 11.1). Also, Reef pups were heavier at birth than pups of the same sex at Staraya Artil [12], but the inter-island differences were smaller than the difference between the sexes. The interaction term sex-by-island was not significant.

Birth weight varied randomly throughout the season at Staraya Artil, but at Reef heavier pups of both sexes were born earlier in the season (St. Paul slopes significant in table 11.2).

Males and females were born in nearly equal proportions at all times of the season at both sites (table 11.3). On a daily basis, one sex was often much more numerous than the other, but the next day the other sex might predominate [13]. These brief "runs" show the importance of basing sex

Table 11.2
Seasonal Trends in Birth Weights of Northern Fur Seal Pups
by Sex and Island, 1980 (regression analysis)

Island and Sex	r^2	Y Intercept (kg)	Slope	Significance (p Value)
St. George Island[a]				
Females	0.0034	5.03	0.0032	0.1942
Males	0.0018	5.55	0.0029	0.2774
St. Paul Island[b]				
Females	0.0387	5.76	−0.0138	0.0030
Males	0.0556	6.49	−0.0169	0.0002

[a] Staraya Artil breeding area.
[b] Reef breeding area.

ratio estimates on large sample sizes and over long time series. Our sample size of 1,064 pups [14] was too small to conclude that the sex ratio (51% males) differed significantly from 50:50.

The seasonal decline in birth weights at Reef did not result from pups of different sex or size being born in different parts of the breeding area. The catwalks at Reef and Staraya Artil were divided into five and three equal sections, respectively, and differences in birth weight and sex were compared among sectors by ANOVA [15]. No significant differences were found in any sectors. Therefore, birth weights were not clumped along the catwalks.

Growth Rates

Frequently weighed individuals, those with 6–16 samples each, did not predominate in the data set and did not have a disproportionate effect on the results [16].

Overall, growth for pups of both sexes at both islands between birth and 50 days of age was generally linear with moderate variability. Figure 11.1 shows the typical appearance of the raw growth rate data. Note that after 30 days the number of weights recorded per day decreased, a result of the increasing wariness of pups to being captured.

Weight gain did not increase progressively in a strictly linear fashion, but instead it oscillated. Both sexes at both islands showed this oscillation in plots of their daily mean weights (example, fig. 11.2). The oscillation appeared to break down after 30–35 days of age, possibly because the number of daily weight samples decreased (as pups matured they moved away or were more able to elude capture), or because the feeding cycles of mothers became more individualistic (chapter 10).

Table 11.3

Seasonal Trends in Birth by Sex of Northern Fur Seals at St. George
and St. Paul Islands, 1979 and 1980 (χ^2 contingency table).[a]

	Quarters of Birth Season				
	1	2	3	4	Totals
Females					
Observed	186	184	104	51	525
Expected	185.0	187.5	101.2	51.3	
χ^2	0.01	0.07	0.08	0	
Males					
Observed	189	196	101	53	539
Expected	190	192.5	103.8	52.7	
χ^2	0	0.06	0.08	0	
Total observed	375	380	205	104	1,064[b]
Total χ^2					0.30[c]

[a] Includes data from Staraya Artil, St. George Island, 1979 and 1980, and Reef, St. Paul Island, 1980, combined because no differences were found among groups.

[b] Overall sex ratios = 51% males, 49% females. Sample size is too small to detect a significant difference at p = 0.05.

[c] p value = 0.9600.

The sine function we added to the linear regression of pup weights gave an improved fit in all cases, somewhat more for female than for male pups. A Fourier transform analysis confirmed that the cycle length (L) we had derived empirically was valid. It also suggested other cycle lengths that were not obvious by inspection.

The sine waves for female pups at Reef and Staraya Artil had periods of 12 and 7 days, respectively, which were about equal to the maternal foraging trip length at the two islands (chapter 10). These results support the hypothesis that the cyclic rise and fall of pup weights reflected the mother's attendance patterns. The first cycle included a 7–8-day perinatal stay on shore followed by a brief trip to sea.

The results for male pups did not fit the foraging patterns of mothers as well as for female pups. Both our empirical and Fourier analyses suggested multiple periods for males. The main period was 17–20 days, which was not realistic [17], and the secondary one was around 4–8 days. Unfortunately, the data were not adequate for a more detailed comparison of islands and sex. Nevertheless, the data demonstrate that growth in both sexes was not strictly linear, but resulted from weight oscillations that may have predictive qualities.

Male pups gained more actual weight per day than females, as we would expect for a dimorphic species. The range of growth rates (slope of weight

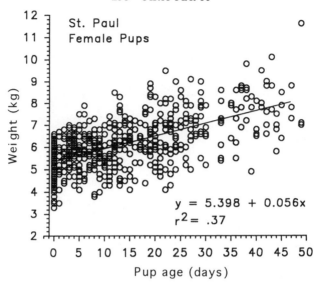

Fig. 11.1. The change in actual weight with age in female pups at Reef, St. Paul Island, in 1980. The line, fit by linear regression, shows that the growth rate was 0.06 kg/day. The total sample size was 651 weights for 161 individuals.

vs. age) for the sexes was largely overlapping, except at the very fastest rates where males predominated (fig. 11.3). Growth rates differed significantly by sex but not by island [18]. A separate analysis [19] showed that pups at Reef weighed significantly more during the first 20 days of life, but thereafter the difference between sites was not significant.

Despite differing in actual weight gain, pups of different sex on the two islands grew at about the same relative rates [20]. Males at Staraya Artil and Reef gained 1.3% and 1.2% of birth weight per day, and females gained 1.2% and 1.1% per day, respectively.

Age, sex, island of origin, age by sex, and age by island all contributed significantly to predicting pup weight. Age and sex contributed the most to the model fit. Island of origin had only a marginal effect [21].

DISCUSSION

Small but consistent inter-island differences in female attendance patterns and diet differences were associated very weakly with differences in pup growth rates in the first 50 days of life. St. George Island mothers had shorter trips to sea and a greater number of days ashore for suckling than St. Paul Island mothers, and they fed on fish and squid while St. Paul females fed primarily on fish (Antonelis et al., 1997). Presumably St. George Island

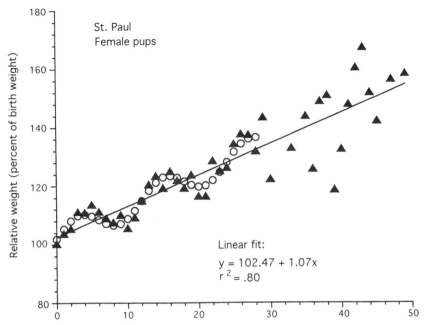

Fig. 11.2. Daily averages (triangles) of relative weight (actual weight/birth weight) calculated from the data points in figure 11.1. The model fit ($r^2 = 0.97$) a sine wave (circles) having a period of 12 days. The equation for the graph was:
$y = 0 + 1.174t + 4.833 \,(sin(57.2958 * ((2\pi t)/p)))$, where y = relative weight gain as a percent of birth weight, t = pup age in days, p = 12 days. The increased variance after 30 days of age accompanied a decrease in daily samples available.

pups, which were significantly smaller at birth during our study, burned less of their acquired energy in maintenance metabolism because they fasted for shorter periods between visits of their mothers than did St. Paul Island pups (Costa and Gentry, 1986). Apparently none of these differences were sufficient to cause significant or consistent inter-island differences in growth during the first 50 days of life. It is possible that the effects were cumulative and that growth rate differences developed later in the season. Such a possibility cannot be discounted given that growth rates sharply increase after the molt (Boltnev et al., 1997).

At both islands male pups gained more weight than females; during the first 20 days, St. George pups gained more weight than St. Paul pups. However, the relative growth rates (weight relative to birth weight or to the previous day) did not differ, either between the sexes or between islands. That is, over the first 50 days of life the growth increment was not much affected by the sex of the pup or the island from which their mothers foraged.

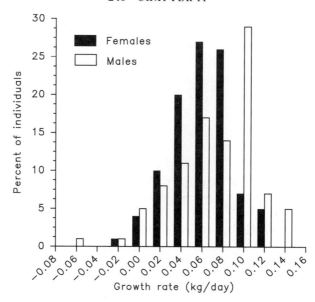

Fig. 11.3. Individual slopes for the regression of actual weight on age for male and female pups showing displacement of the means and general overlapping of the ranges.

The question of how small a difference in environment can cause a measurable change in growth rate is still open. Doidge et al. (1984) looked for inter-island differences in growth where the attendance patterns of antarctic fur seal mothers differed, but found none. The only large intraspecies differences in growth rates reported to date come from a comparison of subpolar and temperate populations of antarctic fur seals (Kerley, 1985).

Northern fur seals show growth characteristics similar to other fur seals studied to date. Absolute growth for males and females (84 and 66 gm per day, respectively) were significantly different, as they were in virtually all other fur seal species reported to date (antarctic fur seal: Payne, 1979, Doidge et al., 1984, and Doidge and Croxall, 1989; New Zealand fur seal: Crawley, 1975; subantarctic fur seal: Kerley, 1985; Galapagos fur seal: Trillmich, 1986; Cape fur seal: Rand, 1956). Also, relative growth rate was not different between the sexes in the present study, or in New Zealand fur seals (Mattlin, 1981). Relative growth has not been systematically reported in other fur seal studies.

Our results suggest that weight does not increase linearly, but increases when the mother is present and decreases when she is absent, at least for female pups. Our data extended only to 50 days of life, but from the attendance patterns we would expect the period of this oscillation to increase with time of season. Also, we would expect the amplitude of this oscillation to be greatest in populations where females make long trips to sea (St.

Paul Island) and least where mothers are on shore frequently (Commander Islands). The results suggest that growth rates of males were more variable and did not fit the mothers' trips well, but those of female pups did. However, the data collection system (capturing pups from fixed points) was too imprecise to resolve these cycles. This subject should be pursued using a better method of pup recapture.

Disagreement exists on whether the sex ratio in newborn northern fur seals deviates from 50:50. Our small sample suggested a ratio at birth that was consistent with Fowler's (1997) estimate (50.65% males, based on 127,849 animals pooled from different locations, years, and sampling methods). This ratio differs significantly from 50%, but not by as much as would be expected for an animal of such extreme sexual dimorphism. Antonelis et al. (1994) report a more strongly male-biased ratio (54.7% males, n = 8,290 dead and live newborns). However, York (1987b) reported that sex ratio in fetuses did not differ significantly from 50:50 over the months of gestation, which implies that the sex ratio at birth should equal that among fetuses (based on 7,000 fetuses collected pelagically).

Trivers and Willard (1973) predicted that where extreme variance exists in male and female reproductive success, females should selectively invest in the more variable offspring, resulting in deviation from a 50:50 ratio at birth. For species like fur seals, males should numerically predominate, be born to larger mothers, and be born first in the season. They cite supporting evidence from three pinniped species (Coulson and Hickling, 1961; Stirling, 1971).

If the Trivers and Willard prediction fits any mammal it should fit the northern fur seal because of the high value of males implied by the life history traits. We found that males were not born first in the season, nor were they born in particular parts of the breeding group. Absolute growth was greater in males than in females, as it is in other fur seals. Mothers supported the growth of male and of female pups with no measurable difference in foraging behavior (chapter 10). The prediction that larger mothers should produce male offspring and smaller mothers females was not met in either antarctic (Costa et al., 1988) or northern fur seals based on large sample sizes (ca. 7,000; York, 1987b; Trites, 1991). The differential cost of male and female fetuses to antarctic fur seal mothers was only 0.3% of total metabolic cost (Trites, 1992a; based on results of Boyd and McCann, 1989). This same conclusion probably applies to the northern fur seal. Finally, it is arguable whether sex ratios at birth fit theoretical predictions (above).

In conclusion, the Trivers and Willard (1973) model fails to predict the cost, timing, growth characteristics and possibly the sex ratio of the northern fur seal. As suggested in chapter 10, perhaps mothers can use a single foraging strategy to rear male or female offspring because weaning at 120 days places the energetic cost of dimorphic growth onto the offspring.

SUMMARY

We measured the weights at birth for 1,064 northern fur seal pups at study sites on two islands, and reweighed 712 of them at intervals to 67 days of age in 1980 to determine whether growth rates of pups differed on islands where maternal attendance patterns and diet were known to differ. These study sites did not necessarily represent the islands on which they were found because sites on various parts of an island could differ. All work was completed before the autumn molt, when growth rates are known to increase.

The sexes were born at equal rates throughout the season and were not geographically clumped. The sex ratio, 51% males, was not significantly different from 50:50 due to small sample size. Pups at Reef (St. Paul Island) were heavier at birth than same-sex pups at Staraya Artil (St. George Island). The heaviest pups were born earliest in the season at Reef but not at Staraya Artil. Pups were not geographically clumped by birth weight. At both sites, males averaged 0.6 kg heavier at birth than females.

Male pups gained more actual weight per day than female pups (84 vs. 66 gm, respectively). Same-sex pups at the two sites did not differ from each other in weight gained per day. The sexes did not differ in growth relative to birth weight at either site; females gained 1.1–1.2%, and males gained 1.2–1.3% of birth weight per day, a nonsignificant difference. Growth rates did not vary by weight at birth or date of birth. A multiple linear regression showed that age and sex were the best predictors of pup weight, and that island of origin improved the prediction very little.

Growth was generally linear from birth to 50 days of age with no inflections. However, in the first 35 days the mean daily weight oscillated in a sinelike pattern. Statistical tests showed that for female pups these oscillations matched the attendance pattern of the mother. They did not match for male pups, which oscillated with multiple periods. To test whether attendance pattern causes this oscillation, a difference in sine wave period and amplitude should be sought at islands where females have very long and very short foraging trips.

The general conclusion of this study was that pup weight and growth rates differ only marginally and weakly between two islands where maternal attendance behavior and diet are known to differ. The sex ratios at birth, growth rates, and foraging and suckling behavior of mothers did not support the theoretical predictions that, in a species such as this, male offspring should be born first, predominate, and have faster relative growth as a result of differential parental investment.

CHAPTER 12

Female Foraging Behavior: Inter- and Intra-Annual Variation in Individuals

WITH MICHAEL E. GOEBEL

A MAJOR question when the St. George Island Program started was whether a changed relationship between the northern fur seal and its prey base caused the herd to fail to recover from the herd reduction program of 1956–68 (Anon., 1973). A pelagic research program was to have addressed this question, but it was terminated in 1975. As a substitute, we initiated the use of Time-Depth Recorders (*TDR*s) (Kooyman et al., 1976, 1983a) to measure maternal diving behavior, arguably the most important link between fur seals and marine food resources. These instruments could not reconstruct pre-1956 diving behavior, but they could document the important sources of variability in diving behavior as the basis for future comparisons. Previous papers have described northern fur seal diving in general terms (Gentry et al., 1986b), defined its spatial variability (Goebel et al., 1991; Loughlin et al., 1993), compared variability with that of other otariids (Gentry et al., 1986a; Antonelis et al., 1990a), and looked at El Niño events as a source of variability (Gentry, 1991). Most of these studies used data from different individuals from year to year. But the individual as a source of variability has not received enough attention.

Since 1981 there have been numerous studies of diving behavior on many species of pinnipeds. Information on foraging patterns now exists for eight of the nine species of fur seals (Northern fur seals: see above; antarctic fur seals: Boyd and Croxall, 1992, Boyd et al., 1994; South American fur seals: Trillmich et al., 1986, and A. York [pers. comm.]; Galapagos fur seal: Kooyman and Trillmich, 1986, and M. Horning [pers. comm.]; Juan Fernandez fur seal: J. Francis et al., 1997; Guadalupe fur seal: Gallo, 1994; Cape fur seal: Kooyman and Gentry, 1986; New Zealand fur seal: Mattlin, 1993.) These studies show that fur seals generally feed at shallow depths (<30 m), primarily at night, on diurnally migrating species of fish, cephalopods, and crustaceans.

Some northern fur seals foraging from the Pribilof Islands appear to be an exception to this general pattern. Some females dive both day and night to depths of 75–200 meters (Gentry et al., 1986b; Goebel et al., 1991).

Gentry et al. (1986b) categorized the variability into three general patterns: shallow, deep, and mixed. The shallow diving pattern has only nighttime dives less than 30 meters, and depth changes with time as the prey approach and descend from the surface. The deep diving pattern has dives in excess of 75 meters both day and night, and depths do not change with time because the prey are near the bottom and do not migrate vertically. The mixed category has both shallow and deep dive bouts, usually on different days.

Given this diversity in diving patterns, the question of individual variability in the different seasons or years becomes more specific. Do females shift among dive patterns as season changes? Aside from average dive depths and changes in depth with time, what aspects of diving change with season? Which ones show no change? Do females at St. George and St. Paul islands show the same kinds of seasonal changes, given that their foraging environments, diets, attendance behavior, and pup growth all differ somewhat?

In this chapter we compare the foraging patterns of nine female northern fur seals that were instrumented with *TDR*s at least twice. Five of these females contribute to an intra-annual comparison (different seasons of the year), and four of them contribute to an inter-annual comparison (same season in different years) [1].

METHODS

All foraging records were obtained from females breeding at East Reef (St. George Island) or Zapadni Reef (St. Paul Island; see fig. 1.4) from 1981 through 1986. Data for the intra-annual comparison were collected from 13 July to 11 August and again from 7 to 25 October. Data for the inter-annual comparison were collected in July and early August of different years (four females from East Reef). We report separately the results of one female instrumented in July 1981 and October 1983 at East Reef. In all, we report on twenty records from nine females (table 12.1).

We captured females on their second day on shore following a trip to sea. We tagged, weighed, and instrumented them, and released them to their capture sites [2]. All instrumented females departed to sea within 24 hours of receiving a TDR. After recapture, film from the TDRs was developed, printed on paper, and digitized to create a data file [3]. Each record was analyzed for bouts of diving using the same dive bout criterion used by Gentry et al. (1986a) and Goebel et al. (1991) for this species (i.e., five or more dives with less than a 40-minute surface interval between each dive).

We developed four indices of the foraging patterns (duration of diving bouts, number of dives per bout, percent time below the surface, and number of dives per hour) and compared these values within and between years for the same individuals as the basis of our conclusions about diving patterns. All statistical comparisons of means were made with analysis of variance (ANOVA) or Student's t-test.

RESULTS

Intra-annual Comparisons

ST. PAUL

All three females instrumented at Zapadni Reef were similar in mass in July and all three had gained mass by October [4].

Many aspects of overall diving differed between July and October. The October trips lasted much longer than July trips for each female (table 12.2) [5]. Dives in July occurred primarily at night (92%) whereas dives in October occurred throughout day and night and were more evenly distributed by hour (fig. 12.1 *top*). Mean dive depths were about twice as deep (45 vs. 93 m) in October as in July (fig. 12.1, *bottom*) [6]. These changes in the depth and timing of dives are evident in three-dimensional plots for each female (fig. 12.2). All three females showed a mixed diving pattern in July (predominant shallow diving <30 m at night but with some deep diving day and night). By October, all three had shifted to a deep diving pattern (fig. 12.2). The predominant depth of dives in October varied for each female [7].

Some dive indices increased, some showed no change, and some decreased between July and October diving. Specifically, the total number of dives made per trip to sea and the number of diving bouts increased in October compared to July. No seasonal change occurred in the rate of diving (dives/h) for the entire trip [8], the number of foraging bouts made per hour [9], or the mean duration of foraging bouts [10]. A decrease occurred in the mean number of dives within foraging bouts [11], with the result that the mean rate of dives (dives/hr) within bouts decreased [12] in October compared to July.

ST. GEORGE

The seasonal change in timing and depth of dive that were seen in the St. Paul records (shallow nighttime diving in July and deeper diving day and night in October) was not observed at St. George Island [13]. There, diving was primarily at night in both July and October (fig. 12.3, *top*), and the bimodal distribution of shallow and deep dives existed in both seasons (fig. 12.3, *bottom*). The mean dive depth for one female (P9) increased in October, as at St. Paul [14]. But for the other female (no. 1789), mean dive depth decreased in October and the percentage of her shallow dives increased (fig. 12.4)

Table 12.1

Summary Information from Twenty Diving Records of Nine Lactating Female Northern Fur Seals Foraging at the Pribilof Islands, 1981–86

Female ID	Record Initiation Date	Time	Duration (h)	No. of Dives	Depth (m)	Mean Depth of Dive (m)	Dives/h	No. of Dive Bouts
J8	5 Aug 1984	15:59	162.14	200	168.1	74.5 (36.8)	1.23	7
J8	18 Jul 1986	11:07	122.00	155	169.1	48.6 (34.5)	1.27	7
J8	24 Jul 1986	14:25	121.35	135	177.3	54.5 (39.2)	1.11	9
P9	5 Aug 1983	10:50	137.54	175	205.7	127.5 (46.6)	1.27	7
P9	14 Oct 1983	12:15	183.16	301	242.2	170.3 (72.4)	1.64	19
1208	7 Jul 1981		231.00	302	139.0	48.0	1.31	14
1208	13 Oct 1983	12:56	264.50	1,453	233.0	64.3 (47.3)	5.49	41
1789	9 Jul 1982		192.00	365	98.7	44.9 (13.7)	1.90	18
1789	13 Jul 1983	22:54	124.00	148	146.6	69.1 (23.4)	1.19	11
1789	10 Oct 1983	08:54	250.00	1,221	230.7	56.5 (25.7)	4.88	22
2775	27 Jul 1984	22:01	106.50	250	100.9	49.7 (17.4)	2.35	7
2775	6 Aug 1986	15:15	156.00	218	136.6	51.3 (21.1)	1.40	11
592	5 Aug 1983	03:43	149.35	154	101.3	34.0 (9.7)	1.03	5
592	19 Jul 1984	22:47	188.00	477	133.0	48.3 (15.3)	2.54	19
308*	24 Jul 1985	20:42	242.20	832	158.9	60.6 (42.2)	3.44	25
308*	11 Oct 1985	09:06	297.00	660	126.1	90.5 (14.8)	2.22	31
767*	22 Jul 1985	00:41	114.44	431	103.0	33.3 (27.7)	3.77	7
767*	13 Oct 1985	09:55	288.01	1,075	110.7	63.6 (20.7)	3.73	21
854*	2 Aug 1985	22:08	91.01	312	96.4	40.6 (28.3)	3.43	9
854*	7 Oct 1985	19:38	305.14	590	206.8	124.9 (32.6)	1.93	27

Note: Standard deviations for mean depth of dive are given in parentheses.
*Females resident on Zapadni Reef, St. Paul Island. All others are from East Reef, St. George Island.

Table 12.2

Comparison of July and October Foraging Data on Six Time-Depth Recorders on Three Female Northern Fur Seals, From St. Paul Island, Alaska

Parameters	July				October				
	Female ID				Female ID				
	308	767	854	Mean[a,b]	308	767	854	Mean[a,b]	p-Value
Trip duration (h)	242	114	91	149 (81)	297	288	305	296 (9)	.041
No. of dives	832	431	312	525 (272)	660	1075	590	775 (262)	.403
Maximum depth (m)	159	103	96	119 (35)	126	111	207	148 (52)	.570
Mean depth (m)[a]	61 (42)	33 (28)	41 (28)	45 (14)	90 (15)	64 (21)	125 (33)	93 (31)	.001
Dive rate (dives/h)	3.4	3.8	3.4	3.5 (.2)	2.2	3.7	1.9	2.6 (1.0)	.160
No of bouts	25	7	9	14 (10)	31	21	27	26 (5)	.070
Dive bouts/h	.10	.06	.10	.09 (.02)	.10	.07	.09	.09 (.02)	.989
Mean bout duration (h)[a]	3.0 (1.8)	4.2 (2.9)	3.0 (1.9)	3.4 (.7)	2.4 (1.5)	5.5 (3.5)	3.1 (2.1)	3.7 (1.6)	.678
Mean no. of dives within bouts[a]	32 (37)	60 (55)	34 (24)	42 (16)	20 (15)	49 (39)	19 (22)	29 (17)	.009
Mean dive rate within bouts (d/h)[a]	10 (9)	13 (7)	13 (10)	12 (2)	9 (2)	9 (4)	6 (3)	8 (2)	.049
Nonbout dives (%)[b]	4.9	3.3	2.6	4.0	3.9	3.5	12.2	5.7	—
Time in bouts (%)[b]	30.9	25.8	29.4	29.3	24.8	37.2	27.2	29.7	—

Note: Each record is for a single foraging trip to sea.

[a] All mean values have sample standard deviation in parentheses.

[b] Overall percent values are reported in the mean column for nonbout dives and time in bouts.

St. Paul - 1985

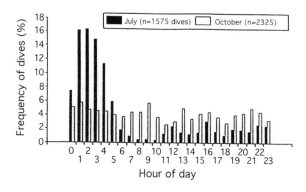

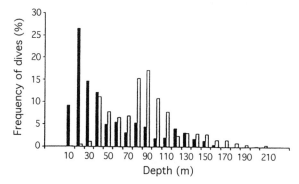

Fig. 12.1. *Top*: The distribution of dives by hour of day for three lactating northern fur seals, St. Paul, Alaska, in July and October 1985. *Bottom*: The distribution of dives by depth (0–210 m) for three lactating northern fur seals, St. Paul, Alaska, in July and October 1985.

As at St. Paul Island, some indices of foraging increased with season, some showed no change, and some decreased between July and October diving. Increases occurred in trip duration and total number of dives per trip for both females (table 12.3), the number of dive bouts per trip (they doubled) [15], and the percent time spent in foraging bouts (they doubled) [16]. In addition, there was a suggested increase in the dive rate (dives/h) for one female, but the difference was not significant [17]. No seasonal change occurred in the number of bouts per hour [18], duration of dive bouts [19], number of dives per bout [20], or the dive rate (dives/h) within bouts [21]. A marked decrease occurred in the percent of nonbout dives [22].

In summary, the aspects of diving that changed with season were generally related to the environment (hours of darkness or depth of prey), such as trip duration, number of dive bouts made, total dives made, depth of dives, increased feeding in daytime, and change in dive pattern. Those aspects that did

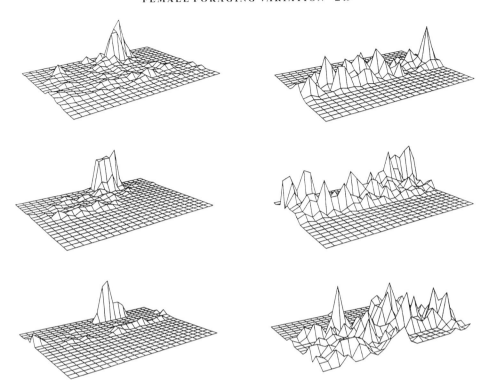

Fig. 12.2. Three-dimensional plots for each female northern fur seal instrumented with time-depth recorders in July (*left*) and October (*right*) 1985 on St. Paul, Alaska. Plots for female 308 are the top two figures, for female 767 the center two, and for female 854 the bottom two. See table 12.1 for number of dives plotted in each record. The X axis is hour of day, with midnight in the center of each plot, the Y axis is depth (m) from 0 to 210 m (depth increases as the Y axis approaches viewer), and the Z axis records percentage frequency of dives.

not change were generally related to diving physiology (recovery from dives), including bouts per hour, dives per hour within bouts, and bout duration. Two seasonal changes that were seen at St. Paul but not at St. George Island (number of dives per bout and dives/h within bouts), may have been related to the much deeper dives that St. Paul Island females made in October.

Inter-annual Comparisons

All inter-annual comparisons were made from records of females foraging from East Reef [23]. Most aspects of diving did not differ between the two years of instrumentation. These included the duration of the foraging trip (table 12.4) [24], number of dives per trip [25], dive rate (dives/h) for the

St. George - 1983

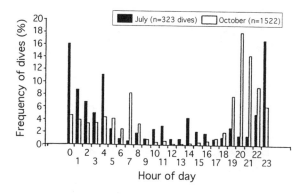

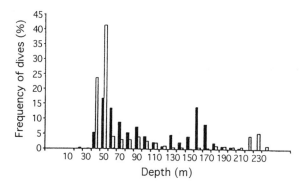

Fig. 12.3. *Top*: The distribution of dives by hour of day for two lactating northern fur seals, St. George, Alaska in July and October 1983. *Bottom*: The distribution of dives by depth (0–210 m) for two lactating northern fur seals, St. George, Alaska, in July and October 1983.

trip [26], number of dive bouts [27], number of dive bouts per hour [28], and number of dives per bout [29]. Other aspects of foraging changed for some females but not for all of them. These included the mean depth of dive (changed for 3/4 females, although not so much that their overall pattern changed) [30], duration of dive bouts (changed for 1/4 females) [31], and dive rate within bouts (changed for 1/4 females) [32].

Diving did not change from a nighttime to a day and night activity for any of the four females (fig. 12.5). Dives were primarily made during nighttime hours. Only one female showed a detectable change from the first to the second season (♀ 592 shifted the peak frequency of dives from shortly after dusk, at 2200 hours, to just before dawn, at 0100 hours; fig. 12.5, *bottom left*).

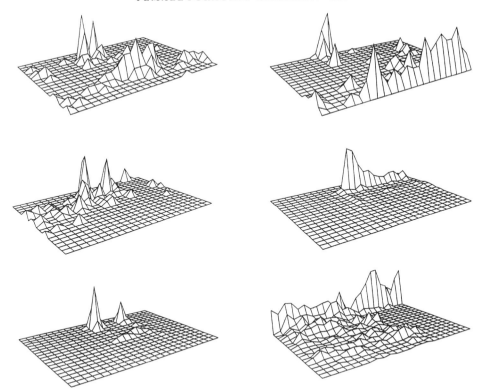

Fig. 12.4. Three-dimensional plots for female northern fur seals, P9 and 1789, instrumented with time-depth recorders in July (*left*) and October (*right*) 1983 on St. George, Alaska. Plots for female P9 are the top two figures and for female no. 1789 the center two. Plots for female no. 1208 instrumented in July 1981 (*left*) and October 1983 (*right*) are at the bottom. See table 12.1 for number of dives in each plot. The X axis is hour of day, with midnight in the center of each plot; the Y axis is depth (m) from 0 to 210 m (depth increases as the Y axis approaches viewer); and the Z axis records percent frequency of dives.

The general depth pattern of dives did not change from the first year to the second year (fig. 12.6). Female J8 had more dives at 110–140 meters in the first year that instruments were attached but still maintained a greater frequency of dives below 60 meters in both years and the basic bimodal distribution of dive depth.

None of the St. George females showed a fundamental change in dive pattern with season as the St. Paul females did. Female nos. 1789 and 2775 were predominantly mixed divers (shallow nighttime dives on some days and deep diving on the first and last days of the trip; fig. 12.7), and female no. 592 was a shallow diver in both 1983 and 1984 (fig. 12.7). Female J8 was a shallow diver with some deep dives on all three trips to sea (fig. 12.8).

Table 12.3

Comparison of July/August and October Foraging Data from Four Time-Depth Recorders
on Two Female Northern Fur Seals from St. George Island, Alaska, 1983

	July/August			October			
	Female ID			Female ID			
Parameters	1789	P9	Mean[a,b]	1789	P9	Mean[a,b]	p-value
Trip duration (h)	124	138	131 (10)	250	183	216 (47)	.282
No. of dives	148	175	161 (19)	1221	301	761 (650)	.426
Maximum depth (m)	147	206	206	231	242	242	.242
Mean depth (m)[a]	69 (23)	127 (47)	98 (41)	56 (26)	170 (72)	113 (81)	.687
Dive rate (dives/h)	1.2	1.3	1.2 (.07)	4.9	1.6	3.2 (2.3)	.448
No. of bouts	11	7	9 (2.8)	22	19	20.5 (2.1)	.028
Dive bouts/h	.09	.04	.07 (.03)	.09	.10	.09 (.01)	.500
Mean bout duration (h)[a]	1.3 (0.6)	3.7 (1.5)	2.3 (1.6)	3.3 (2.5)	3.1 (1.7)	3.2 (2.1)	.154
Mean No. of dives within bouts[a]	12 (8)	20 (16)	15 (12)	55 (74)	15 (11)	36 (57)	.872
Mean dive rate within bouts (d/h)[a]	9 (6)	5 (3)	8 (5)	14 (7)	5 (2)	10 (7)	.068
Nonbout dives (%)[b]	13.5	21.7	18.0	0.7	6.0	1.7	—
Time in bouts (%)[b]	11.6	19.1	15.5	28.8	32.0	30.2	—

Note: Each record is for a single foraging trip to sea.

[a] All mean values have sample standard deviation in parentheses.

[b] Overall percent values are reported in the mean column for nonbout dives and time in bouts.

Table 12.4

Inter-annual Comparison of Diving of Four Female Northern Fur Seals, from East Reef, St. George Island, Alaska, 1982–86

| | Female ID/Year | | | | | | | | |
| | J8 | | | 1789 | | 2775 | | 592 | |
Parameters	1984	1986 (18 July)	1986 (24 July)	1982	1983	1984	1986	1983	1984
Trip duration (h)	162	122	121	192	124	106	156	149	188
No. of dives	200	155	135	365	148	250	218	154	477
Maximum depth (m)	168	169	177	99	147	101	137	101	133
Mean depth (m)[a]	74 (37)	49 (34)	54 (39)	45 (14)	69 (23)	50 (21)	51 (21)	34 (10)	48 (15)
Dive rate (dives/h)	1.2	1.3	1.1	1.9	1.2	2.3	1.4	1	2.5
No. of bouts	7	7	9	18	11	7	11	5	19
Dive bouts/h	.04	.06	.07	.09	.09	.07	.07	.03	.1
Mean bout duration (h)[a]	3.1 (1.8)	2.9 (1.8)	2.1 (0.5)	2.6 (1.8)	1.3 (0.6)	2.9 (2.2)	1.5 (0.6)	3.1 (2)	1.7 (1.1)
Mean no. of dives within bouts[a]	29 (16)	19 (23)	12 (6)	19 (15)	12 (8)	31 (42)	18 (11)	23 (15)	22 (14)
Mean dive rate within bouts (d/h)[a]	19 (15)	6 (3)	6 (3)	8 (5)	9 (6)	8 (5)	13 (5)	10 (5)	15 (11)
Nonbout dives (%)[b]	6	12.3	23	7.9	13.5	52	8.3	24	13.8
Time in bouts (%)[b]	13.6	16.4	15.3	24.8	11.6	19.4	10.3	10.3	17.4

Notes: Each record is for a single foraging trip to sea. All records were collected sometime between 13 July and 13 August.

[a] All mean values have sample standard deviation reported in parentheses.

[b] Overall percent values are reported in the mean column for nonbout dives and time in bouts.

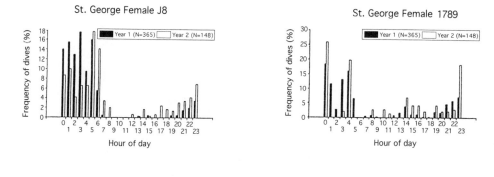

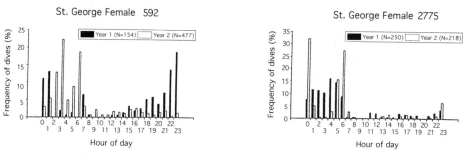

Fig. 12.5. The distribution of dives by hour of day for four lactating northern fur seals, St. George, Alaska, instrumented in July in different years.

DISCUSSION

Studies of foraging and diving behavior of fur seals have focused primarily on the early lactation period, within the 2 months following parturition (Gentry et al., 1986a; Goebel et al., 1991; Boyd and Croxall, 1992; Boyd et al., 1994; Trillmich et al., 1986; A. York [pers. comm.]; Kooyman and Trillmich, 1986; Francis et al., 1997; Gallo, 1994; Kooyman and Gentry, 1986; Mattlin, 1993). Females of migrating subpolar species, environmentally constrained to a 4-month lactation strategy, are faced with two important selective pressures that make foraging patterns of the late lactation period important. The first is feeding the pup prior to weaning and the abrupt termination of maternal investment. The second is the need to retain their own condition while feeding the pup. Given these selective pressures, and particularly the need to feed an ever growing pup, it might be expected that females will increase their foraging effort as the season progresses. Pups in fact have an accelerated growth rate after approximately 80 days of age (Boltnev, 1991). Females may increase their rate of feeding or pups may partition maternal resources differently after the molt.

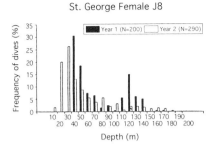

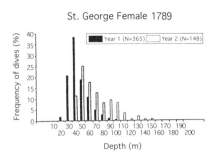

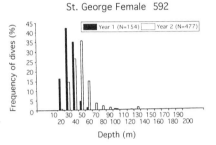

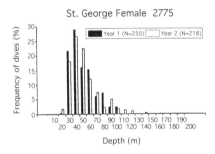

Fig. 12.6. The distribution of dives by depth (0–200 m) for four lactating northern fur seals, St. George, Alaska, instrumented in July in different years.

For females that dive predominantly at night, prey availability may increase simply because photoperiod changes greatly during the course of lactation. At the latitude of the Pribilof Islands, day length (sunrise to sunset) decreases from 17.8 hours on 1 July to 8.1 hours on 15 November (NOAA tide tables). For the period of our study, nighttime hours more than doubled (from approximately 0600 h to 1300 h). Therefore, vertically migrating prey may spend more hours in surface waters.

In our intra-annual comparison, all females showed differences in foraging from early to late lactation. At Zapadni Reef, St. Paul, all three females exhibited the same general foraging pattern of diving primarily at night to depths of less than 50 meters during early lactation. By October all three had made a shift in their foraging patterns to diving throughout night and day to depths in excess of 50 meters. By the classification of Gentry et al. (1986b), all three females switched from a mixed (though predominantly shallow) to a deep diving pattern. Parameters that showed differences (mean depth, mean number of dives within bouts, and mean dive rate within bouts) were associated with a shift to a deeper foraging pattern. The main difference however, besides an overall shift in foraging pattern, was an increase in trip length.

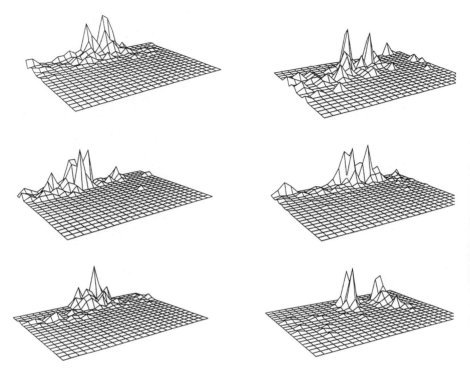

Fig. 12.7. Three-dimensional plots for three female northern fur seals instrumented with time-depth recorders in July of different years on St. George, Alaska. ♀ 1789, *top*; ♀ 592, *center*; ♀ 2775, *bottom*. The X axis is hour of day, with midnight in the center of each plot; the Y axis is depth (m) from 0 to 210 m (depth increases as the Y axis approaches viewer); and the Z axis records percent frequency of dives.

Though females foraging from East Reef, St. George, did not show a switch in their overall foraging patterns as did females at Zapadni Reef, they did show differences, mainly associated with an increase in trip duration. The percent time in foraging bouts doubled and the percent of non-bout dives decreased. Most parameters we measured, however, did not change between early and late lactation. The most consistent change in behavior between July and October appears to be the proportion of time that females spend at sea. This is confirmed by studies of attendance patterns that have much larger sample sizes and show an increasing duration of trips to sea as the pup-rearing season progresses (Gentry and Holt, 1986). It suggests that females tend to forage near their individual optimum capability given their foraging location and prey availability.

Females may forage in the same area from trip to trip (Loughlin et al., 1987) although sample sizes to confirm this are limited. Recent data for females that were instrumented with satellite transmitters for nearly the

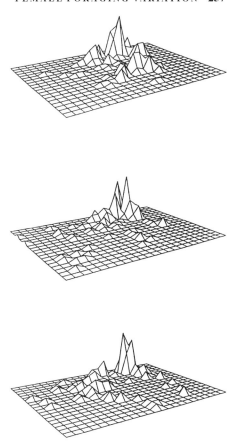

Fig. 12.8. Three-dimensional plots for female J8, instrumented with a time-depth recorder in August 1984 (*top*) and again in July 1986 for two successive trips. The X axis is hour of day, with midnight in the center of each plot; the Y axis is depth (m) from 0 to 210 m (depth increases as the Y axis approaches viewer); and the Z axis records percent frequency of dives. See table 12.1 for number of dives in each plot.

entire lactation period confirmed that there is a degree of site fidelity in foraging (Loughlin et al., 1993). The work of Goebel et al. (1991) and Loughlin et al. (1993) suggests that the Zapadni Reef females, because of their deep diving pattern, were foraging over the continental shelf.

Though our study was not designed to compare St. George and St. Paul islands, we would expect the proportion of the different foraging patterns to vary with island. Because St. George is nearer the shelf break, we would expect a greater proportion of animals at St. Paul to forage over the continental shelf, and therefore a greater proportion of deep divers there.

Information from diet studies supports this idea. Antonelis et al. (1993)

showed that animals foraging from St. Paul had a greater occurrence of walleye pollock and other shelf-associated species in their diet than those at St. George. St. George animals consumed, by proportion, more species associated with the oceanic domain off the continental shelf.

Little is known about how prey species of northern fur seals move seasonally. However, there are data on the types of predominant prey species and their locations relative to the continental shelf. Kajimura (1984) summarized the variation in principal forage species for fur seals by habitat. Fur seals feeding in the Bering Sea beyond the continental shelf over deep water fed on oceanic squid of the family Gonatidae (primarily *Gonatus* spp., *Berryteuthis magister*, and *Gonatopsis borealis*) or the bathylagid deep-sea smelt, *Leuroglossus schmidti*, of the family Bathylagidae. These prey species, and fish with swim bladders, exhibit diel vertical migration and inhabit relatively shallow depths at night as they move vertically in synchrony with the deep scattering layer (Roper and Young, 1975; Pearcy et al., 1977). During the night they are fed upon by fur seals that rarely dive beyond 200 meters (Gentry et al., 1986b). Fur seals foraging over the shelf were likely to feed on walleye pollock, Pacific herring, and capelin (Kajimura, 1984; Sinclair et al., 1994). Each of these prey items is distributed throughout the water column over the shelf, depending on the sex and age of the individual and time of day; however, they are principally found near the bottom (Bakkala and Wakabayashi, 1985). Even when prey are near the bottom over most of the shelf floor, they are shallower than the maximum diving depths observed for most fur seals and are accessible during all hours of the day.

It would thus appear that all three of the Zapadni Reef females preyed upon a species (or species) that changed its behavior and location within the water column from July to October. The apparent shift in foraging behavior may also be caused by a shift in the prey species taken. Little is known about whether individuals shift prey items. Fur seals forage on a wide range of species, but the number of species found in any one stomach tends to be small (1–2 species). Sinclair et al. (1994) conclude that fur seals specialize in taking prey of a particular size. Foraging pattern and prey may therefore be more a result of foraging location than an individual's specialization.

Considerable variability exists among lactating females in their diving patterns. Our inter-annual comparison, however, shows that for a given time of year females do not change their dive pattern. The degree to which they specialize is likely a result of the prey characteristics of particular foraging locations and a tendency to site fidelity. Intra-annual changes may be reflective of changes in prey at particular foraging locations.

This study shows that females while at sea are expending similar amounts of effort in July and October as measured by dive bouts per hour and bout duration. But because overall proportion of time at sea increases, the total amount of effort increases toward the end of lactation. This is important because subpolar fur seal species are environmentally constrained to a 4½-month lactation period. Pups are abruptly weaned in November (Ragen et al., 1995) and must begin foraging for themselves just as they start their migration south and winter begins. Pups in better condition may stand a better chance of surviving the transition from nutritional dependence to foraging on their own. Females should optimize their lifetime reproductive success by balancing the conflicting demands of investment in their pup and maintaining their own condition for survival and future reproductive output.

SUMMARY

Intra-annual (summer to autumn) and inter-annual (summer to summer) comparisons in diving behavior were made for five and four females, respectively, to determine the extent to which individuals change their diving habits over time. The intra-annual comparison included females from both Pribilof islands, while the inter-annual comparison included St. George Island females only.

The inter-annual comparison showed that summer diving (early lactation) patterns tend not to change from the broad categories of shallow, deep, or mixed diving in different years, suggesting that females for a given time and location prefer a particular prey.

The intra-annual comparison showed that, compared to summer diving, autumn diving (late lactation) was associated with increases in trip duration (table 12.3), number of dives per trip, number of dive bouts, mean depth of dive, proportion of time at sea, and absolute amount of time spent diving, and with a decrease in the number of exploratory dives. It also showed that there were no seasonal changes in the dive rate within bouts, dive bout duration, number of dives per bout, and maximum dive depth. That is, the aspects of diving that relate to environmental factors (such as day length and depth of prey) change seasonally, but those that are related to physiological capabilities of individuals (such as recovery from diving) do not change. This suggests that seals dive at an optimal rate near their individual physiological capabilities at all times of the pup-rearing season, and that the primary mode of increasing effort with increasing pup demand for resources is to increase the proportion of time spent at sea.

The results suggest that females from the two islands may differ in the way they change their diving patterns with season.

CHAPTER 13

Female Foraging Behavior:
Effects of Continental Shelf Width

WITH MARK O. PIERSON AND VALERY A. VLADIMIROV

THE FINAL aspect of the northern fur seal maternal strategy considered here is the effect that the local foraging environment has on diving behavior (using diving as a proxy for foraging). Foraging behavior has been broadly compared across taxonomic lines (Costa 1991a,b, 1993), in different maternal strategies (Antonelis et al., 1990a), and in a search for global trends (Gentry et al., 1986a). In these comparisons, the taxonomic status of the seals, their diet and maternal strategies, climate, and other factors may differ. The recent trend is toward comparing the same species in different environmental settings (examples, Costa et al., 1989; Trillmich, 1990; Trillmich et al., 1991; Boyd et al., 1994). The value of this approach is that the reduced number of variables makes it easier to assign causes to observed differences.

This chapter reports on a comparison of diving behavior of northern fur seal females at St. George Island and at Medny Island, Russia. The islands are at about the same latitude and both are affected by the Bering Sea climate regime. However, the underwater environments differ. St. George Island is atop a broad continental shelf, while Medny Island is part of an island arc system and has a narrow shelf. This distinction of islands being on either a wide (St. Paul Island, St. George Island, Robben Island) or a narrow shelf (Kurils, Bogoslof Island, Bering Island, Medny Island) applies to most sites where northern fur seals breed. It may not apply to San Miguel Island, which is atop a shelf unlike most other shelves in the world (it is a series of basins and ridges termed the "southern California continental borderland," Kennett, 1982).

The width of the local continental shelf may affect fur seals in several ways. The swimming distance from shore to the nearest point of deep water (shelf break) can be much greater at wide shelf islands (table 10.1). The prey species taken also may differ as this distance changes (chapter 10). Females take mostly fish where the distance is great (St. Paul Island), mostly pelagic squid where it is short (Commander Islands), and a mixture of pelagic squid and fish where it is intermediate (St. George; Antonelis et al., 1996). Finally, the width of the shelf may be a general index to the duration of foraging trips there, but it may not determine this duration (chapter 10, table 10.1).

Wide and narrow shelf islands also tend to differ in their past population trends. Most of the wide shelf colonies declined from 1956 to 1981 (St. George, St. Paul, and Robben islands; table 1.1). The narrow shelf colonies (Bering, Medny, the Kurils; table 1.1) remained stable or increased in those years (Bosoglof and San Miguel are in a different category because they were new colonies that grew largely from immigration). This difference in trend is usually not considered in the search for causes of the population's failure to recover from the herd reduction program of 1956–68. That search has focused more on single, simple factors such as disease and entanglement (chapter 1) than on complex environmental causes. A notable exception is the work of York (1991b, 1995).

In 1990 we addressed the problem of shelf width by comparing female diving at St. George Island and at Medny Island, where past population trends and female attendance behavior were known to differ. The questions this research addressed were the following: Which aspects of diving are the same or different in these two environments? Do females take prey at different depths and water temperatures? Do they use different dive patterns (change of depth over time; chapter 12)? Given that trips to sea are of different length, does dive effort differ? Is transit time to and from foraging areas a constant fraction of the trip to sea? Do dive shapes differ between islands? Is the correlation between dive depth and dive duration the same at both islands? These questions are important because differences in foraging behavior, acting through energy delivered as milk, and differential pup growth and survival rates could form the basis of long-term population trends.

METHODS

We collected data at East Reef, St. George Island, in July and August 1988 and 1989, and at Urilie, Medny Island (54 36.9′ N lat., 167 46.2′ E long.), in July and August 1990 (fig. 13.1). At both sites we used identical capture techniques and measured dives using the same electronic Time-Depth Recorders (TDRs) [1]. The TDRs we used at Medny Island made frequent measures of light level as the basis for calculating a daily approximate location at sea (termed "geolocation"; Hill, 1994).

To conform with present practice in diving studies, we treated all shallow dives (≤ 5 m in depth and ≤ 0.5 min in duration) as if they represented traveling rather than foraging. Dives of the type we excluded are known to occur during daylight hours when northern fur seals do little surface feeding (Kajimura, 1984; Gentry et al., 1986b). We conducted separate analyses that included and excluded these "traveling dives."

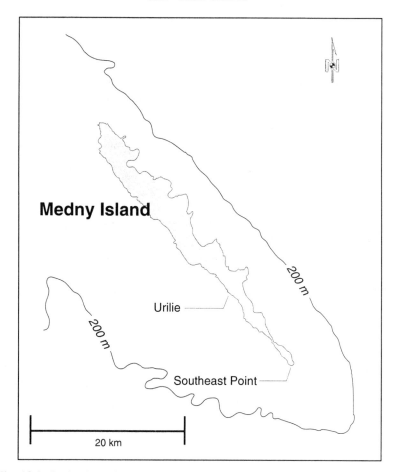

Fig. 13.1. Study site at Medny Island, Russia, approximately 1,400 km west and 0.5 south of St. George Island, Alaska.

The dive records were analyzed for most of the variables mentioned in chapter 12. In addition, we analyzed all dives of 20 meters for descent and ascent rate and for shape [2]. Dive shape was analyzed to test the hypothesis that sudden reversals of depth (fig. 13.2) correlate with some aspect of the physical environment, specifically water depth or temperature. The analysis of depth reversals was important because they may relate to prey capture or physiological necessity.

The two islands were compared for the "dive patterns" that occurred there [3]. St. George Island females are known to use three different dive patterns termed "shallow," "mixed," and "deep" (definitions in chapter 10). By assigning dive patterns to females at Medny Island we were testing the hypothesis that deep and mixed diving occur only when a wide continental shelf is available.

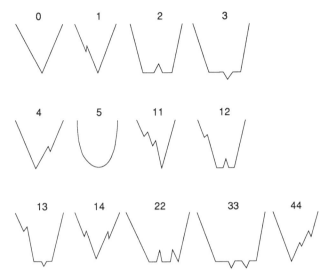

Fig. 13.2. Dive shape categories used in analyzing dives of   20 m depth. Assignment to categories depended on the location and number of depth reversals that exceeded the depth resolution limit (± 2 m) of the instrument.

Diving is not continuous in fur seals, as it is in sea lions, but occurs in bouts. Various statistical techniques have been used to define the ends of bouts with a view to producing one more trait by which fur seal species can be compared. The method we used [4] was modified from that used by Boyd and Croxall (1992) to establish bout-ending criteria (BEC) for antarctic fur seals.

All statistical tests were made on the entire data set except for a comparison of maximum vertical distance moved within bouts, which was a measure if dive effort. For this test we selected short segments of the deepest dives available at the two islands [5]. Our rationale for this selection was that if foraging effort differed at all, it was most likely to differ in the extremes. Total vertical distance moved was calculated as:

sum of maximum depths ∞ 2/trip or set duration.

Attendance behavior of noninstrumented females was inferred from observing suckling in marked pups.

RESULTS

Our analysis was based on twenty-seven trips to sea by twenty-three females (St. George, ten trips for nine females; Medny, seventeen trips for fourteen females).

Some of our inter-island comparisons were confounded by the fact that

Table 13.1

Foraging Trip Characteristics for Northern Fur Seal Females
at the Pribilof and Commander Islands

Female ID	Start Date	Weight (kg)	No. Dives Recorded	Hours of Diving[a]	Dives in 24 Hours[b]
MB	7/90	28.3	353	56.4	6.3
MS	7/90	29.4	568	101.9	5.6
MT	7/90	40.4	742	99.7	7.4
M9	7/90	28.9	909	130.2	6.9
M13	7/90	28.9	606	107.3	6.2
M16	7/90	32.4	1,551	152.3	10.2
M17	8/90	35.4	924	219.5	4.2
M26	7/90	—	159	29.1	5.5
MG2[c]	8/90	29.3	817	129.7	6.3
MG3	8/90	43.2	226	82.9	2.7
MG4	8/90	28.6	413	112.5	3.7
MG5[c]	8/90	35.6	730	110.3	6.6
MG6	8/90	31.7	723	69.0	10.5
MG8[c]	8/90	31.7	313	123.7	2.5
S1687	9/88	36.6	575	178.3	2.2
S2067	7/87	—	551	92.6	5.9
S2084	10/88	35.1	463	172.8	2.7
S2093	7/87	—	831	147.7	5.6
S2115	8/89	39.0	368	117.7	3.1
S5243	10/88	42.9	436	171.6	2.5
S6000[c]	7/89	35.4	1,721	302.8	5.7
S6001	7/89	44.0	936	188.4	5.0
S6002	7/89	42.6	1,148	142.3	8.1

Note: M = Medny female; S = St. George female.

[a] Time in hours between the first and last dives to more than 5 m.

[b] Does not include dives to 5 m or less.

[c] Combines data from two successive trips to sea.

instruments and/or harnesses had a much greater effect on females at Medny than at St. George Island. They increased trip duration by only 0.6 days at St. George and by 7.7 days at Medny [6]. Compared to the females at St. George Island, those at Medny Island spent more time ashore after being instrumented, and more days at sea without diving. Initially females at Medny Island had aberrant diving behavior (short bouts with irregular surface intervals and inconsistent dive depths) unlike any we saw at St. George Island (Gentry et al., 1986b). After 2–3 days this behavior ended and diving became regular, suggesting that they had learned how to dive with the instruments.

Females at the two islands did not differ in the proportion of their dives that were considered to be "traveling dives" [7]. On average, 32.3% of dives were removed from analysis because they were ≤ 5 meters in depth and ≤ 0.5 minutes in duration.

The proportion of trips to sea that females spent in transit to and from foraging areas [8] did not differ significantly between the two islands despite the fact that trip durations were very different. That is, transit time seemed to comprise a constant fraction of the trip to sea despite absolute trip length. However, the sample size for inbound transit times was small at Medny Island (n = 10) [9].

Females spent more hours diving on a foraging trip (first to last dive per record) at St. George than at Medny Island (table 13.1; means = 168 and 109 hours, respectively) [10]. Given the exaggerated response of Medny females to the instrument or harness, the difference was undoubtedly greater than this comparison indicated.

Our analysis did not produce an unambiguous BEC, so we could not compare the islands for dive bout duration. We found a sharp and distinct inflection point in only two of the twenty-four probability plots. The others produced smoothly continuous curves with no obvious inflection points. This finding does not imply that diving was aperiodic. At Medny Island, virtually all foraging dives occurred between dusk and dawn (fig. 13.3, *top*). The same was true at St. George Island (fig. 13.3, *bottom*), although all females there showed some clusters of deep dives (100–190 m) during daytime. However, these clusters did not cause obvious breaks in the probability plots of interdive intervals.

Foraging effort at the two islands was different using some measures and similar using others. Medny Island females had a greater dive rate (dives/hr from first to last dive on the record) than St. George island females when shallow traveling dives were included [11]. Also, Medny Island females spent a significantly greater proportion of their time at sea foraging (under water) than St. George Island females (table 13.2), again only when traveling dives were included [12]. Dive effort was not different at the two islands when measured by the total number of dives per trip to sea [13], nor by the vertical distance moved per hour, either averaged over the entire trip or within selected dive bouts [14]. Finally, dive effort was not different at the two islands when measured by the total vertical distance moved on a foraging trip [15]. The latter result was unexpected because St. George females tended to make a few deep dives per hour (average 2.9 dives/h to 100–170 m) whereas Medny females made many shallow dives (average 15.9 dives/h to 20–63 m).

The two populations did not differ in the mean depth or duration of dives (table 13.2) [16]. However, females at St. George Island made single dives with much greater maximum depths (fig. 13.4, *top*) and durations (fig. 13.4,

Medny female G6: n = 908 dives

Medny female 17: n = 1914 dives

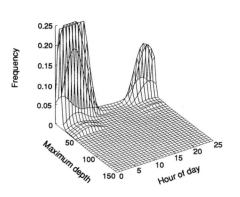
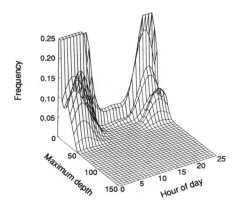

St. George female 2115: n = 578 dives   **St. George female 5243:** n = 478 dives

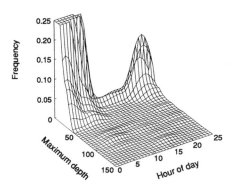
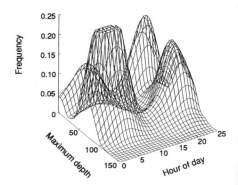

Fig. 13.3. Frequency (Z axis) of dives by hour of day for representative females at St. George and Medny islands.

bottom) than females at Medny [17]. These maximum performances by St. George Island females were too few in number to affect the overall population means.

Time/depth profiles suggested that females at the two islands used different search strategies. At St. George Island, the longer the females were down (4–5 min) the deeper they went (100–200 m). That is, dive duration

Table 13.2
Comparative Dive Characteristics of Northern Fur Seal Females
at the Pribilof and Commander Islands

Female ID	Mean Max Depth[a]	Mean Dur.[b]	Dur./ Dep. Corr.[c]	Slope[d]	% time under Water[e]	Mean Bottom Time	AID[f]
MB	14.4	1.5	0.567	0.063	19.7	0.63	20.6
MS	13.6	1.1	0.552	0.077	11.4	0.69	7.6
MT	15.8	1.5	0.412	0.073	22.7	0.92	13.2
M9	15.9	0.9	0.533	0.059	12.8	0.51	6.1
M13	16.2	1.1	0.657	0.075	12.1	0.59	13.4
M16	16.9	1.2	0.521	0.063	21.9	0.88	17.6
M17	20.6	1.5	0.487	0.066	13.1	1.01	14.7
M26	17.1	1.4	0.382	0.050	12.2	0.82	3.0
MG2[g]	17.9	1.5	0.648	0.058	17.6	0.80	19.1
MG3	13.8	0.8	0.540	0.046	5.1	0.67	18.0
MG4	19.5	1.2	0.794	0.064	7.3	0.75	23.3
MG5[g]	18.3	1.0	0.486	0.046	11.9	0.40	6.3
MG6	11.3	1.1	0.335	0.046	16.3	0.64	13.9
MG8[g]	13.8	1.3	0.344	0.043	9.9	0.41	9.5
S1687	24.8	1.6	0.605	0.024	8.7	0.71	29.9
S2067	29.1	1.4	0.854	0.028	14.4	0.48	19.5
S2084	14.3	0.7	0.676	0.031	4.5	0.55	43.3
S2093	23.7	1.4	0.714	0.037	14.2	0.54	17.5
S2115	16.2	0.8	0.831	0.040	5.0	0.45	31.2
S5243	61.1	2.6	0.778	0.035	11.2	0.78	52.4
S6000[g]	14.8	0.8	0.815	0.041	8.9	0.45	20.7
S6001	19.0	1.0	0.841	0.039	9.6	0.47	27.1
S6002	9.9	0.5	0.686	0.045	7.7	0.37	26.7

Note: M = Medny female; S = St. George female.
[a] Mean of the maximum depth attained on each dive.
[b] Mean durations of all dives.
[c] r^2 value for regression of dive duration on dive depth.
[d] Slope of the same regression.
[e] Including all dives > 2 m.
[f] Ascent Inflection Depth = mean depth at which inflection (depth reversal) occurred during ascent.
[g] Combines data from two successive trips to sea.

was a good predictor of dive depth (table 13.2, all r^2 .61). But at Medny Island, many long dives were shallow (ca. 20 m); females descended to a particular shallow depth and remained there for 4–5 minutes, with the result that dive duration did not predict depth well. The inter-island difference in whether duration predicted depth was significant [18]. Because of these long-duration shallow dives, females at Medny Island spent significantly more time at the bottom of a dive than those at St. George Island (table 13.2) [19].

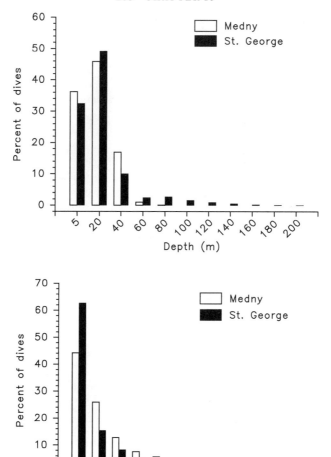

Fig. 13.4. Comparison of maximum dive depths and maximum dive durations for female fur seals at St. George and Medny islands.

Considering the data as a single pool, females ascended faster than they descended [20]. However, we could not compare the rates of descent and ascent between the islands. Scatterplots of ascent against descent rates, and ascent rate against dive duration for each individual, were so varied that an inter-island comparison would have been invalid.

The analysis of dive shape showed that simple, spike-shaped dives with no depth reversals (shapes 0 and 5; fig. 13.2) made up half (50.7%) of all dives we categorized [21]. This finding contradicts a previous statement (Gentry et al., 1986b) that all dives in this species are spike shaped [22]. About 18.6% of all dives had a single reversal on descent, 5.5% had a single reversal on ascent, 17.6% had either two or three reversals, and 7.6% had a reversal only at the bottom of the dive. That is, spike-shaped dives and dives with a single reversal on the descent were the predominant shapes.

Females showed individual variation in the dive shapes they used. The single most frequently used shapes (for twenty-two of twenty-three females) were 0, 1, and 5; one St. George female made mostly shape 14 dives. Taken as a group, St. George females used dive shapes 0 and 4 more extensively than Medny females [23], and Medny females used dive shapes 1, 12, and 13 more extensively than St. George females [24]. Inter-island differences in the other shapes were not significant.

Given that females from the two islands differed in the depth and durations of the most extreme dives (fig. 13.4), do the dive shapes in which this difference occurs suggest a consistent difference in foraging strategy? Analysis showed that St. George females reached significantly deeper depths than Medny females in dive shapes 0, 1, 2, 4, 5, and 14 [25]. These shapes are quite diverse, which leads to the conclusion that shape is of secondary importance to maximum depth attained.

Depth reversals occurred across a wide range of depths at both islands (descent, 2–124 m; ascent, 2–120 m). Interestingly, reversals never occurred in the deepest dives we recorded (124–170 m). The two islands did not differ in the depth at which reversals began on the descent, but they did differ on the ascent (depths at St. George were deeper; table 13.2) [26]. An analysis of reversals during descent, the most numerous kind, showed the majority clustered between 10 and 40 meters at both islands.

The temperature ranges at which depth reversals occurred were narrower than the depth ranges just discussed. On either descent or ascent, reversals began within a fairly narrow range (4 C) for most (60%) females. The others had wider ranges (to 7.3 C) that were probably attributable to the deeper depths they attained (water temperature decreases with depth). Reversals during descent clustered within a 2 C range for most individuals (fig. 13.5).

These results led to the conclusion that the depths and temperatures at which reversals occurred were predictable for individuals, especially within brief bouts of sequential diving. But individual variation was so great that the two populations could not be separated on these bases. Generally, depth reversals on descent occurred between 10 and 40 meters at temperatures between 7.5 and 11.5 C.

Medny females had a uniformly warmer foraging environment than St. George females (table 13.3). Water temperatures were significantly warmer at the surface, at the start of depth reversals, and at maximum depth at Medny than at St. George Island [27]. The temperature difference at maximum depth

TEMPERATURE AT DEPTH REVERSAL

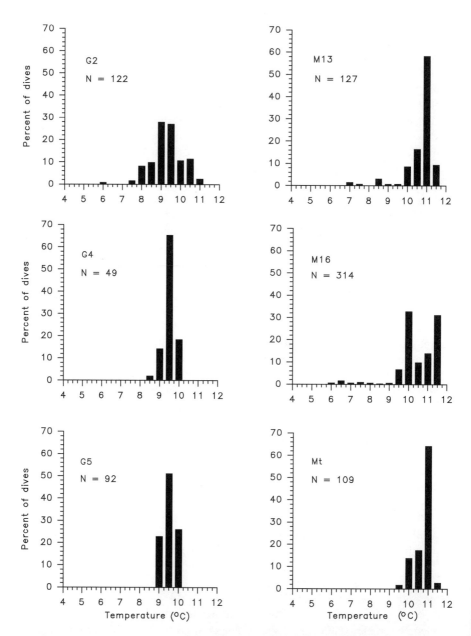

Fig. 13.5. The temperature at which depth reversals began during the descent phase of the dive for six Medny Island females (nos. G2, G4, G5, M13, M16, and MT).

Table 13.3

Water Temperatures at Selected Points in the Dive Profiles of Female
Northern Fur Seals; Comparison of Medny Island and St. George Island

Female ID	Mean Surface Temp.	Mean DIT [a]	Mean Bottom Temp.[b]	Mean AIT [c]
MB	10.6	10.1	6.6	7.5
MS	10.3	10.3	7.5	8.7
MT	10.5	10.6	8.5	8.8
M9	9.8	9.7	7.8	8.6
M13	10.6	10.4	6.9	8.0
M16	10.3	10.2	7.1	7.7
M26	9.7	9.8	7.9	8.5
MG3	9.4	9.3	6.2	7.1
MG4	9.1	9.3	7.6	8.5
MG5[d]	9.1	9.2	8.2	7.6
MG6	10.5	9.9	6.1	8.1
MG8[d]	10.2	9.3	7.1	8.6
S1687	7.5	6.8	6.8	7.1
S2067	9.3	8.6	5.7	6.7
S2084	7.5	7.1	5.9	5.9
S2093	8.5	8.1	6.1	7.0
S2115	9.3	9.2	6.7	6.7
S5243	7.8	7.9	3.6	4.1
S6002	8.9	8.6	6.6	6.4

Notes: M = Medny female; S = St. George female.

[a] DIT = Descent Inflection Temperature, the temperature at which inflection (depth reversal) occurred on the descent.

[b] Temperature at the maximum depth recorded.

[c] AIT = Ascent Inflection Temperature, the temperature at which inflection (depth reversal) occurred on the ascent.

[d] Combines data from two trips to sea.

was partly explained by the fact that maximum depths were greater (hence colder) at St. George Island (fig. 13.4, *top*). The temperature differences on the shallowest dives (≤ 20 m) were about 2 C.

Medny females moved from cooler into warmer water as their trips to sea progressed, but St. George females did not. The evidence for this conclusion is that temperature at the surface and at maximum depth increased during the trip (fig. 13.6 a,c). This result cannot be explained by a shift to shallower dives (hence warmer water) because depths did not change over time (fig. 13.6e). Unfortunately, the TDR memories were filled by day 6 and therefore did not show a return to colder water at the trip's end. The records of St. George females did not show this change in temperatures.

The approximate locations at sea, derived from the TDR's geolocation feature, supported the suggestion that Medny females moved into warmer water

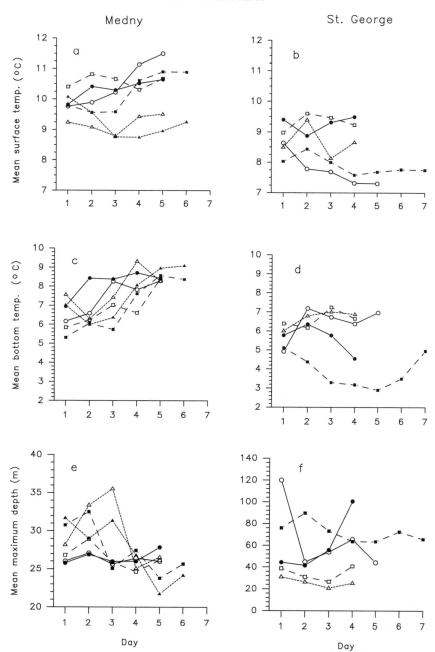

Fig. 13.6. Water temperature at the surface and at maximum depth, and maximum depth attained by day for six Medny and five St. George Island females. Note progressive warming for Medny Island females (panels a and c).

as the trip progressed. These locations were all in a southwest direction from Medny Island [28]. Maps of sea surface temperature for the month these measures were made, August 1990 showed that temperatures gradually increased going southeast from the island, but not in other directions (NOAA, 1991).

DISCUSSION

The first question asked by this research—which aspects of foraging behavior change with shelf width where taxonomic status is identical and latitude and overall environment are similar—was answered clearly. Females from Medny Island (narrow shelf) expended more dive effort (dives/h) per day at sea, spent a greater proportion of their sea time under water, uncoupled dive duration from dive depth (had long time at the bottom on shallow dives), and linked time at the bottom with water temperature more than females from St. George Island (near the edge of a wide shelf). Some of the results from Medny Island may have been affected by instrument or harness problems and should be repeated. Attendance behavior also differed; females at narrow-shelf islands cycle faster (chapter 10). Related work (Antonelis et al., 1996) showed that the diet also differed at these two islands.

The larger question that these differences relate to—whether foraging differences affect population trend through energy flow—cannot yet be addressed. Measures of energy flow and pup growth rate were made (unpublished data). But the most critical comparison—pup survival rate—has not. Such a study seems desirable.

This study suggests that narrow continental shelves do not support a biomass of prey that induces fur seals to dive for it. Of the three dive patterns described for northern fur seals (Gentry et al., 1986b; Ponganis et al., 1992; also see chapter 12), only the "shallow" pattern was found at Medny Island. A comparable sample size at a wide shelf island would produce numbers of "deep" and "mixed" divers that forage on the continental shelf (Goebel et al., 1991). The absence of these patterns at a narrow-shelf island suggests that females bypass the shelf, which they would not do if abundant prey were present.

Contrary to the way it has been used previously, the correlation between dive duration and dive depth is not a characteristic of the species (Gentry et al., 1986b). The correlation differs among populations of the same species feeding in different environments. Therefore, this correlation may reflect the local foraging environment rather than the physiology of the diver.

The finding that the vertical distance moved per hour did not differ in deep and shallow dive bouts suggests that the animals may dive to some physiological limit whenever they are foraging. Deep, long-duration dives were followed by surface intervals of about 20 minutes and shallow, brief dives by intervals of less than 0.5 minutes. The difference may relate to the need to clear lactate after anaerobic dives (Kooyman, 1989; Kooyman et al., 1983b). Different divers appear to cover an equal amount of vertical distance per hour and to vary their surface intervals according to the depths they are exploiting. This finding supports the conclusion in chapter 12 that females dive to some physiological limit whenever they dive.

The dive contours of female fur seals seemed more varied (thirteen types) than reported for elephant seals (five types; Asaga et al., 1994; Crocker et al., 1994). If this difference is not an artifact of the scoring methods used, it may reflect differences in body size, maneuverability, speed of the prey, or energetics of foraging (Costa, 1993).

The finding that about 50% of dives have at least one depth reversal on descent or ascent refutes a previous statement that fur seal dives are simple spikes (Gentry et al., 1986b). The superior resolution of modern TDRs revealed these reversals. The function of depth reversals, if any, is unknown. They probably do not accommodate lung collapse because collapse is virtually complete by 50–70 meters (Kooyman, 1972, 1988), whereas depth reversals occur from 2 to 124 meters and do not occur at all on the very deepest dives (124–170 m).

Depth reversals may be associated with foraging. The majority of reversals occurred between 10 and 40 meters at temperatures from 7.5 to 11.5 C, a relatively narrow range given the maximum depths and temperatures recorded. Most reversals occurred within a 2 C range for each individual (fig. 13.5). If they reflected exploration, play, or other nondirected behavior, they might be expected to occur over a wider range. The hypothesis that depth reversals indicate prey detection is testable using TDRs that measure prey capture attempts.

Unexpectedly, identical TDRs and harnesses affected foraging trips differently at the two islands. The ratio of instrument to animal cross section area (Wilson et al., 1986) were the same in both cases. More days were added to trips where females experienced warmer water temperatures, dove less than 70 meters deep, had long times at the bottoms of dives, and took mostly squid in their diet than where they performed some deep dives (> 100 m), had short times at the bottom, and took a diet of mixed fish and squid (Antonelis et al., 1996). A difference in swim velocity for females from the two islands could explain the difference because instrument drag increases nonlinearly with swim velocity (Feldkamp et al., 1989).

The instrument effect prevented an adequate comparison of trip length, proportion of trips spent in transit to and from foraging areas, and overall dive effort. The Medny measurements should be repeated using smaller TDRs. Although a formal comparison was not possible, transit seemed to occupy a similar proportion of sea trips at both populations.

Evidence that Medny females went southwest on foraging trips was not surprising. Stejneger (1896) reported that most of the animals he collected at sea near Medny Island were found at the 100-fathom contour southwest of the island. Our animals went considerably farther than that point, perhaps because the TDR lengthened the trip to sea.

SUMMARY

We instrumented nine females at St. George Island and fourteen females at Medny Island (Commander Island Group, Russia) with Time-Depth Recorders to compare maternal foraging behavior at two islands where the continental shelf was wide and narrow, respectively. All measures were made during summer. The comparison was based on thirty variables.

Where the continental shelf was wide, females had longer trips to sea, performed deeper and longer single dives into colder water, and expended less foraging effort (measured as dives per hour and proportion of sea time spent under water) compared to females where the shelf was narrow. Where the shelf was narrow, females cycled between land and sea more quickly than where the shelf was wide. At the narrow-shelf island, all females used a shallow dive pattern that is associated with nighttime foraging on vertically moving prey, whereas females from the wide-shelf population used mixed and deep dive patterns that are associated in part with shelf feeding. This difference suggests that narrow continental shelves do not support a biomass of prey that can sustain a fur seal population.

Medny Island females performed many long-duration (4–5 min) dives to a shallow (20 m) depth, apparently in association with a warm water layer. This foraging strategy was not seen at St. George. Because of this difference, depth could not be predicted from dive duration at Medny Island as it could at St. George Island, and females spent more time at the bottom of a dive at Medny compared to St. George.

The two populations differed in minor ways in the shapes of dives individuals preferred and in the depth of dives in a given profile. Depth reversals on descent and ascent, which may indicate prey capture attempts, occurred on about 50% of dives. Reversals occurred within fairly predictable depth (10–40 m) and temperature (7.5 –11.5 C) bounds that differed slightly between islands.

These results show that for the same species breeding at similar latitude, local environmental differences may create diving/foraging strategies that are as dissimilar as those seen for different species feeding at different latitudes. The extent to which these differences in maternal foraging contribute to differences in population trend is still unknown.

PART FIVE

Summary, Comparisons, and Conclusions

———————————

In the final section of this book I try to synthesize the information
in the preceding chapters at two levels. The first chapter in this
section draws a comprehensive picture of the northern fur seal by
linking the findings of separate chapters and interpreting them.
The second considers the implications of the present work for
otariids in general. The latter is not intended as a comprehensive
review of all findings about otariids; rather, it focuses on
those concepts in behavior and ecology that the
present work adds to, offers alternatives
to, or contradicts.

CHAPTER 14

Synthesis

THIS CHAPTER summarizes the natural history, mating system, and maternal strategy of the northern fur seal from information presented in earlier chapters. It reviews the research project that produced this information and answers some of the key questions posed by the St. George Island program, largely from the behavioral standpoint. It also reviews some possible causes for the long-term decline in fur seal numbers that started the research project reported here (including some information that does not appear in earlier chapters). Finally, this chapter discusses some implications of the present work for those that manage fur seal populations.

THE BEHAVIORAL RESEARCH PROJECT

In this book I report the results of a 19-year study (1974–92) on the behavior of northern fur seals. The study was part of a broader program of fur seal research that was funded and conducted by the U.S. government to meet its obligations under the Interim Convention on Conservation of the North Pacific Fur Seal (in force from 1957 to 1985).

In the periods 1820–67 and 1911–38, northern fur seals recovered after having been reduced by sealing to nearly the brink of extinction. Both times, stopping the killing of females allowed the species to grow to at least 2 million animals. However, starting in 1956, 330,783 females were killed intentionally to reduce the herd, but this time the herd failed to recover. Based on the 1911–38 period, an 8% per year recovery rate was expected. Ten years after the desired herd size had been reached and recovery still had not occurred, a research program into possible causes was initiated.

The research program was to focus largely on the effects of human activities on fur seal survival and reproduction. With the agreement of treaty signatories, the U.S. government suspended the commercial kill for fur seal pelts in 1972 on one of the Pribilof Islands (St. George) to create a research reserve. Coordinated land and sea research on several topics, including behavior and population dynamics, was to have been conducted for at least 15 years as the population changed in size and sex ratio (Anon., 1973). The effort was known as the St. George Island Program. This book is a final report of the behavioral portion of the St. George Island Program only; it does not include a detailed analysis of results from the population project.

The general questions to be addressed by the St. George Island Program were listed in the document that established the research preserve (Anon., 1973). The overriding question of why the herd failed to recover from intentional reduction still cannot be answered fully.

The specific questions to be asked in the behavioral project, as well as experimental design and field methods, were left to the discretion of field personnel. These consisted mainly of various colleagues under contract, field assistants, scientists in other parts of the St. George Island Program, and me.

Historically, the fur seal research program on population dynamics used a different approach than that used by the behavior project. It collected a long time series of population parameters (largely estimated) and inferred behavior from population trends rather than by measuring behavior directly. It considered only those behavioral topics that could directly affect population dynamics, and only in terms that the models could accommodate. The behavioral project used data on many subjects to investigate the effects of age, population size, sex ratio, environment, and other factors on the behavior of individuals. Its approach was to show how variations in individual behavior combine to create trends observable at the population level. The two approaches are complementary.

The behavioral project was empirical. It began with measurable components of behavior and tried to build a profile of the species' behavior through induction. Assumptions and preconceptions of a behavior's evolutionary value were kept to a minimum. Although the data were not intended to address specific aspects of theory, they were sometimes useful for that purpose.

The behavioral topics selected for study were initially those that might change with population size and sex ratio. Other topics were added later. One discovery—that nursing mothers moved reliably to sea and back and might be able to carry instruments to measure diving—led to a new way of measuring foraging behavior (Kooyman et al., 1976, 1983a). Time-Depth Recorders (TDRs) described the northern fur seal's pelagic behavior and permitted comparison of foraging in fur seals worldwide (Gentry and Kooyman, 1986a). These results partly satisfied the program's original goal of showing the relations between fur seals and other living marine resources through pelagic collections. Although the instruments have changed, the use of TDR's to study foraging has become a standard in marine mammal biology.

Behavioral data were collected every breeding season from 1974 through 1992 on established study sites. Some types of data were collected daily every year. Other types were collected only before and after the herd recovered from the effects of the former kill of males, and still other data were collected during experiments that changed from year to year. The core of the program was the large number of known females (1,051) and males (1,542) in the population.

Two major events occurred during this study. A climate regime shift occurred in the late 1970s and lasted until this project ended. Also, the strongest El Niño of the century occurred in 1982–83. Measures of fur seal behavior during these hemispheric events gave some new information about how fur seals relate to their environment.

Most of the program's questions about behavior were answered. But the behavioral project had two notable failures. One was that extreme tag wear and breakage prevented us from getting good estimates of philopatry, individual survival, and kinship relations. The second was that after the kill of males was stopped in 1972, the population continued to decline. Because of this decline, we could only compare behavior between a medium and a small population, not between a medium and a large one as the St. George Island Program had conceived.

After the start of the herd reduction in 1956, the Pribilof population declined until 1970, increased briefly until 1976, then continued declining until 1983 when the decline stopped on St. Paul Island (York and Kozloff, 1987; York, 1990). The decline stopped at St. George Island in 1996. The total world population is still about 60% smaller than it was in 1956. That is, recovery toward the 1956 level appears to have barely begun, if it is occurring at all.

ANSWERS TO ST. GEORGE ISLAND PROGRAM QUESTIONS

The following is a list of the key questions posed by the St. George Island Program (Anon., 1973) and partial answers mostly from the standpoint of behavior. Most of the program's questions, large and small, were answered at least in part.

1. *Did the killing of juvenile males for pelts prevent the herd from recovering as predicted*? It clearly did not. The St. George Island population stopped declining in 1996 despite the fact that the kill there ended in 1972. On the other hand, the St. Paul Island population stopped declining four years *before* the kill ended there in 1985. Thus, stopping the kill for pelts did not coincide with a change in population trend at either island. Furthermore, the kill for pelts did not stop the herd from increasing markedly between 1918 and 1956 when the kill of females began (1,896,623 males were killed in this period, data in Engle et al., 1980).

2. *Did the relationship between survival and abundance of juveniles change between the 1930s and the 1950s*? The behavioral project did not address this question about population dynamics. But, it showed that there is good reason to question whether behavior on land is a likely agent in

density-dependent mortality. Pup mortality on land is density dependent (Fowler, 1990). But this mortality seems more likely to result from disease than from the number of adult animals. Female density on land is independent of population size because individual females retain the same spacing despite the total number on shore. They nurse and mate at the same local density in females per square meter whether animals are sparse or numerous. "Crowding" is not likely because females expand or contract the area occupied as their numbers fluctuate. Male densities are limited (minimum territory size is one body length in diameter; chapter 4), and males actively avoid trampling pups. Further evidence of density-dependent mortality should be sought through competition at sea or in diseases, not in behavior that affects reproduction.

3. *Did human disturbance on breeding areas (for management purposes) affect survival by breaking nursing cycles?* Brief, infrequent human disturbances are not likely to affect fur seals through breakage of the maternal bond within a season. Mothers are so strongly bonded to their young and to their parturition sites that occasional intrusion of humans has little lasting effect. Disturbed females go to sea briefly, then return to their young. Most females (67%) will tolerate being physically moved to a different breeding area or island with their young, without the bond being broken (chapter 7). However, the effects on maternal bonds within a season represent only short-term effects. It is highly likely that human disturbance has a cumulative, long-term effect that can be measured in terms of population redistribution and productivity (see next question).

4. *Did human disturbance (as above) cause a change in herd behavior or activity cycles that could reduce survival?* The kind of human disturbance during the present program (management disturbance of breeding areas once or twice per year; occasional visits by hunters after 15 October) did not seem linked to a change in behavior. Many aspects of behavior did change with herd size and sex ratio, but usually following, not preceding, the population change. For example, the frequency of male territorial defense increased, but only after male numbers increased. All other aspects of male behavior that showed measurable changes were not directly related to reproduction.

The activity pattern on shore was also little affected by these occasional disturbances. Northern fur seals are active at night while on land. This pattern did not change with population size or sex ratio in this study. Weather and social disturbances, such as fights among adult males, caused temporary increases in activity cycle. But infrequent disturbances by humans (once or twice a year) were not sufficient to cause a change in activity cycle.

Human disturbance correlates with long-term redistribution and selective declines of the breeding population. The breeding areas on St. George

Island that had road access declined faster from 1914 to the present and contributed less to overall pup production than breeding areas that had no road access (Ream et al., 1994). The conclusions about the effects of human disturbance on fur seal behavior in this work do not apply if the disturbance level increases.

5. *What is the optimum sex ratio?* Soon after the kill of males stopped, the sex ratio on shore was twenty-seven to forty-six females per breeding male (depending on the site). The ratio rapidly declined to 9:1 at two sites and remained there for 9 years. There is no evidence that pregnancy rates increased as the sex ratio decreased. Even at high sex ratios, pregnancy rates may exceed 85% in some age groups.

The ratio of 9:1 females per breeding male on shore may be considered the optimum sex ratio because the undisturbed population maintained this ratio for many years. This ratio affects reproduction through the difficulty males have detecting estrous females. Since about 66% of the female population is at sea on any given day, the 9:1 ratio on shore translates to an adult sex ratio of 26.5 females per breeding male for the population at large. The turnover rate in adult males is about 34–40% per year.

6. *Did the killing of males act as an artificial selection pressure that altered their behavior or survival?* Probably not. There was only a slight decline in juvenile males counted on landing areas during the weeks when kills were usually made (Gentry, 1981b). Year-to-year survival was about the same before (1964 = 80%; Chapman, 1964) and after (1977–79 = 75–78%; Gentry, 1981b) the kill of males.

7. *What effect did human disturbance have on reproductive and survival rates?* Behavioral studies failed to answer these questions because of marking problems. Thousands of plastic tags abraded or broke in much less time (< 5 years) than animals lived. Metal tags were not easily read from a distance.

8. *Have disease or pollution increased such that survival rates were affected?* Various single causes for the decline were considered, such as disease, pollution, entanglement in fishing debris, and competition with fisheries (Fowler, 1982, 1987, 1990). However, none of these has been universally accepted as the probable cause of the decline (Trites, 1992b; York and Kozloff, 1987; also chapter 13).

None of these causes can explain all the fur seal population trends. Only two fur seal populations (Pribilof and Robben islands) declined. Those at the Commander Islands, Kuril Islands, Bogoslof, and San Miguel Island did not (fig. 1.10). Any disease that affected the Pribilof and Robben Island populations would eventually have spread to the other populations because juvenile males intermix on land (chapter 7), and females and juveniles intermix at sea (North Pacific Fur Seal Commission, 1984). Marine pollution cannot have been responsible unless animals from the Commanders,

Kurils, Bogoslof, and San Miguel somehow found unpolluted waters, an unlikely event given intermixture at sea. Entanglement in marine debris alone is not likely to have been responsible (Trites, 1992b) because the entanglement rate at the Commander Islands was, if anything, higher than at the Pribilof Islands [1]. Finally, fishing is not believed to have been the cause of the Pribilof decline (Baker and Fowler, 1990; Fowler, 1982; National Research Council, 1996). Fishing was so heavy around the Commander Islands (data unavailable) that a 20 km fishery exclusion zone was necessary. Nevertheless, the fur seal population there remained stable. In short, no single factor, acting alone, is likely to have driven down two populations but not the others.

9. *Did the availability of food change between the 1930s and the period when females were expected to recover from being reduced*? Probably. See next question.

10. *Did the environment change between the 1930s and the period when females were expected to recover from being reduced*? Yes. The climate regime when fur seals were at their peak in productivity (late 1930s) was distinctly different from that in the 1960s to early 1970s when they were expected to recover from the reduction. Climate regimes in this century are reported as 1900–24, 1925–46, 1947–76, and 1977–present (National Research Council, 1996). Apparently a regime shift occurred in 1947, between the time the fur seal population peaked in productivity (late 1930s) and the time it was intentionally reduced (1956–68) with the expectation of recovery at 8% per year.

Although the full effects of climate regime shifts on northern fur seals have not been studied, it is clear that the survival of juvenile fur seals correlates positively with El Niño events (York, 1991) and with air and sea surface temperature trends over decades (York, 1995). Trends in herring and perhaps squid and mackerel correlate with sea surface temperature and may mediate the survival of juvenile fur seals. York (1995) advances the hypothesis that warmer temperatures in the year before birth increase the prey available to 4-month-old fur seals leaving the Bering Sea in November and December.

The kind of analysis York has done should be expanded to include Pacific salmon landings [2] and fur sea population trends in the context of regime shifts. Specifically, northern fur seals reached their peak in productivity during the 1925–46 climate regime (National Research Council, 1996) when the troposphere was warming (Graham, 1995), atmospheric pressures over the North Pacific were decreasing (Trenberth and Hurrell, 1994), and salmon landings were large (Beamish and Bouillon, 1993). In the next regime, 1947–76, northern fur seals were reduced by 330,783 females and were expected to recover at 8% per year; the Robben Island population began an unexplained decline in 1964 (fig. 1.9). In the 1947–76 regime, tropospheric temperatures were declining (Graham, 1995), atmospheric pressures were mostly high [3], and salmon landings were greatly reduced (Beamish and Bouillon, 1993). The most recent regime (beginning in 1977) was like the 1925–46 regime in that tropospheric temperatures were rapidly increasing (Graham, 1995), atmospheric pressures were very low (Trenberth and Hurrell, 1994), and the landings of many fish species were high (Lluch-Belda et al., 1989; Beamish and Bouillon, 1995; Bakkala, 1993; see notes 19–25 in chapter 1 for details). Hawaiian monk seals and their prey increased (Polovina et al., 1994). Birds in the Bering Sea had greater foraging success (Decker et al., 1995), and the feeding trips of northern fur seals became shorter, suggesting more food was available (chapter 10). Most importantly, the decline in northern fur seals that had begun in 1956 stopped in 1983, six years after the regime shift occurred. Interestingly, the time from birth to recruitment for female fur seals is about five years (York, 1983). It is possible that increased survival of juveniles started by the 1977 regime shift improved recruitment 5–6 years later and ended the long-term decline [4].

A testable hypothesis from the above is that in regimes with increasing tropospheric temperatures and low atmospheric pressures (termed the negative state; Wallace and Gutzler, 1981), changes in wind fields and circulation pattern enhance primary productivity in many parts of the fur seal's range, increase overall productivity of the food web, and enable growth in the fur seal population through increased survival of juveniles. In regimes when air temperatures are declining, atmospheric pressures are high, and fish populations are lower (positive state; Wallace and Gutzler, 1981), primary productivity is not enhanced, growth conditions for organisms in the food web are less than optimal, and the survival of fur seals decreases such that rapid population increases may not be possible. Regional differences in productivity (North Pacific versus Bering Sea, or eastern versus western Bering Sea) exist in any climate regime. These differences could account for fur seal populations at the Pribilof, Commander, and Robben islands following dissimilar trends over the past 40 years. A comparison of historical trends in circulation patterns, air and sea temperatures, fish landings, and fur seal population trends should be done.

CONCLUSION OF THE ST. GEORGE ISLAND PROGRAM

The St. George Island Program came to a virtual end when the Interim Convention on the Conservation of the North Pacific Fur Seal lapsed. Public opposition to the killing for furs, an objective of the convention, caused the United States to withdraw from the agreement in 1985. Without U.S. support the convention ended, research budgets decreased, and international cooperation to find the causes of the decline ceased. The behavioral project continued to collect data until 1992. The population dynamics project changed its emphasis and now collects data related to a conservation plan (National Marine Fisheries Service, 1993). Understanding the population trends from 1956 onward is still important. Without knowing why the reduced herd failed to increase we cannot predict whether, and for how long, the present stability in northern fur seal numbers will continue.

NATURAL HISTORY

Northern fur seals are the oldest of the living otariid genera, with ancestors dating back 5 million years. Fossil deposits suggest that the species evolved within at least part of its present range. That range has undergone multiple glacial cycles that radically altered the fur seal's physical and biological environment. The prey on which modern fur seals depend, the areas where they feed and mate, their behavior, and their migratory habits are all cumulative expressions of this history.

The northern fur seal is primarily adapted to exploiting small fish and squid. Myctophid fish and squid are the main dietary item in the broad oceanic areas at the southern end of the range (Taylor et al., 1955; Perez and Bigg, 1986; Walker and Jones, 1990), but a more diverse diet is taken in coastal areas (Perez and Bigg, 1986; Sinclair et al., 1994; Antonelis et al., 1996). Fur seals are usually solitary at sea, perhaps because their prey items are dispersed. For most of the year (327 days for females, about 305 days for males) individuals are simultaneously pelagic, largely solitary, and nocturnal. Diving records of females in the summer suggest that most animals feed in the top 100 meters of water.

The species' distribution is effectively bounded east and west by continental shelf breaks. Presently their distribution ends at the transition zone between subtropic and subarctic water masses in the south, and somewhere near 60 north latitude in the north. The continental shelf does not limit distribution in the Bering Sea apparently because of the vast food resources it supports. Thus, the species' present range consists of neritic, epipelagic, and mesopelagic areas (to 200 m) from temperate to subpolar latitudes, including a few islands. Climate within this range is largely driven by changes in strength and position of the Aleutian Low.

Most females and juveniles migrate annually, north in spring and south in early winter. Adult males do not follow this pattern. Based on dive records, we have learned that most females in summer forage on vertically migrating prey at shallow depths, sometimes along shelf breaks. Where conditions permit (mainly atop broad shelves), some females feed on prey that do not move vertically. The southern migration begins coincidental with weaning and the advent of freezing weather, although the event that triggers the migration is unknown.

The northward migration gives fur seals access to a more abundant, predictable, and concentrated prey base than when they are in their winter range. Northern areas may be less affected by unpredictable El Niño/Southern Oscillation (ENSO) warm events than more southerly areas where seal populations are sometimes devastated (Trillmich et al., 1991). Fish abundance in the North Pacific Ocean and Bering Sea is not constant but fluctuates on a decadal scale with climate change (Beamish and Bouillon, 1995) and on longer cycles (Baumgartner et al., 1992). Even so, in the past 200 years these fluctuations have not caused periodic crashes like those reported for ENSO events among more southerly pinnipeds.

Fur seals, like most pinnipeds, are constrained to give birth and rear young on land. Like other otariids, females must feed while lactating (Oftedal et al., 1987). These dual needs force lactating females to become central-place foragers, commuting between suitable birth substrate and highly productive marine foraging areas. At any given sea level, few land areas meet these criteria. Most that do are islands near shelf breaks in the northern parts of the range. The paucity of suitable sites concentrates fur seals onto a few land areas for rearing the young. Breeding on a few islands established the conditions for the evolution of otariid polygyny (Bartholomew, 1970).

Migration and annual mating may be entrained by some predictable environmental signal, such as day length (Temte, 1985). The St. George Island population peaked during the same calendar week (ending 13 July) for 15 years despite changes in local temperatures, a major shift in the Aleutian Low, and a large ENSO warm event. Breeding at the northern and southern ends of the range is offset by about 2 weeks, which implicates day length as a determinant. Day length probably affects the individual's arrival date through its effect on implantation of the embryo in the previous autumn. The physiological mechanism by which this occurs is not known, nor is the mechanism by which individuals navigate to the breeding islands.

Individuals are highly predictable in time and (on some breeding areas) location of arrival. Most individual females arrive in an 8-day span unique to them, as in antarctic fur seals (Boyd, 1996). On one shoreline breeding area (East Reef, St. George Island), females used parturition sites on average 8.3 meters from the site in previous years, and suckled within 11 meters of the parturition site. This kind of specificity could explain why some breeding areas have not changed for at least 250 years. It could also account for the apparent resilience of breeding sites to occasional disturbance by humans.

Predictability in time and place of female arrivals may provide the basis for male competition. Males arrive a month before the earliest females and compete for sites that females will soon use for parturition, and on which they will mate a week after arrival (peak 10–12 July). The large body size of males allows them to fast for 30–50 days (range to 87 days) before abandoning their territory, usually about 6 weeks after the first females arrive. Large body size (relative to females) is an epigamic character that evolved along with polygyny in response to male competition (Bartholomew, 1970). Large body size contributes to fighting success, territorial location, and mating. But without proper timing, skillful use of terrain, ability to detect and respond to estrus, and other factors, males will not have high reproductive success.

Social relations among animals on shore are parsimonious in the extreme, as befits usually solitary animals gathering for a brief period (males 50, females 38 days per year) to mate, bear, and feed young. Males divide the available space among them while reducing aggression to the lowest possible level (by remaining in fixed sites, habituating to known neighbors, centering boundaries along topographic features, and reducing male-male relations after females arrive). Females form an aggregation in which individuals are closely spaced but lack social bonds with, or social status relative to, others. It is an open aggregation that females can repeatedly enter and leave (for feeding) with little aggression or disruption. This openness is appropriate to a highly dynamic group in which members maintain individualistic foraging schedules. They may enter and leave it free of the constraints that social status places on more permanent groupings.

Mature females are individualistic with a narrow range of preferences in many aspects of their terrestrial lives, and some preferences in their pelagic lives. On land they use a small area and arrive on it near the same date each year, as mentioned. Their behavior at sea also suggests some site fidelity and more repetition, individuality, and specialization than was once thought. At sea, females may use a similar dive pattern and foraging location on several trips to sea, may reach those areas by similar routes, and may alternate between land and sea on a predictable (for each individual) time schedule. Individuals specialize in taking prey of a particular size at a given location (Sinclair et al., 1994; Antonelis et al., 1996). Each individual probably takes much less than the full range of prey types recorded for the species (seventy-five known prey species; Kajimura, 1984; Perez and Bigg, 1986; Sinclair et al., 1994).

Females appear almost rigid in their use of land sites. But experiments show that if they are translocated, most females (67%) will rear young on breeding areas (even islands) different from the ones they prefer. Perhaps the environment does not usually change enough to make females demonstrate the flexibility in site use of which most of them are capable. However, this flexibility is conditional. If other females do not use the translocation site, females moved there will abandon their young, which suggests that the locations of other females may be more important to a mother than the location of her own young. Only young females make any attempt to rear their young away from established female groups.

Extreme site fidelity within a lifetime seems incongruous given that the species probably moved among many breeding sites over evolutionary time. Their history of being forced to change breeding sites due to glacial cycles seems to have shaped a tendency for more, not less, rigorous site fidelity than in other species of otariids.

SOCIAL BEHAVIOR

As stated above, social behavior in this species is marked by extreme simplicity. Individuals have no group behavior, form no social bonds (except mother/young), join no coalitions of several against one, and form no hierarchies of one individual over others, although size dominance exists. Female groups are structureless aggregations of individuals without status, except that young females tend to be on group edges. Groups can be readily entered or left, even by nonresidents. Females exhibit one set of behavior toward all adult males without regard to individual identity, and a second set toward all females. That is, behavior is class, not individual, specific. If neighbors know each other as individuals, the difference between their behavior and that of any two strangers is not detectable by the methods used in this study. A simpler form of society is difficult to imagine, given the necessity of individual foraging at sea.

Because behavior is not subject to many conditions or contingencies, the behavioral repertoire is not large, and social signals are clear and redundant compared to social mammals like primates (Dunbar, 1988). Subtle variations exist within a type of behavior, but not to the extent that the types of behavior blend continuously one into another (contra Lisitsyna, 1973).

The behavior of northern fur seals on land might be interpreted as showing that they cannot learn from experience; for 200 years they used the same land sites where humans conducted frequent kills for pelts. But individual captives learn complex behavior related to taking food about as quickly as other otariids. Their failure to abandon risky land areas probably results from their extreme fidelity to these sites, the use of which appears to be imperative to them.

Males

Male development involves a shift from frequent play bouts that are based on size-related dominance and involve much biting to infrequent threats that are site specific and involve no biting or play. In the final stage before the onset of territoriality, males defend rest sites around the periphery of breeding areas. Territorial defense is highly site specific and involves the use of threat displays that include biting, visual and vocal signaling, and (infrequently) fighting.

Juvenile males make frequent runs, here termed forays, across breeding areas during the breeding season. Sometimes several males do so at once. Multimale forays are not coordinated group action (Campagña et al., 1988). They are nearly simultaneous but nevertheless they are individual responses of juveniles to momentary lapses in attention to territorial boundaries by adults. They are triggered by any small disturbance.

Forays are not a reproductive strategy in the sense of an attempt to establish an immediate territory and mate. Most males making a foray are sexually or socially immature. About 97% of forays end in neither territorial establishment nor mating. Many adult males defend territory on areas across which they made forays when they were juveniles. This finding suggests that forays represent a brief visit to a favored site, perhaps the former natal site.

Forays are disruptive and may cause a redistribution of females and the deaths of some pups. But these effects are limited to times when the herd is recovering from the effects of juvenile males being killed for pelts. Forays should not occur often when the population breeds at its optimum sex ratio.

Males establish a territorial system on the same grounds where females later bear young. From the female standpoint, male territoriality is temporary. Males rigidly adhere to the boundaries in this system, but females largely ignore them. The areas males maintain are nonoverlapping territories with definable boundaries that are defended using aggressive behavior and within which males have exclusive reproductive access to females. Only those males that defend territories on sites traditionally used by females for parturition will encounter enough estrous females to make a major contribution to the next generation. Territory size varies inversely with the number of males present.

Males do not openly compete for any territorial site available, but instead are highly site specific. New males do not challenge territory holders at random, only those at a particular location. Once a male establishes a territory, he will not move to a completely new location in subsequent years, although he may not occupy exactly the same boundaries again. About two-thirds of adult males on a breeding site return the next year. Extreme site specificity and repeat occupancy create stability and continuity in the male territorial system. If disrupted by humans, the system reforms in its original configuration with little aggression because males compete for and defend only specific sites.

In a declining population, males continue to defend territories that females have abandoned with the result that large inland areas may be populated by males only. These males obviously cease copulating. Their reluctance to change locations suggests that considerations other than increasing their reproductive success shape their behavior.

Early in their breeding lives, males defend as large a space as possible. As they age they defend smaller, previously defended areas, arrive on territory earlier in the year, and increasingly use alternatives to fighting (such as skillful use of terrain) to defend their territories. No prior residence effect was measured (Braddock, 1949). But because older males arrive earlier and established males seldom lose fights, such an effect is likely. The type and amount of aggressive behavior that males use to defend territory depends to a large extent on factors such as time of season, proximity and number of other males, and presence of females. No single level of male aggression typifies the species.

Fighting is a fairly uncommon means of territorial defense for the individual. Long-distance threats, calls, and stereotyped threat displays are more frequently used. Habituation among neighbors, a change of attention from other males to females early in the season, and possibly a prior residence effect all reduce fighting to a low level.

Females

Females are gregarious but asocial (in the sense that they form no social bonds except with their young). Each is attracted to a specific, small area of land as described. Each is also attracted to groups of other females, but not to individuals within these groups. Experiments with captives show that residence outside a group is strongly aversive to most females. Once in a group, however, they show no communal, helping, or play behavior but instead are aggressive toward all other females and pups. Female pairs do not repeatedly rest together, make simultaneous movements to sea and back, or inhibit aggression toward one another in ways that suggest a social bond. Adults that behave toward each other as mother and daughter are not apparent.

Female social behavior is limited to the exchange of simple, redundant, aggressive signals, mostly open-mouth lunges. Female aggression is linked directly or indirectly (through spacing) to reducing pup injury. Female aggression is frequent, of low intensity, and usually lacks an identifiable cause or definite outcome, such as one animal moving away. Aggression has these traits because females use it preemptively to keep neighbors from moving close enough to harm the young. Young need protection because trauma (mostly female induced) is the main cause of death in 17% of the pups that die on land. Threats act to position females such that young are protected, but not to the extent that a social hierarchy forms or that females segregate into different groups by age or reproductive status the way males do. Aggression is fundamental to the way females relate. It is the same on the breeding area, in captivity, in large groups, or between two isolated individuals.

Male herding behavior is not the immediate cause of females forming groups because captives form groups even when no male is present, and space is not limited. Therefore male herding behavior on the breeding area reinforces a preexisting female tendency to form groups. The density within groups declines seasonally from an initial high of about two females per square meter to less than one. This decline is not a response to a cessation of male herding because captives that lack all contact with males undergo the same decrease. Aggression declines with intragroup density.

Several aspects of grouping seem incongruous. Why should females that are usually solitary voluntarily form groups? Why should residence outside of groups be aversive, given that female aggression occurs only within them? Why is density highest and aggression most frequent during the days when neonates are most vulnerable to attack from foreign females? That is, why do females not separate from the group at parturition like sea lions do (Gentry, 1970; Heath, 1990)? Why do density and aggression decrease over the season?

One possible answer to these questions is that grouping is less a proximate reaction to male herding behavior than an adaptation to avoid injury from males. Extreme sexual dimorphism for size and the tendency of males to suppress females with force make all male-female interactions potentially injurious or lethal to females. Females reduce the risk of injury through a suite of traits, two of which are forming groups and seeking their centers when males approach (Francis, 1987). The risk of injury is greatest before and during estrus (early in the season) because males selectively interact with pre-estrous and estrous females. A declining risk of injury in a population that has an increasing proportion of post-estrous females may explain the seasonal decline in female density and aggression. Mothers do not protect their young from aggression from foreign females by residing outside of groups possibly because, in so doing, they increase the risk of male-induced injury to themselves. Females do not avoid male proximity, only male-female (and male-male) interactions.

Female groups remain discrete or coalesce, grow or shrink, depending on the size of the female population, not on the behavior of males. Males are usually ineffectual at determining discreteness, location, or size of female groups. Groups persist until mid-November, whereas adult males abandon their territories in late July. Therefore, female groups are the most consistent feature of the terrestrial part of the northern fur seal's annual cycle. All the important characteristics of these groups are determined by females.

Estrus occurs at a predictable interval after the female's arrival on the island, which is itself predictable. Therefore, individuals tend to mate at a given location on a fairly predictable date (4 days either way of a date specific to the individual). Mating in the population is less a chaotic scramble of competing females than an orderly progression of individuals with highly specific tendencies in space and time. Estrus ends after 34 hours if the female fails to copulate, and it does not recur later in the year.

Fully estrous females will mate with a male of any size, including juveniles. Once they are in estrus they have no apparent mechanism for rejecting potential mates based on their fitness. They present sexually to adult males, pups, other females, or humans, as if they are unable to discern differences among them. Decisions on where to mate and with which male are largely subsumed within a long-standing preference for a given site. They mate at a given site for longer periods than the average reproductive life of an adult male (1.5 years).

Males can affect the onset and termination of estrus more profoundly and rapidly than males in most other species of mammals. Male aggressive behavior can induce the onset of full receptivity (the Whitten effect; Whitten, 1956a,b) within minutes. Coitus may terminate receptivity even before the male dismounts. Estrus for such females may be as brief as 15 minutes per year. Together, these traits reduce the number of males that females contact and the duration of these contacts, and thereby decrease the risk of female injury or death that impregnation carries. Furthermore, brief estrus allows females quickly to dispense with mating and begin the foraging cycles that lead to weaning at about 120 days.

Females voluntarily enter male territories not because they are attracted to males but because such entry is unavoidable if they are to reach preferred parturition sites. Male territories cover all the ground where females bear young, and more. Most males adopt an inconspicuous posture (lie prone) and allow a female group to grow in their area through the natural attraction of females for other females. If females do not arrive quickly, some males actively appropriate them until a nucleus of about ten forms, whereupon herding becomes counterproductive and decreases. The presence of females in a territory causes a sudden and marked decline in the aggressive encounters used by the resident male to defend his space.

Arriving females focus on the terrain and treat males as impediments to reaching specific destinations. Females avoid contact with males by not crossing open spaces and not entering a territory that lacks other females. They reduce contact with males further by joining female groups and avoiding any form of behavior that would attract the male's attention, such as calling, making eye contact, or moving quickly. Male avoidance behavior creates the false impression that females are docile and compliant. Their aggressive tendencies toward males emerge after estrus when males approach them less, and when such aggression toward males is not likely to trigger male retaliation and injury.

Most females return to the same parturition site, or close to it, for many years. Some of these returns represent philopatry (return to the natal site). At least 78% of females mate and rear young somewhere on the central breeding area of their own birth. In a small number of known cases, parturition occurred on or within 10 meters of the exact site where females were themselves born. This precision may be typical of most females. Other returns represent site fidelity (immigrant females that adopt a nonnatal site). Those showing site fidelity return to their adopted sites as faithfully and precisely as females returning to their natal sites.

Philopatry has a measurable ontogeny. In the first 30 days of life, pups form an attachment to a site that is so strong they will return to it from great distances if displaced. Experiments show that suckling, not birth or contact with peers, is the experience upon which this initial attachment is based. A testable hypothesis is that this attachment lasts until adulthood and determines the parturition site most females use throughout their lives.

The data showing philopatry are weaker for males than for females, but male philopatry may nevertheless exist (Chelnokov, 1982). Between 60% and 80% of juveniles use landing areas near the central breeding area of their birth. Forays may represent visits to a natal site (a testable hypothesis). Males seem to compete for only a single site of many available, and they are reluctant to abandon territories when females do. Philopatry may be a common denominator in all these aspects of male behavior.

Breeding groups form in the same locations over years largely because individual females repeatedly use favored parturition sites. In declining populations, females will go inland only as far as the last group of females because females avoid being the only female in a male's territory. Therefore declining populations shrink in inland areas first, not because the terrain is inhospitable, but because male avoidance takes precedence over reaching the preferred site. If females are abundant and not declining, individuals can move group to group and reach their preferred parturition site without being along with a male.

FUR SEAL MATING SYSTEM

The mating system should be analyzed from the behavioral tendencies of individuals, not from the appearance of social groupings. Social groupings are compromises among several conflicting behavioral tendencies of individuals and change with population size. Small groups of females attended by a single male each (formerly called harems) may predominate whenever females are scarce but are uncommon when females are numerous. Therefore, these single-male groups are not an important endpoint in this mating system. They are only one form of compromise among conflicting female tendencies to reach preferred parturition sites, join other females, and reduce contact with males. The underlying behavioral tendencies are more important to the mating system than are the single-male groups in which they sometimes result.

The primary behavioral tendency on which the mating system depends is predictability in where and when females arrive on land. Predictability sets the stage for vigorous male competition. Males learn which areas females occupy through experience and compete to occupy those areas before females arrive. To mate with the most fit males in the population, females need only arrive predictably and mate with any male near their usual parturition sites, because all males resident at such sites are winners of the competition their own arrival triggers. To mate with males that lose this competition, females would need to arrive at an atypical time or place and have an unusual preference to reside outside of female groups. That is, to mate with males of low fitness, females would need many behavioral traits that are outside the norm for females.

Within an established breeding group, any mate selection the individual female could make would be small compared to the selection process imposed by male competition (McLaren, 1967). Females can select mates only from among males that have first been selected by other males (more exactly, that other males have failed to exclude from the breeding area). For that reason, females derive small benefit from direct mate choice, and none is apparent in this species.

An absence of direct female mate choice is suggested by several lines of evidence. They mate at the same site over years regardless of the male defending it. They do not selectively mate with the same male in different years despite the opportunity to do so. They usually mate with the largest males available, not out of choice but because male competition has excluded all other potential mates. Given the opportunity, they will mate with a male of any size or age. When fully in estrus, females are universally receptive. Females do not base their behavior toward males on contingencies such as size, color, behavior, or any other phenotypic trait that would index fitness. Juvenile and senescent males cannot trigger rapid onset or termination of receptivity as can robust, territorial males. However, this difference does not equate with female mate choice because ovulation is spontaneous, irrespective of how estrus begins or ends.

Some northern fur seal females in estrus wander away from the parturition site, moving randomly in a nearly comatose state with their eyes closed, not responding to animals around them. Wandering is probably not a search for a fit male (Boness, 1991). Wandering females will sexually present when touched on the perineal area by a human using a stick. If they do not discriminate between species when they sexually present, it is not likely that they can discern subtle differences in fitness among males of their own.

Indirect mate choice is not a useful or valid inference for this species, because females choose a site based on features other than male quality. Females use a site because of long-term attachment to it (possibly philopatry), their pup resides there, or it is occupied by other females. It is the absence of these three conditions that prevents females from landing on areas used by juvenile males, not the lower fitness of the males there. One-third of translocated females will reject a site that has prime males, their pup, and other females on it, apparently because it does not contain their preferred site. Therefore, male quality does not seem to be a criterion for site selection in northern fur seals.

Some secondary aspects of female behavior contribute measurably to the degree of polygyny that exists in the mating system. No more than one-third of the females are on shore daily because of individualistic arrival dates and feeding schedules. Because of absent females, the species practices serial polygyny. Also, individual site preference spreads estrous females geographically and prevents excessive sex ratios from developing locally. The characteristics of estrus (receptivity to all males, a single, brief copulation, and estrus that is triggered by the male and terminated by coitus) increase the number of females a male may attend per unit time. By forming groups to avoid contact with a male, females increase the number of females to which each male has access.

No single aspect of male behavior has the same organizing effect on the mating system that female predictability has. Male behavior is much more contingent on female behavior than the reverse. For example, individual males are only slightly less predictable in time of arrival and site use than females. But the territorial system they create depends upon, rather than shapes, the distribution of the later-arriving females.

Male competition reduces the number of males that mate at any one time. Male competition divides a predictable resource (female parturition sites) among those males that are the hardest for other males to exclude from those sites. The age and size of breeders are not constant but change with the age of the colony and the age structure of the male population. The mating system may be classified as serial resource-defense polygyny, recognizing that parturition sites are not a critical resource.

Reproductive success is more variable for males than for females because it depends on a balance of many attributes. Females are almost assured of mating with a fit male by arriving predictably. They need not compete with other females for access to mates, and need only mate with a single male in each year of their life to realize maximum lifetime reproductive success. Males must time their arrival to match their fast with the availability of estrous females. They must defend territory, suppress this defense after females arrive, herd females only briefly, and interact with females often enough to detect estrus but not often enough to be a source of harassment. Males that have high copulation frequencies are successful at all these and more factors. However, their copulation frequency cannot be predicted from their standing in any one of these categories. Low copulation frequency can result from an inadequacy in only one item on this list, whereas high success results from competence in all of them simultaneously. Finally, male reproductive success changes markedly with the sex ratio, whereas that of females does not.

Much remains to be measured before the number of young that each sex produces in a lifetime can be compared well. Methodological problems prevented us from making these measurements. Male and female reproductive output should be compared at a stable sex ratio, after the effects of kills for pelts have lapsed.

As predicted for species in which females do not transfer among breeding sites, the delay between birth and first reproduction for females exceeds the average male reproductive lifetime (Clutton-Brock, 1989b). About 20% of females fail to mate on their natal breeding area (above). Females bear their first young at age 4 or 5 years (3 in the western Pacific populations), but adult males last on average only 1.45 seasons. Thus, most fathers are out of the breeding aggregation before their daughters mate and inbreeding is avoided. However, given philopatry, it is possible that mothers could mate with their sons or half-brothers; the extent of male transfer is still not known.

MATERNAL STRATEGY

The main maternal strategy involves seasonally migrating from southern (mostly) epipelagic areas where food is widely distributed to northerly waters where more abundant, concentrated food can support the switch to central-place foraging and the increased energy demand that accompany lactation. The cost of exploiting rich subpolar waters in summer is that weaning must occur relatively early in the pup's life (120 days) due to the onset of harsh weather and a resultant migration. Females at the southern end of the range (San Miguel Island) need not migrate but still wean by 142 days (DeLong and Antonelis, 1985). Thus, time to weaning is an adaptation to short summers and harsh winters and may be timed by photoperiod (Temte, 1985).

Females provide the neonate with nourishment and protection from other females. Except for the 5 days after a pup is born, mothers spend less than half of each visit with their offspring. Mothers neither teach their young nor play with them. The young are precocial, requiring no maternal guidance to assemble adultlike behavioral patterns from elements that are present near birth. Females invest in a fat layer that will sustain the young briefly during its transition to solid food. Most young wean themselves and enter a highly variable marine environment still naive about foraging. This apparently risky strategy can produce populations of large size.

The transfer of maternal energy to the young depends in part on environmental factors, such as time of season, distance to food, prey species available, and the way females dive for prey. How females develop their dive pattern is not known.

Foraging mothers reveal several general principles about diving that may apply to other species: (1) the proportion of a feeding trip spent in transit seems to be a constant across sites (absent El Niño conditions); (2) the aspects of diving that vary by season or island are related to the environment (e.g., day length), whereas those that remain constant are related to physiology (ability to recover from dives); (3) the relationship between dive depth and duration depends on local foraging opportunities—it is not a characteristic of the species as previously thought (Gentry et al., 1986a); (4) foraging trip duration is determined by local foraging environments (including prey patchiness) and not latitude as previously suggested; (5) depth reversals occur on about 50% of dives and are not related to lung collapse, but they may relate to prey capture; (6) the diving of populations should be characterized by the variations shown by its members; population averages obscure important variations.

Pup growth is an expression of the maternal strategy. Most of the past conclusions about northern fur seal pup growth rate are provisional because they did not take into account the recent finding that growth rate increases sharply after the molt (Boltnev, 1991; Boltnev et al., 1997). The following conclusions are based on growth before the molt: (1) male pups are larger at birth and gain more weight than females; (2) most pups with the fastest growth rates are males, but growth of the sexes relative to birth weight and birth date are the same; (3) growth is only weakly associated with island of origin and diet where these differ; growth in the first 35 days does not result from an incremental adding of weight, but is an average of an upwardly sloped wave function (fig. 11.2), with a period that corresponds to the mother's feeding cycle for female but not for male pups.

According to parental investment theory as applied to northern fur seals, male pups should predominate, be born first, be born to larger mothers, and cost the mother more to raise (Trivers, 1972; Trivers and Willard, 1973), and parent-offspring conflict should occur at weaning (Trivers, 1974). Only the first of these predictions is still uncertain for northern fur seals (see chapter 11). Males are not born first in the season nor do larger mothers carry larger fetuses (Trites, 1991). Males receive 61% more energy than female pups (measured by isotope dilution; Costa and Gentry, 1986), but the difference is not detectable in the mother's diving or attendance behavior. That is, mothers raise male and female young with the same maternal behavior pattern. The difference in energetic cost to a mother of raising a male or a female pup is likely to be only 0.3% (Trites, 1992a). Finally, parent-offspring conflict is weak (most pups leave the island before their mothers, implying self-weaning). Therefore, northern fur seals fit few if any predictions of Parental Investment theory, even though they are the type of species (sexually dimorphic with differential reproductive success) to which this theory applies.

RECOMMENDATIONS FOR NORTHERN FUR SEAL MANAGEMENT

Some misconceptions about animal behavior have led fur seal managers to poorly predict how the herd would respond to various management plans. This section outlines some of the difficulties.

Behavior of Individuals

Unless population models consider the behavior of individuals, they cannot be used to predict individual behavior or even some broad aspects of herd behavior. Present population models treat the herd as a supraorganism with behavior that is described by broad averages (average pupping date, average pregnancy rate, average survival, and so on). The intent of these models is to predict population-level processes such as productivity. They work well in that context, but they fail at the individual level because individuals have behavioral tendencies that are narrower and more specific than that of the population.

The herd is a collection of individuals each of which has a narrow range of preferences for how and when they use land. These preferences are so specific that one animal is not likely to follow the pattern of any other. A changing population will conform to the narrow range of preferences of individuals rather than to the broad population averages that population models use. Individual habits give the herd a structure that cannot be seen but nevertheless determines how it will collectively act during any population change. Predictions that are not based on these preferences will fail.

Many herd changes are attributed to extrinsic factors (predators, temperature, disturbance), when in fact intrinsic factors (individual preferences) are often at work. For example, it was predicted that juvenile males from St. Paul Island would move to St. George Island after 1972 to avoid the kill for pelts (an extrinsic factor) on their home island. No such movement occurred. This prediction failed because it assumed that a site having no kill was preferable to a site that had a kill. This is a uniquely human perspective. Most fur seals are driven by fidelity to a specific site, whether a kill exists there or not. It is the location of that site, not the location of kills, that makes a site favorable to fur seals. Fur seals appear not to choose sites by comparisons; any prediction that they will move among islands to avoid human activities is likely to be wrong.

As another example, abandonment of inland areas during population declines has led to the belief that beach fronts are "good" locations, and inland areas are inherently "bad" ones. Actually, females treat all parturition sites as unsuitable except the one they prefer. If they abandon a site it is because they are unable to reach it and still avoid males (see above), not because some physical quality of the site is repellent. Finally, the population should not be expected to colonize new beaches or islands unless the environment changes enough to overcome the extreme fidelity to specific, small sites that females have.

Interpretation of Data

Data that are based on a behavioral trait of individuals should not be interpreted as showing a broad, general tendency of the herd. For example, a curve showing the number of females on shore by date does not describe the arrival of an average herd. It is a frequency distribution of the individuals that have specific preferences for arrival on each date of the season. The shape of the distribution can change without affecting the inclusive dates within which arrivals occur.

Basic Population Unit

It is probably incorrect to treat whole islands or island groups as the basic population unit in which dynamic changes occur; central breeding areas probably serve that purpose. Each central breeding area has a unique history, age structure, and ties to the environment. Site fidelity of individuals almost completely isolates breeding areas from one another; there is no behavioral mechanism by which changes on one site can be transmitted to other sites. There is no evidence that breeding areas in close proximity are similarly affected by extrinsic factors, such as a change in prey at sea. Central breeding areas are grouped by island for convenience only. The danger of doing so is that grouping may obscure important local processes that drive population trends.

Density-Dependent Mortality

The suggestion that density-dependent mortality manifests itself through behavioral processes on land is probably incorrect. The concept of density dependence has played a central role in many management decisions. This study shows that intragroup density on shore is independent of the number of animals in the population (population density) because of individual spacing tendencies. Therefore, evidence for density-dependent mortality should not be sought in behavior on land.

Detrimental Behavior

Fur seal social behavior tends to be self-regulating. The fear that some form of behavior will become so aberrant that it will jeopardize the population is probably unfounded. Specifically, increases in the male population do not lead directly to increased pup deaths because as the male population increases, males themselves create a buffer zone around breeding areas that protects pups from the large mass of males in the population. This concern was twice used to argue for continuing the kill of males for pelts (for St. George Island in 1972, for St. Paul Island in 1985).

Future Population Reductions

The results of the 1956–68 herd reduction program show clearly that we still know far too little about northern fur seals and their environment to strongly interfere with their population size or structure. The reduction was highly conservative, as shown by comparing it against the present calculation of potential biological removal (or PBR; Barlow et al., 1995). The PBR estimates the number of animals that can be removed annually from a population while still allowing it to grow to optimum size [5]. The PBR for the Pribilof population for 1995 was 20,846 animals from a minimum population size of 969,598 (Small and DeMaster, 1995). Removals from 1956–68 averaged 25,444 females annually (330,783 in 13 years; Kajimura et al., 1979; Engle et al., 1980) from an estimated population of 1,850,000 (North Pacific Fur Seal Commission, 1958; or see table 1.1). Juvenile males were also killed from 1956 to 1968, but these kills may have had no effect on population trend [6]. Clearly, the kill of females was conservative enough to allow the population to recover unless the carrying capacity of the environment changed. It may have done so because in 1996, 40 years after the kill of the first females, the population had only just stopped declining (York and Kozloff, 1987; York, 1990).

The failure of northern fur seals to recover is even more surprising given that the period 1956–81 was an era of unparalleled recovery and growth for other otariid populations worldwide (Merrick et al., 1996). Managers would have had no hint from studying other otariids that northern fur seals would fail to recover from such a conservative reduction. The reasons for the failure to recover are not known; perhaps a change in environment (above) was involved. Certainly most of the otariid species that increased from 1956 to 1983 were outside the area that is affected by climate in the North Pacific.

The lesson from the past 40 years of northern fur seal management is that unforeseen environmental events can prevent otariid populations from recovering from herd reductions, even when the parameters used in population dynamics are well known, and even when the reductions are conservative. Management tools, such as PBR, seem designed for stable environments, not for those that include profound and unexpected changes. Too little is yet known about environmental changes, marine mammal trophic relationships, and human effects on both to confidently reduce a population on the expectation that it will rebound. To do so in the modern world seems extremely risky.

CHAPTER 15

Implications for Otariid Studies

A BRIEF history of research on the eared seals may be useful to those who are unfamiliar with this field, and as a prelude to the following summary of the present work's contributions to the field. Prior to 1953, researchers tended to make qualitative observations on single species (example; Hamilton, 1934, 1939). Many species were recovering from sealing; they were at low numbers and poorly known. Bartholomew (1953) introduced the use of marked animals to address particular questions and thus ushered in the modern era of research. Throughout the 1950s and 1960s, studies of increasing sophistication were conducted (examples: Rand, 1956, 1967; Paulian, 1964; Vaz-Ferreira, 1950; Peterson, 1965, 1968) on a few species. Some compendia compared the species, mostly on anatomical grounds (examples: Sivertsen, 1954; Scheffer, 1958).

In 1970, Bartholomew published another watershed paper which included a model of pinniped polygyny that identified key concepts for otariid research. These ideas began to be incorporated into field studies, but research questions and study methods were still largely diffuse (see Ronald and Mansfield, 1975, for examples). By the end of the 1970s, a few species were known well, and a few (especially Juan Fernandéz fur seal and Hooker's sea lion) were largely unstudied. By the early 1980s, basic behavioral and population data were available for most species of otariids (Ridgway and Harrison, 1981). The antarctic (Payne, 1977) and Cape fur seals (Shaughnessy, 1984) had staged impressive recoveries from sealing.

Researchers working on otariids began to unite and gather momentum in the 1980s. The first international fur seal symposium was held in Cambridge, England (Croxall and Gentry, 1987). Then, a worldwide comparison of maternal strategies in fur seals (Gentry and Kooyman, 1986a) introduced some new technology and outlined some organizing principles. One result of this work was that researchers began combining methods from the fields of behavior, physiology, and ecology, such that the distinctions among these fields became blurred. Also, researchers began to use standard methods to carefully compare different species and sites. As a result of standardization, the first comparison of pinniped responses to a major environmental event of hemispheric proportion, the 1982–83 El Niño, was possible (Trillmich et al., 1991).

In the 1990s, new analytical tools became available for investigating trophic relationships through fatty acid analysis (Iverson, 1993). Also, microsatellites (Gemmell, 1996) and DNA sequences (Lento et al., 1996) began to be used to resolve taxonomic issues. In 1996 the first all-otariid symposium was held in Washington, D.C. (Boness and Majluf, in prep.). Like other modern researchers, those working on otariids now are in daily contact through electronic bulletin boards.

The present study was a long-term (19 years) empirical study of the behavior of known individuals. It used direct observation, selected comparison, field and laboratory experiments, and dive instruments as its main tools. Its main contribution to studies on other otariids is a body of comparative data on a variety of behavioral subjects. These data were presented in the previous chapters. It also makes a contribution by posing some key questions about otariids that only further research can resolve, and by making some observations about the methods, approaches, and theories that are being applied to these animals. These questions and observations comprise the rest of this chapter. They are presented in approximately the order in which they appear in this book.

BEHAVIOR AND ANTIQUITY

A major unresolved question in otariids, to my mind, is whether their behavior reflects their taxonomic affinities or, more generally, their antiquity. From my observations on other species (eight), the northern fur seal seems more channelized and predictable in its behavior than the others. It also seems more tightly bound to smaller breeding areas than other species. Other otariids, especially the sea lions, seem flexible in their behavior. Is this difference real or only apparent? If it is real, does it reflect the antiquity of the genera (northern fur seals are oldest, sea lions are youngest)? Alternatively, does it result from the complexity of environmental changes they have faced (northern fur seals share a complex history with glacial cycles)? If so, why should an older species that has changed locations often over the long term evolve more narrow behavior in the short term? We need behavioral data that relate to the evolutionary history of the family (Gentry, 1975). Such data might help resolve recent conflicts between the stratigraphic and the molecular biology interpretations of otariid evolution.

SOCIAL BEHAVIOR: GENERAL

All otariids are gregarious, but are any of them social in the sense of forming complex social bonds among adults or in maintaining kinship ties? Some aggregations of otariids seem semipermanent. But in my experience, adults act toward each other like tolerant strangers. Northern fur seals have behavior that is appropriate to another animal's age-sex group but none that is specific to another individual (except the pup). Otariids may be gregarious because the energetics of central place foraging restrict them to only a few islands in their range (chapter 1). But why has the ability to form social bonds not evolved in these aggregations as they have in many other groups of mammals? Does the need for solitary foraging inhibit the formation of social bonds? Do any of the low-latitude otariids with prolonged breeding seasons form complex social bonds?

A related question is: Do any otariids display true, coordinated group actions? It has been suggested that "group raids" of juvenile southern sea lions are coordinated (Campagña et al., 1988). Group action requires animals to make conditional alliances, use complex signaling, and suppress selfish behavior to foster group cohesion. True group action is not likely in the northern fur seal, even though the species makes "forays" that superficially resemble group raids. The play of northern fur seal juveniles and the territorial system of adult males both involve skills that are related to individual combat. It seems unlikely that half-grown males making forays would possess these group skills and that younger or older males would not. Similar dashes by juvenile males across breeding areas probably occur in all species of otariids. Do any of them show evidence of coordinated group action, or do they more resemble the northern fur seal?

Why do otariid aggregations seem unable to disband and quickly reform in another location when they are frequently disturbed? For example, many southern fur seals were driven almost to extinction without abandoning their traditional breeding grounds (Busch, 1985). Northern fur seals used the same landing sites for 200 years while literally millions of them were killed for pelts. Steller sea lions in British Columbia withstood a 30-year attempt to exterminate them (Pike and Maxwell, 1958). Judging from tests with California sea lions, otariids seem to have complex cognitive abilities, learn quickly, and retain information well (Schusterman, 1966; Schusterman and Thomas, 1966; Schusterman and Krieger, 1986; Schusterman and Kastak, 1993). Why then do social groupings of otariids not learn to abandon disturbed areas, as we would expect? Intense site fidelity may explain this contradiction for northern fur seals. But how can it be explained for otariids in which site fidelity is less specific?

SOCIAL BEHAVIOR: MALES

Failing to abandon dangerous land areas is an example of apparently maladaptive behavior in northern fur seals that has a parallel in the ways some adult males defend territorial sites. Miller (1975) found that male New

Zealand fur seals would defend more than one site, and argued that such lability should be advantageous in increasing the frequency of copulation. While this argument is logical, northern fur seals and Steller sea lions (Gisiner, 1985) seem to defend only a single site, and northern fur seals will not change sites even when females abandon the one they hold (chapter 4). Apparently males of some species are more labile than others. Do differences among otariids result from the extent to which they base the locations of territories on philopatry? Even if they do, how can some species maintain behavior that appears to be maladaptive?

SOCIAL BEHAVIOR: MALE-FEMALE

What accounts for the marked differences among otariid species in the extent to which females form groups, move among groups, reside outside of groups, and act submissively toward males? This study suggests that the relative risk of injury in male-female relations may affect many of these aspects of female behavior. Clearly, the risk of injury varies among species. In some, females are rarely bitten or harmed (e.g., Hooker's sea lion), while in others (northern fur seals) females are at risk of injury or death in virtually any interaction with a male. Why does the risk of injury vary among otariids, and to what extent can this risk explain other differences in otariid social groupings? Comparative data are needed from many species to answer these questions.

Male herding behavior is probably not an adequate measure to use in addressing these questions. In captivity with no males present, female northern fur seals form dense groups, avoid leaving the group for parturition, and try to rejoin groups if separated from them. In the field, females form groups that are larger than any male could herd together and that last longer into the year than male herding lasts. Therefore, herding by males seems not to be the proximate cause of grouping behavior of females, but only reinforces a preexisting tendency in females to do so. A more likely hypothesis is that grouping is an evolutionary adaptation that serves to reduce the risk of injury from male-female interactions in general.

A test of the hypothesis lies in the following predictions about species in which males do not injure females. The risk of injury should increase with size difference between the sexes, just as the extent of polygyny does (Boness, 1991). (Herding apparently does not vary with size difference [Boness, 1991], but herding and the risk of injury are different, as noted above.) Females should not avoid being alone with males, avoid interacting with them, or act submissive in their presence. Females should not be as attracted to other females as in the northern fur seal, and should form groups less readily in captivity. The density within female groups should not peak when most females are in estrus, but should be more uniform throughout the season. Females should leave the breeding group at parturition to protect their pup from attack by other females. (This seems to be the case for Steller [Gentry, 1970], California [Francis, 1987], and Hooker's sea lions [Gentry and Roberts, in prep.]).

MATING SYSTEM

The history of thought about otariid mating systems from Steller (1749) until now leads to one conclusion: we need a more balanced approach in this area. Past researchers placed undue emphasis on the male contribution to the system. There are signs that modern researchers are tending to place undue emphasis on females. While the reason for these differences make for interesting speculation, the fact is that neither sex drives the mating system. The mating system is an intersection of two distinct sets of tactics that are under different sets of selective pressures. While female tactics have received less attention than male tactics, this should not lead to the conclusions that females drive the system. A more balanced approach would be to look for the complex ways that male and female tactics interact to create a mating system in a given environment. This approach shifts the emphasis from the two sexes to the processes that regulate their pairing.

Stressing processes also shifts the emphasis away from classifying mating systems, which is not particularly fruitful. Researchers probably agree that most otariids follow a kind of resource defense polygyny (Emlen and Oring, 1977; Rubenstein and Wrangham, 1986), the resource being sites that females use for parturition. This is a soft resource because these sites are not essential to the females' survival and can be changed in her lifetime. Also, whether any otariid uses a lek system seems a matter of semantics. In all, the fit between otariid mating systems and existing classifications of mating systems seems somewhat strained. The benefit of belaboring this fit is unclear. By contrast, investigating the link between mating systems and the environment for different species offers unlimited possibilities.

As a basis of comparison, I offer this outline of male and female tactics in northern fur seals. The female tactic is to migrate, arrive predictably in time and space, mate with any male on her preferred site, and quickly get on with pup rearing. The male tactic is to compete for only a single preferred site in specific years of his life and balance a great number of conflicting variables in a brief season while maximizing contact with females. Females reach maximum reproductive success by persisting in a fairly set pattern and mating with any winner of the male competition that their arrival triggers. Males reach maximum reproductive success by minimizing failure in any of the variables alluded to. The male tactic has more possibilities for error and therefore produces more variable reproductive success than the female tactic. Also, for comparison, the northern fur seal mating system hinges on predictability in time, precision in space, and persistence in both. This is a system in which season length is paramount. What are mating systems based on at latitudes where season is less important?

FEMALE MATE CHOICE

This work searches for evidence of female mate choice by many lines of evidence, not because it seems important in this species but to demonstrate the kinds of evidence that can be brought to bear on the question. Several conclusions are apparent. One is that the idea of indirect mate choice (females reside where the best males are found, thus implying an indirect choice of them) is a tautology (Peters, 1976), not a hypothesis. An indirect choice can never be measured and therefore can be neither proved nor disproved. Indirect choice is only an inference based on the simultaneous occurrence of the two sexes on one site. Whether it is an apt inference for a given species depends on whether the females choose that site on some basis other than male quality. The basis of site choice must be known from empirical tests before it is concluded that indirect mate choice occurs.

In a similar vein, the proximity of a female to a given male on a breeding area does not necessarily imply an active choice of that male. Northern fur seal females make an active choice of a parturition site based on many factors (e.g., preexisting preference, time of season, population size, location of other females). The presence of a male and the male's qualities are secondary consequences of females using the site predictably, because males compete for and cover all such areas. The evidence that the site is primary is that females will mate at the same site for many more years than the average male breeds. This simple metric should always be applied before concluding that juxtaposition implies choice.

The central idea behind female mate choice is that through their direct actions, females can increase their chances of mating with a male of high fitness. These actions imply the existence of an entire suite of female capabilities aimed at detecting and responding to differences among males. To date there is no evidence that these capabilities exist in eared seals. Rather than assuming that they do (thereby violating the law of parsimony, or Occam's Razor), simple empirical tests should be performed. Estrous females should be given free access to males of different size and age, and the extent to which they choose among them should be measured (chapter 8). If females are incapable of discriminating among males in captivity they are unlikely to do so in the wild.

There are ways that females can mate with males of high fitness without actively choosing individuals. Northern fur seal females need only be predictable and persistent in using their parturition sites, and intense male competition for those sites will result. This competition assures that females, through no individual mate selection, have available as mates only the winners of male competition. If females mate with any winner they therefore mate with a male of the most fit class. A claim that females choose individuals must be supported by evidence that selection by females is more stringent than that by male competition. Predictability, persistence, and universal receptivity are appropriate for species that are highly seasonal and in which injury from males is likely. What traits have evolved in species that differ in these regards?

Care must be used in interpreting the strange behavior of females at estrus as showing mate choice. Estrous Steller sea lions walk with an unusual gait, look sleepy, drool, and sometimes run as if to attract attention (Gentry, 1970). Female Hooker's sea lions sometimes wander away from the parturition site (Gentry and Roberts, in prep.). Female northern fur seals look sleepy, pant, make a characteristic sound, and sometimes wander (chapter 8) in a nearly comatose state. Before assuming that this behavior represents a search for a particular mate, the ability of such females to make discriminations should be tested. The present work shows that wandering females cannot discriminate even to species. For them, wandering is an artifact of universal receptivity, not a studied search for a particular mate. This may be the case in other species as well.

In summary, no claim for female mate choice should be made for any species of pinniped without experimental evidence for it in individuals.

MATERNAL STRATEGIES

The next frontier in studies of otariid maternal strategies should be to identify the mother's role in the pup's transition to solid food and its survival. Population trends are very sensitive to the survival rates of juveniles. But very little is known about the factors that affect these rates, including what weaned pups eat (Peterson, 1961; David, 1987b), how they find it, how their mothers contribute to their transition, and how all of these are affected by the local environment. In the northern and antarctic fur seal (Gentry et al., 1986a), all maternal investment occurs before weaning. Apparently, sufficient prey are available to support the pup's transition without a continuing maternal contribution. However, maternal contribution is essential to offset the effects of lunar cycles during the pup's transition to solid food in the Galapagos fur seal (Horning, 1996). When unweaned pups are learning to forage, full moons drive their prey to depths they cannot attain and reduce foraging for a period that is longer for small pups than for larger ones (because dive depth increases with body size). Mothers must continue suckling until their young are large enough to compensate for the effects of the full moon. The other otariids may be intermediate between these two extremes.

The foraging portion of maternal strategies should be studied through long-term trends in individuals, with emphasis on how they acquire their diving traits and how their diving changes with the environment. This approach will reveal more about a species than broad population averages based on the performance of anonymous animals that have been useful up to now. For example, a population average may not show that animals are exploiting more than one foraging environment (chapter 12).

The study of individual diving in northern fur seals has suggested the following principle that may apply to other species: females dive to some physiological limit on the scale of hours, but they dive appropriate to the season or setting on the scale of days to weeks (chapter 12). This finding explains why some aspects of diving change by season or by island while others do not.

The maternal strategy of northern fur seals produces a growth pattern in pups that is not linear but proceeds as a sine function. For female pups, but not males, the period of the curve matches the presence and absence of the mother. If the mother's attendance causes this shape in daily growth, then both the period and the amplitude of the curve should differ at islands where maternal foraging trips are of very different length. In other species of otariids, do male pups also show a sine function in their growth curves?

PARENTAL INVESTMENT THEORY

Parental Investment theory seems weak in predicting pup-rearing practices in otariids. According to this theory, for highly dimorphic species in which reproductive success is more variable for one sex than the other, mothers should selectively invest in the more variable sex (Trivers, 1972; Trivers and Willard, 1973). Otariids fit this description well; mothers should selectively invest in male offspring. However, little evidence for differential investment was found in northern fur seals (chapter 11; also Macy, 1982), the differential cost to the mother of male and female offspring may be only 0.3% (Trites, 1992a), and no parent-offspring conflict occurs at weaning, although it may in species with longer lactation. A more thorough search for differential investment in California sea lions also failed to support theoretical predictions (Ono and Boness, 1996). Perhaps by weaning the young at an early age, otariid mothers shift the burden of dimorphic growth to the neonate. For the future, researchers should carefully review the literature before committing valuable field time and effort to further tests of parental investment theory.

SITE FIDELITY AND PHILOPATRY

The role of site fidelity and philopatry in otariid societies needs more thorough study. In northern fur seals, site fidelity and philopatry in individuals seem to account for the distribution of animals on land, stability of the male territorial system, persistence of breeding areas in the face of human disturbance, and low colonization rate. Individuals use only a small piece of the available land and return to it faithfully year to year with the result that the population shows few changes. Experiments show that the attachment to a site may develop in the first 30 days of life in response to suckling there. Whether this attachment forms within a critical period, as in imprinting, is not known. Are northern fur seals extreme in their attachment to sites compared to other otariids? If so, what factors are responsible for the continuation of breeding sites in the other species?

ESTRUS

The broad pattern of estrus in northern fur seals is typical of otariids but differs fundamentally from that of phocid seals. Otariids enter estrus at the start of lactation (except for California sea lions, which enter at 27 days postpartum; Heath, 1990), and end estrus after a single copulation (Rand, 1955; Miller, 1974; McCann, 1980; Gentry, 1970; Peterson and Bartholomew, 1967). Phocids enter estrus at the end of lactation (Harrison, 1969), which is usually 2–6 weeks postpartum (except in the hooded seal, where lactation ends at 4 days; Bowen et al., 1985). Phocids copulate several times per estrus (up to 18 times in northern elephant seals, Le Boeuf, 1972; average 2.5 times in gray seals, Anderson et al., 1975). Also, behavioral estrus does not last as long in otariids (34 hours for northern fur seals) as in phocids (5 days in southern elephant seals; Laws, 1956).

Whether the details of estrus in northern fur seals are typical of other otariids is yet to be determined. For example, northern fur seals seem to show the Whitten effect (females may be brought into estrus by contact with a male; Whitten, 1956a). Males seem to exert their influence by actively suppressing the aggression that females direct toward them instead of through smell as in other species of mammals. This raises the question of how it is mediated in otariid species in which males are not aggressive toward females. Perhaps the effect does not exist except where the breeding season is short and polygyny is extreme. Another detail of estrus in northern fur seals is that females seem to have a "vaginal code" in which estrus terminates rapidly as a result of coitus (a particular pattern of physical stimulation during coitus ends receptivity; Diamond, 1970). Tests show that most northern fur seal females are nonreceptive immediately after a single copulation, and that coitus has a more rapid and profound abbreviating effect on estrus than in most terrestrial mammals. As with the Whitten effect, does the vaginal code exist in other species of otariids, or is this an adaptation to extreme polygyny and a short, compact breeding season?

ONTOGENY OF BEHAVIOR

At least some individuals in every species of otariid defend a territory during breeding. Therefore, like northern fur seals, they may show a shift from size-related dominance to site-specific behavior as part of maturation. In northern fur seals, the shift involves a reorganization of behavioral elements that are present from an early age. For example, male pups can mount an estrous female appropriately and make coordinated pelvic thrusting motions at 5 days age (fig. 8.6, *left*). A similar reorganization process seems to occur in Steller sea lions (Gentry, 1974) and may in other species as well.

Forays of juvenile males may be better interpreted as a stage in the ontogeny of territorial behavior than as a reproductive strategy. If they are a reproductive strategy, they are highly ineffective because only a few percent succeed. Most northern fur seal males that make forays are too young and small to seriously contend for territory. Instead, their forays may represent visits to a preferred site, such as the former natal site, because adult males defend areas they visited on forays as juveniles. This result suggests that forays may be part of a prior residence effect (Braddock, 1949) by which males exert influence over a site by retaining contact with it frequently until they reach adulthood. This avenue should be explored in other species.

The term "mate substitution" has been used to describe the abduction of pups by juvenile male otariids that sometimes end in the death or injury of the pup. These events should be studied in detail. They seem to be related to the ontogeny of male-female relations, just as play bouts are part of the ontogeny of male-male relations. At least in northern fur seals, juvenile males that abduct pups use many blocking, herding, biting, and mounting movements that closely resemble adult male-female interactions. If abductions are part of a normal ontogenetic pattern, they are probably unavoidable and are not likely to become so frequent that they adversely affect the population.

ENVIRONMENTAL CHANGES

There is a new realization that many vertebrate populations may be affected by long-term, large-scale environmental changes acting through food webs. These effects should be sought in the otariids. To date, the best known environmental effects on otariids are of aperiodic El Niño warm events (Trillmich et al., 1991). Long-term cycles in sardines (Baumgartner et al., 1992) may affect the abundance of some tuna (Polovina, 1996) and perhaps other predators, such as otariids. A climate regime shift in the North Pacific in the mid-1970s may have affected the abundance of many marine vertebrates (National Research Council, 1996), including fish (see details in note 22, chapter 1) and the Hawaiian monk seal and its prey (Polovina et al., 1994). The possibility that climate regimes coincide with some past population trends of northern fur seals was suggested in chapter 14.

Singular, noncyclic, unpredictable environmental events also may affect marine mammals. For example, polar bears weighed more in a 2-year period following the eruption of Mt. Pinatubo in 1992 than before or after it (Stirling and Lunn, 1997). Also, a recent die-off of Cape fur seals in Namibia may have occurred simultaneously with a change in circulation pattern of the Benguela current.

The search for the effects of environmental change on otariids should center on the food web because this is the route by which these effects are mediated. Otariid researchers should carefully track the long-term trends in phytoplankton (example, Venrick et al., 1987), zooplankton (examples, Roemmich and McGowan, 1995, and Brodeur and Ware, 1992), and fish species that are of no commercial value.

EXPERIMENTAL STUDIES

To the greatest extent possible, field experimentation should be made an integral part of any behavioral study in eared seals. With a few notable exceptions, otariid researchers have not taken behavioral studies beyond the level of description and correlation. Without experimentation, the stability of a character cannot be distinguished from rigidity in that character. Without forced choice experiments, researchers can only speculate on the factors that motivate animal behavior; experiments can show the order in which the factors are important. Captive pairings can reveal important details about many behavioral patterns that are too transitory or subtle to be analyzed in the wild. Finally, experiments can quickly resolve theoretical issues that prolonged debate based on observation cannot. Time-Depth Recorders have proven the value of experimental manipulation. Some experiments can be conducted on most age-sex groups, and with fairly minimal facilities. The main tool needed is the conviction that skillful manipulation offers at least as much, if not more, promise than the observation of undisturbed animals.

Appendix

Scientific and Common Names of Species
Referred to in This Work

(alphabetically by common name)

Antarctic fur seal   *Arctocephalus gazella*
Australian sea lion   *Neophoca cinera*
California sea lion   *Zalophus californianus*
Cape fur seal   *Arctocephalus pusillus pusillus*
Capelin   *Mallotus villosus*
Galapagos fur seal   *Arctocephalus galapagoensis*
Galapogos sea lion   *Zalophus californianus wollebaeki*
Gray seal   *Halichoerus grypus*
Guadalupe fur seal   *Arctocephalus townsendi*
Hawaiian monk seal   *Monachus schauinslandi*
Harbor seal   *Phoca vitulina*
Pacific herring   *Clupea harengus pallasti*
Hooded seal   *Cystophora cristata*
Hooker's sea lion   *Phocarctos hookeri*
Juan Fernandéz fur seal   *Arctocephalus phillipi*
Market squid   *Loligo opalescens*
Northern elephant seal   *Mirounga angustirostris*
Northern fur seal   *Callorhinus ursinus*
New Zealand fur seal   *Arceocephalus forsteri*
South American fur seal   *Arctocephalus australis*
South American sea lion   *Otaria flavescens*
Southern elephant seal   *Mirounga leonina*
Steller sea lion   *Eumetopias jubatus*
Subantarctic fur seal   *Arctocephalus tropicalis*
Walleye pollock   *Theragra calcogramma*
Weddell seal   *Leptonychotes weddelli*

Notes

Chapter 1
Introduction

1. Pinniped ancestor: Description is based on *Enaliarctos mealsi*, a late Oligocene or early Miocene proto-pinniped of California (Berta et al., 1989).

2. Mating at sea: Seems to be a response to high temperatures and radiant loads in some California sea lions (Heath, 1990), Juan Fernandéz fur seals (Francis and Boness, 1991), and New Zealand fur seals (Gentry, pers. obs.). It occasionally occurs in northern fur seals in cold climates (Baker, 1989).

3. Strategy: The term "strategy" here refers to an adaptive response of animals that is shaped by natural selection and does not imply cognition.

4. Phocid feeding: The smallest phocid species (harbor and ringed seals) cannot store enough energy to rear young in marginal habitats and supplement their stored fat by feeding during lactation (Bowen et al., 1992), somewhat like otariids.

5. Phocid exceptions: Northern and southern elephant seals and gray seals are as sexually dimorphic and polygynous as some otariids. Also, several polar phocids have reverse sexual dimorphism; females are larger than males (Ralls, 1976).

6. Oldest otariid: Description based on *Pithanotaria starii*, a fur seal of late Middle and early Late Miocene in California (Kellogg, 1925). Undiscovered fossil species probably exist.

7. Migrations: California sea lion males migrate north in winter (Odell, 1981).

8. Population trends: The California sea lion may be increasing because it is not restricted to taking prey from continental shelf areas like other sea lions. That is, it partly feeds like, and is increasing like, a fur seal (Antonelis et al., 1990a; Merrick et al., 1996).

9. Site fidelity: Site fidelity (repeat mating on the same site) has been measured well in antarctic (Lunn and Boyd, 1991) and northern fur seals (Baker et al., 1995, and present study), and somewhat less in other species (examples: Steller sea lion, Gentry, 1970; Australian sea lion, Higgins and Grass, 1993). Philopatry (repeat mating on the original natal site) exists but has not been measured well (chapter 7).

10. Summer breeding: The Australian sea lion breeds in a cycle described as 17.6 months (Higgins and Grass, 1993), or 17–18 months (Gales et al., 1992, 1994) in different parts of the range. Thus, breeding does not always occur in summer.

11. Postpartum estrus: The California sea lion differs from all other otariids in that it enters estrus 28 days after parturition (Heath, 1990).

12. Diapause: The Australian sea lion has both a postpartum estrus and embryonic diapause, but has slower than usual postimplantation development as befits its 17–18 month breeding cycle (Higgins and Grass, 1993; Gales et al., 1992, 1994).

13. Breeding season: Reported as about five months in the Galapagos fur seal and sea lion and South American fur seal (Boness, 1991), and six weeks in the northern and Antarctic fur seals (Gentry et al., 1986a; Boness, 1991). The Australian sea lion has an unusually long breeding season for a temperate species (5 months; Gales et al., 1992).

14. Misclassification: Boness (1991) classified the Hooker's sea lion as a lek breeder based on a statement made by me and a coauthor that some females wander at estrus. His assumption that wandering denoted a search for mates is invalid (Gentry and Roberts, in prep.). The Hooker's sea lion's mating system is as much like resource defense polygyny as that of any other otariid.

15. Lek: The California sea lion is classified as a lek species because some (not all) females mate at sea 28 days postpartum in "milling groups" that include a male (Heath, 1990). Research has not shown whether females are selecting the male, or mating at a favored location irrespective of the male there. Classification as a lek turns on this distinction.

16. Territoriality: Some California sea lions in hot climates defend aquatic territories (Heath, 1990), some Juan Fernandéz fur seal males defend exclusively aquatic territories (Francis and Boness, 1991), and some New Zealand fur seals, Juan Fernandéz fur seals, and Steller sea lions defend land territories that include water areas (Gentry, pers. obs.; Francis and Boness, 1991). Most other species defend territories exclusively on land. California, Galapagos, southern, Hooker's, and some California sea lions defend territory on sand where boundary cues are minimal. Most of those with territories on land remain in the same location over time, but Hooker's sea lion males move the location of their territory as the female population moves (Gentry and Roberts, in prep.).

17. Genetically determined age at weaning: Antarctic fur seals on (temperate) Marion Island wean at 120 days while subantarctic fur seals on the same island wean at about 300 days (Kerley, 1985; Oftedal et al., 1987). Also, northern fur seals on (temperate) San Miguel Island wean at 142 days (DeLong and Antonelis, 1991), while California sea lions there wean at about 300–335 days (Oftedal et al., 1987; Ono, 1991; Melin, 1995).

18. Intermixture: About 5% of the females of Pribilof origin migrate across the Pacific and winter around Japan (Wilke and Kenyon, 1954), comprising up to 27% of animals wintering in the western Pacific.

19. Atmospheric pressure patterns: Climate over the North Pacific and Bering Sea is shaped by changes in atmospheric pressure at four "centers"—Hawaii, Alberta, the Aleutian area, and the Gulf Coast states. On a decadal scale, the pressure patterns at these four centers oscillate between a "positive" and a "negative" state (Wallace and Gutzler, 1981). In the positive state, ocean climate is dominated by a low pressure system at the Aleutian center (called the Aleutian Low, a climatological composite of winter storms centered south of the Aleutian Islands; Ignell, 1990), a weaker low at the Gulf Coast center, and atmospheric highs at the Hawaii and Alberta centers. In the negative state, the pressure patterns at all these centers may reverse, or at least the Aleutian Low may be weak (Wallace and Gutzler, 1981). The change between these states, which may be abrupt, constitutes a regime shift.

20. Wind fields: In the positive state, the intensified Aleutian Low shifts southward and eastward and produces anomalous northerly winds over the central and western North Pacific, which, through increased ocean mixing and thermal advection, act to cool ocean temperatures (Trenberth, 1990; Trenberth and Hurrell, 1994). Simultaneously, this low brings anomalously warm air to the west coast of North America and into Alaska, warming sea temperatures and reducing ice cover in the

Bering Sea (Niebauer and Day, 1989). In the negative state the Aleutian Low weakens and shifts westward, wind directions reverse, and the North Pacific warms while the Bering Sea cools.

21. Regime shift: The four regimes in the North Pacific were 1900–24, 1925–46, 1947–76, and 1977–88 (National Research Council, 1996; Niebauer and Day, 1989; Wallace and Gutzler, 1981). The latter period has been considered to be a single weather regime with unusually well developed features (Trenberth, 1990) in which the North Pacific cooled while the Bering Sea warmed. The 1925–46 regime had similar features.

22. Regime shift and fisheries: Following the last regime shift (1977), groundfish stocks in the Bering Sea increased (Bakkala, 1993; Fritz et al., 1995) by an estimated 15 million metric tons (National Research Council, 1996), and herring increased (Wespestad, 1991). Fish species off the west coast of North America, including salmon (Beamish and Bouillon, 1993), sardines (Lluch-Belda et al., 1989), and many others (Beamish, 1993; Beamish and Buillon, 1995) also showed great increases. Regime shifts may cause these increases through the food web. Chlorophyll a, an index of phytoplankton abundance, was nearly twice as abundant in the North Pacific after 1977 than before (Venrick et al., 1987), and zooplankton biomass in the subarctic Pacific doubled between 1956–62 and 1980–89 (Brodeur and Ware, 1992). Phytoplankton and zooplankton biomass correlates with winter wind speeds (increased by 3–4 m/sec between 1946 and 1987; Venrick et al., 1987), which probably affects the availability of nutrients. Although not as well studied, the large sardine stocks in the 1940s have also been linked to a favorable climate regime (Lluch-Belda et al., 1989).

23. ENSO effect on the Aleutian Low: At the start of an ENSO warm event, atmospheric Walker circulation set up by the rising of warm, moist air shifts eastward along the equator. Because of the earth's rotation, this shift sets up anticyclonic (high pressure) systems both north and south of the equator. The northern anticyclone forms at the Hawaiian center of the pressure pattern (see note 19). Atmospheric teleconnection between this and the other three centers transmits the atmospheric effects of ENSO events to the Aleutian Low (Rasmusson and Wallace, 1983). Strong ENSO events cause the Aleutian Low to deepen and shift eastward, just as in the positive state. An ENSO signal is detectable in the extent of Arctic sea ice coverage (Gloersen, 1995).

24. Pressure/ENSO combined effects: The unpredictability stems from dissimilar timescales and the variable strength and month of onset of ENSO events. Changes in atmospheric pressure at the four centers occur on the scale of decades, while ENSO warm events occur on average every 4 years (vary from 2 to 10 years; Cane, 1983; Fahrbach et al., 1991). Strong ENSO events occur once every 10 years, but their effects may last for 11 years (Jacobs et al., 1994).

25. History of complex changes: Efforts are underway to reconstruct the past 2,000 years of ENSO events and pressure pattern changes using tree rings (D'Arrigo and Jacoby, 1991; Buckley et al., 1992) and marine sediments (Baumgartner et al., 1992). It is hoped that these studies will reconstruct the history of environmental variation in the northern fur seal's range during the present interglacial period.

26. History of northern fur seal breeding and landing areas from 1742 to 1996:

Forty-six central breeding areas are believed to have existed in 1742. Of these, twenty-one disappeared permanently, and four new ones formed on the same islands. The details are as follows:

Bering Island had two known breeding areas (Reef North and South, or Poludion-noye) and one landing area (at Lissonkovaya Bay) when Steller explored it in 1742 (Frost and Engel, 1988). In 1894 there were three central breeding areas on Bering Island: Reef North, South, and Kishatchnaye (Stejneger, 1896). South was small, and the landing area at Lissonkovaya Bay had already disappeared. By 1922 only Reef North remained as the herd decreased due to combined pelagic and land sealing. A new landing area had formed in 1921–22 (Stejneger, 1925), which may have been the beginning of the present Northwest breeding area.

On Medny (Copper) Island, eleven central breeding areas existed in 1894, consisting of the Karabelnoye group (actually one continuous breeding area), and the Glinka group (ten central Breeding Areas on scattered pocket beaches). By 1922, the entire Karabelnoye group and five of the Glinka group had disappeared (Stejneger, 1925). Of these groups, only Urilie exists today. A new landing area formed in 1917 at Southeast Point and undoubtedly gave rise to the present breeding area there, the largest one on Medny Island (Stejneger, 1925).

The central breeding areas on the Kuril Islands were discovered and exterminated before any systematic records were kept. According to Jordan (1898), nearly every island in the chain was rumored by sealers to have had a seal colony at one time. However, Jordan could only document five sites where breeding areas were likely to have occurred: Raikoe Island, Seal Rock, Manauku (Broughton Island), Lovushki, and Srednev Islands. The first three had not recovered by 1894 when Jordan visited there. Recolonization of the latter two apparently began in the mid-1950s (Kuzin et al., 1973).

When the United States acquired the Pribilof Islands in 1867 there were nineteen central breeding areas on St. Paul Island and seven on St. George Island. By 1884 the numbers had been reduced to sixteen and seven respectively with the loss of Nah Speel, Suthetunga, and Maroonitch (Elliott, 1882; Jordan, 1898). After 1914 they had decreased to fourteen and six, respectively, with the losses of Lagoon at St. Paul and Little East at St. George (Osgood et al., 1915; Roppel, 1984; Kajimura, 1990). No breeding areas have been formed or lost at the Pribilof Islands since about 1914.

27. Colonies: A single fur seal pup was born on the Farallone Islands, California, in 1996 (E. Ueber, pers. comm., September 1996). It is premature to conclude that this event will lead to new colony formation.

28. St. George trend: St. George Island pup estimates for 1996 are larger than for either 1992 (Antonelis et al., 1994a) or 1994 (Antonelis et al., 1996), which may imply an end to the decline that began in 1956.

29. St. George Island Program: The 15-year study period was about equal to the maximum life span of males, and about two-thirds that of females (Anon., 1973).

30. Pollock fishery: The pollock fishery began in 1964 at about 100 metric tons annually and increased to nearly 2 million mt by 1972 (Kajimura, 1984). In that period, fur seals increased (1970 to 1976) and then decreased (1977 to 1983). Northern fur seals consume juvenile pollock in midwater, whereas the commercial fishery takes adult fish near the bottom (Sinclair et al., 1994; also Fritz et al., 1995, for Steller sea lions). When this question was formulated, the concept of climate regime

shifts and responses of pollock to them (Bakkala, 1993; Beamish, 1993; National Research Council, 1996) were unknown.

31. 1977 regime shift: From 1973 to 1976 the pressure pattern was negative. The Aleutian Low was shifted far west (Niebauer and Day, 1989), resulting in Pribilof air temperatures that were 2 C below normal, and ice cover that was 10% above normal (Niebauer and Day, 1989). In 1977 this situation suddenly reversed. The Aleutian Low became unusually strong and shifted eastward, allowing southerly winds to warm the Bering Sea surface more than 1.5 C, warm Pribilof air temperatures by 6 C, and reduce ice cover to 10–15% below normal. The peak of the effect in the Bering Sea was in 1979 (Niebauer and Day, 1989). Simultaneously, the mid-North Pacific cooled by more than 0.75 C (Trenberth, 1990), and California coastal waters warmed by 0.8 C (Roemmich, 1992).

32. Study sites: East Reef was a long, narrow, north-facing, rocky-beach breeding area that supported small numbers of seals. Zapadni was a broad, flat field about 100 m inland and uphill from the sea, an extension of a broad shoreline central breeding area. The Zapadni study site initially supported about four times as many fur seals.

33. Landing areas: We studied the behavior of juvenile males on all landing areas, especially from a blind at Zapadni landing area. In 1977 we constructed an elevated wooden walkway at Staraya Artil for studies on the size, growth, sex ratio, and pathology of newborns (Calambokidis and Gentry, 1985).

Chapter 2
Population Changes, 1974 to 1986

1. Female "groups": Groups were considered to be separate only if a space of 2 meters or more separated them.

2. Maps: Available from mid-June to mid-November in some years. However, only the period 23 June–1 August was represented in all years, so comparisons are based on these dates.

3. Season: Early week = 23–30 June, middle week = 7–14 July, late week = 25 July–1 August. For convenience, these 8-day periods are referred to as weeks.

4. Map analysis: Digitized data were analyzed for the mean location (X–Y coordinates at the center) of each female group, area each group occupied, density of that group in females per square meter, and mean X–Y location of each male. The area occupied by the female population during the peak week of each season was derived by adding the total area occupied by all female groups during that week, and dividing by the number of groups present. The X–Y locations of males were used to classify each 10 ∞ 10 meter grid sector according to the density of males there. Five density categories (1–10 males per sector in 2-male increments) were used. Average territory size was estimated by dividing the area of the study site by the number of males present.

5. In-group ratio: Calculated as the mean number of females on shore per day divided by the mean number of adult males with which they had actual contact.

6. Population sex ratio: Calculated as the mean number of adult females on shore per day divided by the total number of adult males present irrespective of male/female contact.

7. Female classification: We used Vladimirov's (1987) criteria, which classes females as "parous" if they were observed giving birth, suckling, calling for, or resting with a pup at least once in the season, as "nonparous" if within five observations these categories were not scored, and as "undetermined" if they were seen fewer than five times total. We excluded the records of females with worn or broken tags that were hard to read.

8. Female counts: Linear regression analysis, using the model $ln(population\ size) = a + b(year)$, showed a significant population decline (p < .0001; coefficients of determination: East Reef = 0.91, Zapadni = 0.88). The Zapadni population declined 2.5 times faster than East Reef (t-test comparison of regression lines: t = 15.97, p < .0001).

9. Male counts: Differences tested using a three-way Kruskal-Wallis test (Zar, 1984) in which the major sources of variation (main effects) were site, week, and year. All possible two- and three-way combinations (interactions) of these variables were also tested. Each weekly sample was composed of five randomly selected counts from census and distribution maps. Too few daily counts were available to warrant including 1985 and 1986 data.

10. Seasonal change and population size: By 1977 the male population had grown enough that early breeders were always replaced when they abandoned their territories in late July. Before 1977 too few males were present to make this replacement complete.

11. Arrivals: Most males arrived in early June and most females arrived by early July; see chapter 3.

12. In-group ratio: The Zapadni data from 1983 onward were unreliable because by then the population was one-tenth of its original size (fig. 2.1) and was behaving erratically as to arrival date (chapter 3).

13. Natality rate: Two-way Kruskal-Wallis test, H = 2.83 for site, 0.71 for years, and 0.00 for the interaction of the two, all values compared to critical $\chi_{(0.05,\ 1)} = 3.841$.

14. Yearly natality rate and environment: Single- and multiyear females from East Reef and Zapadni, plotted for the proportion parous by year from 1974 through 1989, varied from 34% to 100% but with no obvious trend. The two sites did not change in parallel with each other. In 17 of the 29 years of data, the sample size was small (< 50) due to tag loss and poor tag readability.

15. Relative age: The natality classification of 222 multiyear females (811 female-years) was collated for the first through sixth years of tenure, and for years 7–13 combined. The percent parous was calculated for each group. It varied from 70.5% to 79.5%, but a two-way χ^2 test of the association between year of tenure and percent parous showed that they were independent ($\chi_{(0.05,\ 6)} = 6.76$, p = .34).

16. Group density: A two-way Kruskal-Wallis test was run using site and year as the major variables and site-by-year as the only interaction. Data from 1974 and 1985 were excluded because of small sample size (fewer than six daily counts). Only groups of two females or more were considered, and all computed densities that exceeded four females per square meter were excluded (they would have been physically impossible). The final sample size was 132 daily census and distribution maps from which density was calculated for 3,381 groups. The difference among sites was not significant (p > .05). However, difference over years was significant (p < .05), and the difference in site-by-year was highly significant (p < .001).

17. Group density over years: Kruskal-Wallis tests were run for each site using year as the only source of variation. In both tests, differences among years were highly significant (p < .001). The sites differed from each other in that East Reef showed a trend toward increasing density in the later years (correlation analysis, r^2 = .34, p < .001), whereas Zapadni varied without a trend (r^2 ca. 0, p < .86).

18. Seasonal change in density: Density was tested using a two-way Kruskal-Wallis test on data from the early, peak, and late weeks of the 1977–80 seasons at East Reef and Zapadni, the only years for which each week contained at least five daily counts. Site, week, and site-by-week were used as the sources of variation. Density differed significantly between sites and among weeks (p ≤ .001). The site-by-week interaction was not significant (p .05).

19. Density by thirds of season: Means comparison test of ranked data.

20. Method of measuring territory size: Actual territory size was not measured in the field because it required lengthy mapping of agonistic encounters (Gentry, 1970). Average territory sizes were calculated as the size of the study grid divided by total males, and reported as square meters per male.

21. Male density: For each day of the peak week of all seasons, every grid sector was assigned to one of five density categories (1–10 males per sector in 2-male increments; no zeros were included). The number of males in each category was plotted as a percent of the total male population present by week of the season to produce figure 2.8.

22. Female group regression equations: East Reef $y = 5.1 + 0.3x$, $r^2 = 0.57$; Zapadni $y = -.45 + 0.5x$, $r^2 = 0.81$. The intercepts were not significantly different from zero (p < .25), but the slopes were (p ≤ .002). The slope for East Reef was significantly different from the slope for Zapadni (t-test, p ≤ .05). This analysis and figure 2.10 do not include Zapadni 1978 data, which had improbably large values recorded by an inexperienced observer.

23. Females per group: Two-way Kruskal-Wallis test using site, year, and site-by-year as the main sources of variation. The ranks were significantly different in all categories (all p ≤ .05).

Chapter 3
Temporal Factors in Behavior

1. Solitary: Up to 68% of sightings from ships in the Bering Sea were of solitary animals. In California, Washington, and the Bering Sea, groups of 2, 3, 4, and 5 seals comprised on average 26.5%, 13.1%, 6.5%, and 3.7%, respectively, of seal sightings (recalculated from table 1 in Kajimura, 1980). Groups of 20–100 have been seen rarely (Kajimura and Loughlin, 1988).

2. Describing arrival: Analysis used maps from 15 June to 15 August only because these dates included the arrival of all mature females and the peak of reproduction. Map sets were available for East Reef from 1974 to 1989 (excluding 1987), for Zapadni from 1974 to 1986, and for Kitovi from 1976 to 1978, 1983, and 1985. Data for Kitovi males were inadequate and are not presented. Daily counts were expressed as a percent of the highest census made in the same year to compensate for decreases in population size over years.

3. Comparing arrivals: For each year we determined the median arrival date (date on which 50% of the maximum female population had arrived), calculated the

number of days occurring before and after that date, and used these intervals in a χ^2 test to look for yearly differences. Separate tests were performed for the two sites.

4. Juvenile males: Observers counted only juveniles that lacked secondary sex characteristics (large chest girth and a mane).

5. Copulations: Recorded from June to September in 1974 and 1975 to define the mating period for mature animals. Thereafter copulations were recorded only from mid-June to early August. Data were recorded over the range 0500–2300 hours, but only part of that period was covered in a day.

6. Copulation protocol: Included identity of the male and female, hour of day, location, duration and number of mounts, and other data.

7. Changes in mating date: Years were excluded from this analysis if they were missing more than one day of observational data between these dates. The data sets that met these criteria were analyzed for the date on which the median copulation occurred, the number of days prior to and including the median date, and the number of days from the median date to 28 July. These three values were used in a χ^2 analysis under the rationale that if median copulation date changed by year, the intervals before and after the median would reflect the shift.

8. Restrictions: Analysis was limited to records of (1) males on territory between 15 June and 15 August (the period when most copulations occurred), (2) those holding territories within the confines of the study grid (few females were located outside the grid), and (3) males observed copulating and appearing in the copulation records.

9. Special events: These included being sighted for the first time in a season, parturition, copulation, departing or arriving from feeding trips, being stolen or wounded by a male, and others.

10. Angle-branded females: In 1973, five hundred adult females (minimum 4 years old at marking) were tagged and branded with three marks, each a right-angle mark having the angle apex in different orientations representing different numbers (Keyes et al., 1975). These orientations were difficult to interpret because of the animals' loose skins and unusual rest postures. They were recorded only in 1974, 1976, and from 1982 to 1986.

11. Restrictions: Date restrictions to eliminate the occasional nonbreeding visitors after August (see chapter 7) would have removed late-arriving residents. Time restrictions to exclude females seen fewer than 5 days (the duration of the perinatal visit) would have eliminated nulliparous females that mated but failed to survive until the next year.

12. Female movements: the number landing on and leaving the grid by hour of the day was recorded from 24 June through 27 August to include seasonal changes, and from 0400 through 2200 hours to cover hourly changes. Recording stopped if other behavioral events being recorded became so numerous that a good tally was impossible.

13. Zapadni: We observed on a north-facing, grassy slope about 125 meters from the sea abutting a freshwater creek. The site was used by juvenile males that lacked a mane or other secondary sex characteristics. To test for activity patterns, the day was divided into three sections (0530–1200, 1201–1800, and 1801–2200) and differences among them were tested with Kruskall-Wallis, χ^2, and Dunn's multiple comparison tests.

14. Activity data: Data were collected at East Reef in 1974 and at East Reef and Zapadni from 1981 to 1983. Once each hour, sometimes more often, observers at each study site scanned all adult males on territory, and a contiguous group of 50–100 females, and assigned each individual to one of three behavioral categories: actively moving, inactive upright (resting with eyes closed and nose pointed vertically), and inactive on the substrate (usually sleeping). The results were recorded separately by gender on special protocols including date, time, and weather conditions. Kruskall-Wallis one-way tests were used to determine whether the three behavioral categories differed from one another, and for hourly differences within each group. A Kruskall-Wallis three-way nonparametric test was used to determine whether differences were attributable to breeding site, gender, or year. Dunn's multiple comparison tests were used to determine where the differences occurred.

15. Territoriality: Here defined as a defended space within which other males are not allowed, the boundaries of which often follow topographic features and are defended using vocalization, a formal threat encounter, and fighting (Peterson, 1965, 1968).

16. Female behavior: Arriving females would only move from one female group to another. They would not cross open spaces because of their reluctance to be alone with males (chapter 5). Therefore, Zapadni beach had to be filled with female groups before the first females would arrive uphill on our study site. As the beach population declined, arrival dates on our study site came later and the number present by date varied.

17. Population estimate from marked/unmarked ratios: This procedure assumed that unmarked females moved on and off shore like the marked ones—a fair assumption.

18. Median arrival date: This date varied from 28 June to 7 July at East Reef, and 1–12 July at Zapadni (later at Zapadni for the reason given in note 16, above). These differences were not statistically significant (East Reef $\chi^2_{(.05,13)}$ = 3.50, p ≤ .99; Zapadni $\chi^2_{(.05,12)}$ = 11.03, p ≤ .53).

19. Climate regime: The climate regime shift occurred in 1977 in the North Pacific Ocean, but the height of the effect in the Bering Sea was in 1979 (Niebauer and Day, 1989). Using 1979 as the dividing point, the median arrival dates at East Reef for years before (1974–78) and after (1979–88) the climate change fell on 3 and 4 July, respectively (data from Zapadni were not used because arrivals there changed as the population declined).

20. Weather and median arrival date: In the coldest year (1976), Bering Sea ice cover was 10–15% greater than normal and the median arrival date was average (3 July). In the warmest year (1979), Bering Sea ice cover was 10–15% less than normal and the median arrival date was earlier (28 June). (Ice cover data from Niebauer and Day, 1989.)

21. Timing of events on perinatal visit: mean interval from arrival to parturition = 1.3 days (n = 169, SD = 0.9); parturition to copulation = 5.3 days (n = 22, SD = 0.7); copulation to departure on first foraging trip = 1.2 days (n = 24, SD = 1.0).

22. Median copulation date: Medians were calculated for copulations recorded from 1 to 28 July, a period that included most of the copulations for a season (example, 89% of the copulations plotted in fig. 3.6). The median copulation date at East Reef was 12 July (n = 9 years, range 10–14 July), and at Zapadni it was 10 July

(n = 5 years, range 9–11 July). Differences in the median date of copulation at East Reef (1974–84, excluding 1980 and 1981 due to missing data) and Zapadni (1975–79) were not significant (East Reef, $\chi^2_{.05,8}$ = 1.724 vs. tabled value of 15.507; Zapadni, $\chi^2_{.05,4}$ = 0.419 vs. tabled value of 9.488).

23. Activity level: The three behavioral categories were plotted by hour of the day with no separation by gender, site, or year in order to show activity levels in the most all-inclusive way possible. Average standard errors for Inactive Down, Inactive Upright, and Active = 2.4, 3.5, and 0.9, respectively.

24. Differences among categories: One-way Kruskal-Wallis test, H = 4240.1 vs. tabled $\chi^2_{.05,2}$ value of 5.99; Dunn's multiple comparison test showed Inactive Down > Inactive Up > Active. Analysis of hourly differences within group: one-way Kruskal-Wallis test, H = 35.4, 117.2, and 47.4 respectively, vs. tabled $\chi^2_{.05,18}$ value of 28.9.

25. Hourly differences: Significantly fewer animals were Inactive Down from 0600 to 0900 than later in the day; one-way nonparametric ANOVAS, H values for East Reef females and males and Zapadni males were 20.2, 42.8, and 16.1 vs. $\chi^2_{.05,4}$ tabled value of 9.5. Hourly differences for Zapadni females were not significant: H = 2.5.

26. Behavioral differences: No overriding pattern emerged from an analysis of differences by breeding site, sex, and year. Some sites were higher than others, but not in all categories of behavior, and not for both sexes. Females predominated in the Active and Inactive Upright categories, and males predominated in the Inactive Down category. Year had an important influence on results, but males and females did not vary across years in the same pattern, nor did the categories of behavior.

27. Activity over years: for East Reef 1974 (large population size, many females per male) vs. 1981–83 (small population, few females per male), the Active category was not significantly different. Significant differences in the other two categories were not attributable to the 1974 data being the outlier. The mixed results of this analysis mean that variability was partly but not completely masking the underlying influences of site, gender, and year.

28. Juvenile males: Censuses and activity surveys were made independently, so variability in the two curves in figure 3.9 cannot be compared. Morning (0530–1200) and evening (1801–2200) activity did not differ from each other, but both were significantly higher than afternoon (1201–1800) activity (Kruskal-Wallis test, H = 9.037 vs. tabled $\chi^2_{.05,2}$ of 5.99, combined with Dunn's multiple comparison test).

29. Copulations: Figure 3.10 was based on 6,510 hours of observation. East Reef averaged 227 observer hours for each hour plotted (range 2.5–442 hours, SD 133.6); Zapadni averaged 116 observer hours for each hour plotted (range 7–240 hours, SD 72.5). The hours of least effort were early morning, late evening, and noon when observers broke for meals. The dawn peak was seen in three seasons, the midday peak in five seasons at each site (note that it coincided with a slight noon peak in the Active category in fig. 3.8), and the dusk peak was seen in ten seasons at East Reef and in six at Zapadni.

30. Nocturnal copulation: The high percentages from 1982 to 1984 in table 3.2 probably reflect increasing inaccuracies in estimating the population at its small size rather than a switch away from nocturnal copulation.

31. Male disappearance: Males did not disappear because they were hard to identify. Easily identified males from Zapadni that were tagged and branded disappeared at the same rate as the scarred (more difficult to identify) males from East Reef. Also, males did not disappear because they moved to different breeding locations (chapter 4); they would have been seen during censuses.

32. Male restrictions: The time restriction (sightings between 15 June and 15 August) removed only 2% of recorded males at East Reef and 1% at Zapadni. The location restriction (within grid boundaries only) removed only another 1% at each site. The copulation restriction removed 52% of males at East Reef, and 64% at Zapadni.

33. Years of tenure: The mean years of tenure were not significantly different (Mann-Whitney U test, $U = 13$, $p \leq 0.714$) for males in the unrestricted and restricted data sets (unrestricted = 1.45 years at both sites; $SE = 0.03$ and 0.04 at East Reef and Zapadni, respectively).

34. Days of tenure: Analysis was based on all males without restriction. Variability in the percent of males in each of nine tenure categories (1–90 days of tenure in 10-day increments) varied widely and unpredictably from year to year from 1974 to 1977 (large SDs in each category; fig. 3.12), but it decreased markedly from 1978 onward at both sites.

35. Tenure, early vs. late period: For each year-site combination we calculated an average annual tenure using a single value for each male present that year. These means were used to test whether days of tenure in early years when males were sparse (1974–77) differed from tenure in late years (1978–84), when they were numerous (see fig. 2.2). Mann-Whitney U tests, East Reef, $U = 24$, Critical $U = 25$ ($n = 4$ and 7 in early and late periods); Zapadni, $U = 13$, Critical $U = 20$ ($n = 3$ and 7); East Reef and Zapadni combined, $U = 75$, Critical $U = 76$ ($n = 7$ and 14). No results significant at $p \leq .05$.

36. Tenure, comparison of sites: Mean tenure at East Reef = 32.7 days, Zapadni = 30.3 days; Mann-Whitney U test, $U = 55$ vs. tabled value of 84 at $p = .05$ (2-tailed, $df = 10, 11$).

37. Tenure for males that copulated: For 186 males seen copulating that appeared in our records 3 years, mean tenure at East Reef was 40.5 days ($SD = 19.5$), and at Zapadni it was 42.8 days ($SD = 19.6$).

38. Tag loss rate: Single tag losses 2, 3, and 4 years after tagging with plastic Riese tags was 5%, 7.4%, and 21.4%, respectively (Gentry and Goebel-Diaz, 1987–88). In addition, two instances of double tag loss occurred in the same years. By 4 years after tagging, many tags were worn and unreadable. By 5 and more years after tagging, most tags were unreadable. For this reason, most animals tagged as pups and returning as adults at age 4–5 could not be identified.

Chapter 4
Behavior of Adult Males

1. Turnover rate: It was calculated for each year (except the first) at East Reef (1974–84) and Zapadni (1975–84). After 1984 observer effort declined and the estimates became unreliable.

2. Drawbacks of the subsector method of recording location: (1) Since male

body length was about 2 meters, a 25 square meter site was not a very precise location. (2) The number of sites on which the male was reported did not always translate into area covered. Males at grid intersections might move a short distance and be scored in four different subsectors, whereas males at sector centers might move the same distance and be scored in a single subsector. It is hoped that these effects averaged out in the large sample size.

3. Aggressive acts: These were scored when the small male population had not recovered from the effects of the former kill (East Reef 1974, Zapadni 1977–78), and again after it had recovered (1981–84 at both sites). Control data were collected at Kitovi in 1978 and 1983, where the male kill continued. Observers recorded data for only selected a number of "focal" males that were near enough to the observation blinds that all their aggressive interactions could be observed.

4. Male encounter protocol: The protocol listed each focal male and each of his neighbors. Whenever a known pair had an aggressive encounter, the type was recorded. Types were: two-male display, single-male display, multimale display, fight, chase, and unreturned bite. "Display" refers to a stereotyped threat posture— flippers out, head up, mouth open, and feinting movements—which males used to delineate territorial boundaries. "Fight" refers to nonstereotyped pushing, prolonged biting, and shaking. "Chase" refers to one male fleeing from another. "Unreturned bites" involved a male not responding to a bite from a neighbor, usually due to copulating near a boundary.

5. Male-pup: Data were collected at Zapadni from 20 July to 5 August 1984 only. Nine types of male movements were scored, varying in intensity from slow interactions with females to outright fights with neighboring males. For each event the estimated number of pups in the male's immediate vicinity was recorded along with the number of pups touched or contacted by the male. The male's response to pups was scored as "full stop," "hesitation," or "no hesitation," depending on the degree to which he interrupted his interaction to accommodate the presence of pups.

6. Fasting: A program was written that counted the number of consecutive days that entries appeared on each card. Single-day absences were ignored because some of them were not real (observers failed to record animals that were actually present). Those that were real were too brief to allow males to swim to feeding grounds and back, thus breaking their fast. Longer absences might have been sufficient, so the program ended one sequence and began a new count when it detected an absence of 2 or more days.

7. Male weighing: Males were weighed at the start of six field seasons (1976–80, and 1983). The sample included large males on central territories and small peripheral males to represent the full range of males present. In 1974, anaesthetic gasses like Halothane were unavailable, and the known injectable drugs had side effects that were too debilitating to use on territory holders. Therefore, we devised physical capture methods that required 8–10 helpers using ropes, an immobilization cage, a tripod, hoist, and scales (Gentry and Johnson, 1978).

8. Copulations: Recorded as follows: East Reef 1974–88, n = 2,837; Zapadni 1974–85, n = 2,317. First ranking: males were ranked according to the number of copulations they were observed performing in all years. Second ranking: each 5 ∞ 5 meter grid subsector was assigned a rank value based on the total number of copulations observed there in all years (Rank 4 = subsectors that accounted for the

top third of all copulations; Rank 3 = middle third; Rank 2 = lowest third; Rank 1 = subsectors in which no copulations were observed in any year). From male history cards we tallied the number of days that each male spent at each subsector, multiplied by the rank value of that subsector, added the products for each male, and ranked the males by their total scores. Males could only achieve high ranking by spending many days at sites with high reproductive potential. The two ranks were compared by Spearman's rank correlation.

9. Relative age: After pooling males by their first, second, third, fourth, and fifth-plus years of territorial tenure, we calculated a population mean for copulations per tenure year for each site, and compared tenure years by one-way Kruskal-Wallis tests. This analysis included all years of tenure for every male, including years in which no copulations were recorded.

10. Field experiments: Conducted on 3 June at East Reef and 6 June at Zapadni. At East Reef, males were prodded into the sea by two people approaching from uphill using 5-meter-long bamboo poles. At Zapadni ten people using bamboo poles and rolling noisemakers (5-gallon metal cans with gravel inside; Gentry and Holt, 1982) along the ground drove all males into the sea. Distance to the sea was 20 meters and 100+ meters at East Reef and Zapadni, respectively.

11. Data collection: Data included the time, location, identity, route of access, and behavioral means used by males in establishing territories. The types scored, in decreasing order on intensity were "fight," "boundary display," "chasing" (defined in note 3), and "submitting" (tucking the chin onto the chest creating an arched neck, and sidling away from opponent). Boundary displays sometimes included bites. But this was not scored as a fight because the overall interaction was stereotyped and included no pushing, whereas fights are not and do include pushing.

12. Turnover rate: Rate of all males, ranges = East Reef, 43–94%; Zapadni, 54–82%. Johnson's (1968) samples included only established males on territories while ours included established adults plus some identifiable wandering and peripheral males. Turnover rate for our adult males only; East Reef, n = 10 years, 511 known males, mean annual loss rate 34.16%, SD 14.16, range 21–62%; Zapadni, n = 9 years, 467 known males, mean annual loss rate 40.22%, SD 15.02, range 16–67%. Males may have disappeared from our records from year to year because they (1) failed to survive, (2) moved to a different breeding area, or (3) had no marks that we could identify. Given male site fidelity (2) is unlikely, but we could not quantify any of these three possibilities using our methods.

13. Favored sites: A count was made of the number of times each male was observed at each if his rest sites each year. Counts for the first through eleventh most frequently used sites were pooled for all males, and population averages were calculated as a percent of all male sightings. See table 4.2.

14. Ratio method of calculating site fidelity: Ratio was *number of new sites occupied each year/total sites used in all previous years*. The ratio for the first tenure year was always 1. A ratio > 1 indicated a major expansion onto new sites, and ratios < 1 indicated re-use of old sites. The results for all males were collated by year of tenure.

15. Territorial location and copulation frequency: East Reef, n = 869 males, Spearman's r^2 = 0.52, p < .05; Zapadni, n = 658 males, Spearman's r^2 = 0.28, p < .05.

16. Male arrival: All male records were searched for the date on which each individual was seen on territory at the start of each season. The date of the first sighting was plotted by year of tenure for each male (see fig. 4.1). The mean arrival dates of 634 multiyear males at East Reef and 492 at Zapadni (11 and 16 June) were significantly earlier than the mean arrival dates for 646 and 502 single-year males at the same two sites (29 June and 1 July, respectively; two-sample t-test, both p ≤ .0001 with 986 and 992 df, respectively, and all SE <1.1).

17. Young males: Males that were seen for only a single year arrived significantly later than males seen in multiple years (by 18 days at East Reef and 15 days at Zapadni; see note 16). To eliminate these young males, the data set was restricted to males that were seen on territory for 3 or more years. A plot of these data (n = 174) was nearly indistinguishable from figure 4.1. All 174 males were highly variable in arrival date during their first 2 years and narrowed the range of their arrival dates as they aged.

18. Male weight: Mass in June ranged from 130 to 305 kg, normally distributed about a mean of 208.7 kg (n = 176, SE = 32.9). Males seen copulating averaged 215.2 kg (n = 104, SE = 3.2), those not seen copulating averaged 199.2 kg (n = 72, SE = 3.7), a significant difference (t-test, t = −3.3, 174 df, p = .001). Regression analysis for male initial weight and number of copulations subsequently performed, r^2 = 0.05, n = 176, data points widely scattered.

19. Fasting weight loss: All three males were smaller than the population average (mean 155 kg) at the start. The range in their weight loss was 25–38%. The range in duration of fasting was 43–52 days.

20. Average fasts: East Reef mean = 30.1 days, SD = 20.1; Zapadni mean = 27.4 days SD = 20.8.

21. Fasts for successful breeding males: Successful males = those seen copulating and with 3 years on territory (n = 186). Territorial tenure averaged 41 and 43 days at East Reef and Zapadni, respectively, and fasts averaged 39 days at both sites.

22. Fasting and copulating: All males were classified according to whether they were ever seen copulating (yes or no), and for whether they were recorded for 1 or for 2 seasons. Fasting duration for each multiyear male was represented by a single average of all years to weight them equally with single-year males in the statistical comparison. Also see table 4.4.

23. Fasting by relative age: The records of 186 males with 3 years of tenure were sorted by tenure year, and a mean duration of maximum fasting was calculated for tenure years 1, 2, 3, 4, and 5–10 (combined because of small sample size). The means for these years (32.5, 41.4, 41.6, 40.0, and 40.6 days, respectively) showed significant differences (one-way ANOVA, F = 6.4, 4 df, p = .000). A Tukey's test showed that year 1 was different from all other years, and that years 2–10 were not different from one another.

24. Fasting and male population size: Single-year males were excluded from this analysis because of their atypical fasting patterns. A two-way ANOVA test was run on 186 males with 3 years of tenure from 1974 to 1984. Site and year were the main sources of variation, and site by year was the only interaction. Significant differences were found in all categories (F ratio = 15.3, 17.1, and 2.8 for site, year, and site by year, respectively; all p ≤ .002). Fasts were shorter in 1974 and 1977 than in other years (Tukey's test). The two sites did not covary over the years but often

varied in opposite directions. East Reef males consistently fasted longer (by about 5 days) than males at Zapadni (Tukey's test).

25. Fasting and copulation frequency: Days of fasting and number of copulations observed were plotted for 568 multiyear males seen copulating at least once in any year, showing wide scatter in the points ($r^2 = 0.0001$).

26. Long fasts and copulations: Five males seen copulating 51–72 times per year averaged 30 days of fasting (SD = 9.6), whereas nine males that fasted for 80 or more days averaged only one copulation (SD = 1.8).

27. Behavior: The separate components of male behavior were grouped into six larger classes (see "Behavioral Class," table 4.5). To test for the effects of terrain, the percentage of occurrence of these classes was calculated in year-site combinations (terrain at the two sites differed). The effect of population size was sought by comparing behavioral classes at small (1974–77), large (1978–83), and intermediate (1983–88) population sizes. Two-way Kruskal-Wallis tests were run on the ranked yearly percentages in each behavioral class to test for differences by site and population size group.

28. Behavior and territorial defense: The data set on male territorial defense consisted of 14,808 acts among 238 focal males (range 1–478 acts per individual per year) recorded during 2,738 observation hours (1,451 hours at East Reef).

29. Fighting rate: The fighting rate exceeded two fights per male per year only once (Zapadni 1981, mean = 2.3 fights, SE = 0.50). The rate exceeded one fight per male in only two other years.

30. Territorial defense and increasing male population size: As the East Reef male population grew from 20 to >70 males (1974 compared to 1981–84; fig. 2.2, *top*), territorial defense increased from 1.7 to 3.1 acts per 10 hours of observation, a significant difference (Wilcoxon two-sample rank test, ranking yearly means for all individuals and comparing rates between 1974 and 1981–84; $Z = -4.16$, $p < .05$). As the Kitovi male population grew from forty-three to sixty-one males (1978–83), the mean rate of territorial defense increased from 2.4 to 4.8 acts per 10 hours, a difference that was significant at p = .01 (Wilcoxon-Mann-Whitney two-sample rank test, U = 365 vs. tabled $U_{.05,16,34} = 367$).

31. Territorial defense and decreasing male population size: As the Zapadni male population declined from seventy to thirty (1977–78 to 1984, fig. 2.2, *bottom*), the mean rate of territorial defense decreased from 3.6 to 2.3 (average for 1981–84), a nonsignificant difference, although in the expected direction (Wilcoxon test, Z = 0.52, p ≤ .60).

32. Territorial defense and age: For each focal male present for 2 years (n = 51), we calculated a mean rate of territorial defense for each tenure year (n = 160 male-years), pooled East Reef and Zapadni males due to small sample size, ranked individual rates, sorted by year of tenure (1, 2, 3, and 4–10 combined due to small sample size), and compared years with a Kruskal-Wallis one-way rank test. The results (H = 1.89, p ≤ .60) showed no significant differences among tenure years.

33. Territorial defense and presence of females: It was tested for all years in which male territorial defense rate was recorded (East Reef, 1974, 1981–84; Zapadni, 1977, 1981, 1982, 1984). For each year we calculated a mean rate from the start of data collection to the date on which 50% of the females had arrived, and a second mean rate from then until the median date of arrival, and compared the

means by a Wilcoxon-Mann-Whitney two-sample rank test. The rates of territorial defense in the two periods were significantly different at both sites (East Reef, 7.3 vs. 4.3 acts, $Z = 6.18$, $p \leq .0001$; Zapadni, 10.4 vs. 8.5 acts, $Z = 3.78$, $p \leq .0002$).

34. Female effect on individual males: Using only data from the period before the median arrival date of females, we sorted focal males according to whether they had no female in their territories, or had 1 female (range 1–28) when acts of territorial defense occurred. Rates of territorial defense for males were ranked, and the two categories were compared by a Wilcoxon-Mann-Whitney test (see table 4.7 for statistical results).

35. Territorial defense and copulation frequency: Males were ranked by the number of territorial defense acts they performed per observation hour and by number of copulations in the same years, and the two rankings were tested by a Spearman's rank correlation. See note 3 for the years that data are available. A significant (positive) correlation was found for the three early years at two sites (when females were numerous), but not in the later years (Zapadni 1982, and East Reef and Zapadni 1983 and 1984) when females were sparse. None of the r^2 values exceeded .40 for these years. When males with no copulations were excluded from the analysis, the only significant correlation was at East Reef in 1974 ($r^2 = .399$, $p < .001$).

36. Male-pup interactions: $n = 1,019$ male/male interactions in 56 hours of observation, including 370 with no pups present and 649 with 1 pups blocking a path between two adult males.

37. Contact: Contact occurred in 29% of 649 interactions at which pups were present. Hesitation = 71, no hesitation = 351 pups contacted (likelihood ratio $\chi^2 = 9.921$, $p = .02$).

38. Lifetime copulations: Based on $n = 1,541$ known males (869 at East Reef, 672 at Zapadni) observed until they disappeared from the population. All males with zero copulations were included.

39. Copulations and tenure: Single-year males were compared with multiyear males of the same year to avoid temporal biases. The means (1.7 and 3.3 copulations per year for single- and multiyear males, respectively) were the same at both study sites. The differences were significant (Wilcoxon two-sample tests; East Reef, $Z = 8.20$; Zapadni, $Z = 5.15$; both $p < 0.05$).

40. Copulation by age: One-way Kruskal-Wallis tests, comparing ranked values among five tenure year classes, were run for the two sites. East Reef males showed no significant difference by relative age ($H = 3.49$ vs. tabled $\chi^2_{(0.05,4)} = 9.49$). Zapadni males had an apparent peak in the third tenure year (2.8, 3.8, 4.0, 3.4, and 1.8 for tenure years 1–5+), but this was not significant.

41. Copulations and population size: We pooled East Reef and Zapadni data, computed and ranked annual means, and compared ranked data for the periods 1974–77 and 1978–86 using a Wilcoxon two-sample test. The two periods (early = 5.5; late = 1.7) were significantly different ($Z = 3.641$, $p < 0.05$).

Chapter 5
Male-Female Associations

1. Classes of male behavior: Each of the following classes of behavior had one or more diagnostic behavioral components by which it was recognized. Observers were trained to recognize these classes before data collection began:

a. *Brief Encounter*. Adult male and female touch noses; male may sniff female's nose or open mouth to assess her receptivity.

b. *Copulation*. Scored separately as *Precopulation*, in which female allows male to touch her neck or back with his nose, and *Copulation*, including female lordosis and male mount. Categories were grouped in final data analysis.

c. *Female Threatened*. Male threatens female with an open mouth, holding head in oblique (striking) position, strikes female with muzzle, or bites her.

d. *Retain Female*. Male physically blocks female to prevent departure or suppress movement. No threatening behavior used.

e. *Moving*. Male moves around territory, often to investigate females or bar their departure using position only but no blocking behavior. Measured in body lengths moved per hour.

f. *Male Aggressive Encounter*. Scored separately as *Boundary Display*, a stereotyped single- or multimale threat executed from the prone position at the territorial boundary that included biting, *Fights*, nonstereotyped pushing to displace the opponent, and biting and shaking to cause serious injury, and *Chasing*, usually of intruding juvenile males. Scores in the separate categories were sometimes grouped for analysis.

g. *Rest*. Prone or upright posture with eyes closed, possibly sleeping.

2. Male-female interactions: Male identity, day of territorial tenure, day of contact with females, and number of females present were recorded for each sample. The last three were plotted against the male behavior categories (see note 1) to determine whether the extent of male/female contact affected the type of behavior seen. Data were collected during 8 years at East Reef, 5 years at Zapadni, and 2 years at Kitovi.

3. Kleptogyny: It was recorded on a special form that included date, time, identity of the male, number of females in the territory to which the female was moved, length of residence in the new location, and notes on the female's reproductive condition and extent of her injuries. Data were collected between 1977 and 1988 (8 years at East Reef, 5 years at Zapadni, and 2 years at Kitovi).

4. Female mate choice: The subsector in which parturition occurred was identified by searching history card entries for neonate presence (observed birth, suckling, lying next to a young, calling, etc.) during the perinatal visit. Records of single-year females, captives, and females that were hard to identify were removed from the record, leaving 286 females (176 at East Reef) and 858 female-years of observation for comparison from 1974 to 1988. Males using the same subsector as the female during her perinatal dates were identified from male history cards. When several males defended parts of the same subsector, which produced multiple matches for site and date, all were included. For each female we calculated the ratio *number of unique males/number of years in the record* to summarize the likelihood of repeat copulations. The record was visually searched for all instances of repeated matings. No correction was made for the absence of males in years when no match was found.

5. Female/male interactions: In a sample from 1977, East Reef females arrived singly and moved only a few meters to select a rest site. Ten such females entered an average of 1.6 male territories (SD 0.9) containing an average of ten females each (range 3–25), and had an average of 0.6 interactions (SD 0.5) with each male they met. At Zapadni, females entered the grid in groups. The number of males they

passed in the 100 meters from the shore to the grid was not observable. On the grid, twelve females passed through on average 4.4 (SD 0.5) male territories and had an average of 0.2 (SD 0.2) interactions per male. All territories were full of females (50–100 each).

6. Male/female interactions: The data set comprised 1,042 one-hour samples of the behavior of 122 different males, 26 of which were present for more than one year in the records.

7. Behavior related to females: This is a minimum value because movement around male territories ("Moving" in fig. 5.5) was usually female related as well.

8. Male behavior and female contact: For "Brief Encounters," mean of days 1–24 = 11.8, for days 25–50 = 8.9 encounters per hour (one-way Kruskal-Wallis test, H = 10.44, vs. tabled $\chi_{(.05,1)}$ = 3.84). Means for 10-day blocks of "Boundary Display" and "Rest" were ranked and tested with Kruskal-Wallis one-way ANOVAs (H = 35.4 and H = 25.1, respectively, vs. tabled $\chi_{(.001,4)}$ = 18.5) and Dunn's multiple comparison procedure.

9. Abductions: We recorded 403 abductions at three study sites over 12 years, 1977–88 (range 3–48 recorded per site/year combination, too few to correlate with annual changes in the size of the male population). Abductions were recorded for 237 of 1,541 males. Twelve males abducted a female on five or more occasions; one male abducted females eleven times. In 110 cases in which the female's reproductive condition was recorded, 44% were pregnant, 11% had recently given birth, 9% were receptive at the time, and 36% were postestrus. The population mean for 237 males that abducted at least once was 1.6 females (SE 0.08).

10. Mate choice: Sample sizes given in note 4. The mean ratio *number of unique males/years in the record* averaged 1.2 (SE .02). Females with a ratio of 1.0 (a new male each year) dominated the record; 168 such females ranged from 2 to 8 years (average 2.6 years) of being paired with a new male each year. Ratios of > 1 indicate that females shared a subsector with more than one male.

11. Conservative: The turnover rate among all males, not just large adults, averaged 66.5% for the two study sites, not 37%. Using this rate, 190 of the 286 females would have had the opportunity to select the same partner in 2 years, and only 15.2% (29/190) did so.

Chapter 6
Behavior of Adult Females

1. Francis studied at three sites in several years between 1976 and 1985, using a combination of focal animal studies using marked female/pup pairs and scan sampling. He recorded the type and number of aggressive encounters, context of aggression, number of potential opponents (number of noses within one body length), and the outcomes of aggressive conflicts.

2. Frequency: Females averaged 2.8 to 3 encounters per female per 15 minutes.

3. History cards: Entries were made a minimum of once a day when females were on shore and whenever else significant events occurred, such as first arrival of the season, parturition, copulation, suckling, kleptogyny, and movements related to feeding trips.

4. First arrival: It was used to measure the arrival-parturition interval, and

changes in arrival date by age of female. For that analysis, a file was prepared that listed the first observation for each year that each known female was present.

5. Estimated estrus dates: For parous females, see note 21, chapter 3. Nonparous females copulated 3 days after arrival (SD = 1.7, n = 6).

6. Estrus variation: For all females having estimated or observed estrus date in 2 years, we calculated individual mean dates and plotted the SEs of these means as a frequency distribution to characterize lifetime variation in the dates of estrus.

7. Parturition site: The parturition site was either observed and recorded on the history card, or it was taken as the first site at which suckling, resting with a pup, and calling a pup were recorded. A file was prepared that listed the subsector in which each known individual gave birth each year. The mean distance among parturition sites for each individual was based on straight line, center-to-center distances between the subsectors used for parturition in sequential pairs of years. The population mean included a single measure for individuals seen for only 2 years, and a mean of all measures for females seen in 3 or more years. Females usually did not move from the parturition site before mating, so parturition sites were also used as a means of matching females with males as a measure of female mate choice (chapter 5).

8. Behavior: The categories scored are listed in table 6.4. For each female seen ten times or more per season (to eliminate nonmothers), we calculated the proportion of observations in each behavioral category. We calculated a mean score for each category (table 6.4) using only one number for each female and only females that scored > 0 in that category. We then grouped the separate behavioral categories into six broader classes of behavior.

9. Laboratory: A permanent cage complex was constructed outdoors between two sheltering wings of the abandoned Skin Processing Plant in St. George Village. It had a plywood floor and was divided by 2-meter-high wire mesh walls into five, 3 ∞4 meter cages that were interconnected by sliding doors. Each cage gave access to a 5 ∞4 ∞1 meter pool of clean seawater inside the building. A male could be kept in a separate 5 ∞6 meter cage and paired with females in any desired combination. The cages were protected from the street by a 3-meter high plywood fence.

10. Cage experiments: In 1974 and 1983, social behavior was observed among females being used in an estrus study (see chapter 8). In early July 1974 and late July 1983, pregnant or immediately postpartum females were captured on breeding areas by noose and transported to the compound in cages. In 1974, females were kept for one month and fed frozen fish and squid sufficient to prevent weight loss. In all other years, females were held for one week during their perinatal visits but were not fed because they normally fast at that time.

11. Temporary cage: In 1984 a circular cage 13 meters in diameter was constructed on the tundra 110 meters from the sea at East Reef in a place seldom visited by males. Females had free access to all parts of the cage. Six females and their young were placed in the compound and were not approached again for 6 days during which time data were collected from a vehicle parked 35 meters away. Four types of aggressive acts (from vocal threats to outright fights) and five types of outcomes (from no response to physically moving away) were scored. Individual spacing was estimated as <1 m, 1–2 m, or >2 m for each animal during each scoring session. All females were returned to the central breeding area 7 days after capture.

12. Births: A protocol was kept at the East Reef site that included the time of day,

duration, orientation of the young, and other quantitative measures not discussed here. As many of these items as possible were recorded on every birth that occurred in 16 daylight hours per day. Not all items were recorded on every birth due to the number of different protocols being kept.

13. Sample size example: Metal tags often could be clearly read only once per season, whereas plastic tags could be read daily. Therefore, seasonal and daily processes are estimated using different tagged populations. The Zapadni population included 348 survivors from an original population of 500 females that were marked with "angle brands" (chapter 3). The records of these animals were analyzed separately due to irregular sighting effort.

14. First arrival: It was plotted against the year of tenure for each female (year 1 = first year in record; absolute ages were usually not known).

15. Reproductive status: Arrival date was averaged for each female in years when she was a mother and years when she was not. Paired t-tests for differences in arrival date were: East Reef mean = −6.1 (n = 108, SE = 1.53, p = .0001); Zapadni mean = −4.6 (n = 37, SE = 2.16, p = .042). Females lacking paired comparisons were excluded from this analysis.

16. Estrus date: The date of behavioral estrus was established for 128 females (119 at East Reef, 9 at Zapadni) for which arrival, parturition, or copulation were observed in two or more years. The mean of these dates for individuals ranged from 23 June to 31 July, and individual SEs ranged from 0.0 to 14 days. However, 80% of females had an SE of less than 3.9 days.

17. Parturition sequence: Analysis of table 6.1. Sequence 1 (parous in two successive years) is the typical sequence for this species (pregnancy rate exceeds 80%). The sample size was 417 arrival date differences for 121 females from 1974 to 1989. Each female was represented once, either with one difference between two successive years or a mean of several such differences, to avoid repeated measures problems. No females with missed years of attendance were used. The table entry for Sequence 1 is a mean for one subset of thirty females from this sample, selected by computer to equalize the sample size among sequences in a Tukey HSD test. Slightly different means were obtained on each computer run, but no run showed a statistical difference between Sequences 1 and 2. The Sequence 2 (nonparous in 2 years) table entry is a mean based on a single arrival date difference between two successive years for twenty-nine females. Sequences 3 and 4 were based on thirty-one females seen parous in year 1, nonparous in year 2, and parous again in year 3 of a series. The table entry for Sequence 3 is a mean of arrival date differences between years 1 and 2, that for Sequence 4 between years 2 and 3. The computer selected thirty females from this set for statistical comparison. No female was used to represent more than one sequence, except the Sequence 3–4 animals.

18. Parturition site fidelity: See note 7 for method of measuring. Sample size for East Reef (a rocky, narrow beach) = 365 measures for 176 females; sample size for Zapadni (100 m inland and uphill from the sea, flat and broad) = 80 measures for 44 females. East Reef females ranked significantly lower than Zapadni females (one-way Kruskal-Wallis test, H = 38.3 vs. tabled $\chi_{.05,2}$ = 5.99).

19. Number of suckling sites: The number of subsectors used for suckling by females at East Reef and Zapadni was compared using two-way Kruskal-Wallis

tests. Site, East Reef vs. Zapadni females, H = 10.69. Tenure year (single- vs. multiyear females), H = 192.48. Site by tenure year, H = 0.343. All H values vs. tabled $\chi_{(0.05,1)}^2 = 3.841$.

20. Distance from parturition to suckling sites: Zapadni, n = 288 sites for 213 females; East Reef, n =179 sites for twelve females (Gentry and Holt, 1986).

21. Site fidelity and age: Site fidelity was calculated by the ratio method used for males, namely, *number of new sites occupied each year/total sites used in all previous years*. The ratio for the first tenure year was always 1, a ratio > 1 indicated a major expansion onto new sites, and a ratio < 1 indicated re-use of old sites. No statistical test was done on the ratios reported in table 6.2. But the trend is obvious and the SEs are small.

22. Frequency of use: Sectors where suckling occurred were ranked according to the frequency of visits for a given female and year (1 = most frequently used, 11 = least frequently used site). The results for all individuals and years were collated by sector to produce table 6.3.

23. Together/apart: Summing all behavioral scores that indicate the presence or absence of the pup (table 6.4), pairs were together on 55.4% and apart on 44.6% of 10,659 observations.

24. Behavior changes over years: We calculated percentages for the six classes of behavior in table 6.4 for each study site and year, ranked the percentages, and compared them in three time periods—P1 = 1974–78, P2 = 1979–83, and P3 = 1984–88—corresponding to large, medium, and small female populations, respectively.

Differences at East Reef were tested using a Kruskal-Wallis two-way ANOVA using Site, Time Period, and Site by Period as the variables. The *H* values by behavioral class and rankings (a > b > c) of the three time periods were as follows:

Depart/return	57.40	P1a, P2b, P3c
Social interaction	17.36	P3a, P2b, P1b
Moving	37.18	P3a, P2a, P1b
Reproduction	16.11	P3a, P2a, P1b
Pup care	62.83	P3a, P2b, P1c

Critical value, $\chi_{(.05,2)}^2 = 5.991$.

Differences at Zapadni were tested using a Wilcoxon two-sample test because only two time periods were available there (P1 and P2 above). Z values by behavioral class and rankings of the two time periods were as follows:

Depart/return	−4.77	P1 > P2
Inactive	5.49	P2 > P1
Social interaction	−4.59	P1 > P2
Moving	−2.10	P1 > P2
Reproduction	−2.63	P1 > P2

Critical value, $Z_{(.05,\ 2\text{-tailed})} = 1.96$.

25. Behavioral differences between sites: We tested for differences in the six behavioral classes at East Reef and Zapadni for two time periods (P1 and P2; see note 24) using a Wilcoxon two-sample test. Z values by behavioral class and rankings of the two sites were as follows:

	P1	P2
Depart/return	6.03 E>Z	−6.16 E>Z
Inactive	−1.52	7.48 Z>E
Social interaction	−6.72 Z>E	0.93
Moving	−7.13 Z>E	3.37 Z>E
Reproduction	4.46 Z>E	−1.36
Pup care	−2.39 Z>E	−2.36 E>Z

Critical value, $Z_{(.05, \text{2-tailed})}$ = 1.96. In some tests, large sample size differences occurred.

26. Sightability: The flat terrain at Zapadni made female movements harder to see because they were often hidden behind other females. But copulations were easier to see because of the absence of large rocks.

27. Female aggression: On 16 June 1974 I recorded 30 minutes of behavior between the first two females to arrive at East Reef that year, a mother and a newly arrived pregnant female. They exchanged twenty threats (0.7 min). Nineteen of these were open-mouth lunges, twelve were mutual threats, nineteen were initiated by the mother, eighteen had no discernible cause, and none ended with either female moving away.

28. Births: Mean arrival to parturition interval = 1.3 days, SD = 0.9, n = 169 marked females. The duration of births was highly variable, from 5 minutes to several hours. Placentae were either passed with the neonate (nine of twenty-five births) or on average 9 minutes later. In fifteen cases, the female lifted the pup by the nape of its neck and gently dropped it up to fourteen times.

29. Captive interactions: We recorded 535 encounters in 23.5 hours of observation of six females in 1984, or 3.8 acts per female per hour. Low-intensity threatening with an open mouth = 91% of all cases, and high-intensity hard biting and pushing = 1.1% of cases. Neither female moved away from the interaction site in 83.3% of cases, and both females moved away in only 0.2% of cases. Female 6 was involved in 26.4% of the interactions, female 1 in 9.7%, and all the others were intermediate (14–16% each).

Chapter 7
Site Fidelity and Philopatry

1. Herd reduction program data: The ages of all tagged females and the breeding area on which each was born and killed were reported in the Annual Reports of Fur Seal Investigations (National Marine Fisheries Service, 1963–91) for 1956 through 1964 for St. Paul Island, and 1961 through 1964 for St. George Island. Most, but not all, females killed were parturient. Nine years of these tag recoveries are summarized in Baker et al. (1995).

2. Brief visits: Many females make brief visits to central breeding areas other than those on which their pups reside, especially after July when these samples were taken. Many of the females killed on non-natal breeding areas during the herd reduction program may have been making only brief visits there while actually breeding at their natal site. Such deaths would tend to decrease the apparent extent of philopatry. Conversely, some females that had moved to a non-natal site may have been making a brief visit to their natal site on the day of death, which would falsely

increase the estimate of philopatry. The existence of these brief visits was not known until the present study. They may or may not confound previous estimates of philopatry, depending on their frequency.

3. Female dispersal: In 1956 and 1957, the dispersal of some females killed during the herd reduction program was reported, but the sample sizes were too small to draw meaningful conclusions.

4. Philopatry: We recorded the $5 \infty 5$ m^2 subsector in which tagged animals were born and calculated the average center-to-center distance to the subsectors where they subsequently gave birth or mated. A measurement between exact sites would have been more accurate but was not possible. The number tagged was small because capturing neonates at birth was dangerous and highly disruptive. The number of tagged pups that survived to adulthood was small because of high annual mortality, a long interval from birth to mating, tag failure, and possibly emigration.

5. Brief visits: We observed three tagged mothers from Zapadni (nos. 302, 331, and 862) resting at East Reef for 1–3 days each in late July and August of 1980 and 1982. A mother from East Reef (no. 368) made a 3-day visit to East Cliffs in August 1980. Finally, female no. 329 (instrumented with a time-depth recorder) was recaptured at East Cliffs after a prolonged absence from East Reef in October 1983. It is not likely that this unusual landing was a result of being handled; several dozen other females were similarly instrumented over the years and were not observed landing in unusual locations.

6. Pups: In 1975, twenty-seven mother/pup pairs were captured just after parturition on the breeding area and moved to the cage compound at the laboratory where the mothers were used in studies of estrus (chapter 8). In early August, all pairs were released on the breeding area within meters of their respective parturition sites (no pups were born in captivity). Of the four pups that returned to the laboratory, three had been held captive for 30 days and one for 25 days. None of those held 10–24 days returned to the laboratory.

7. Little East: Mature females (i.e., having white vibrissae; Scheffer, 1962) and their young were captured on the day of parturition on various breeding areas of St. George Island and transported by small boat or carried overland to Little East. They were released into a wire cage measuring $4 \infty 4$ meters and held overnight to ensure that they suckled on the site. The next day the cage was removed. Thereafter the site was visited three times daily and notes were made on all females present.

8. Males: Females avoid males in all situations and show no attraction to them. Adult males were present when females abandoned the breeding areas to which they had been moved (Experiment II), and juvenile males were present when females abandoned the unoccupied beach (Experiment III). Therefore, absence of males is not likely to have been the cause of females failing to adopt the foreign sites to which they had been moved.

9. Young females: Eight females with all-black vibrissae (indicating probable age 4; Scheffer, 1962) and their newborn young were captured mostly from East Reef and carried in cages several hundred meters east of the breeding area. They were released onto a cobble beach adjacent to the East Reef landing area at a site occupied by nonbreeding males: breeding had never been recorded there. Only eight females were moved because they were the only ones found on a single day; we did not want to disturb the group by adding other females later. Translocated females

were fitted with a nylon harness (Gentry and Kooyman, 1986a) and tied by a loop of cord to a post behind the beach. They were left tied to the post for 24 hours to ensure that they suckled on the site. Then the cord loop was untied and pulled free without disturbance. Harnesses were held together by a thin plate of magnesium which corroded in seawater due to electrolysis and caused harnesses to fall off during foraging. Attendance of all females was checked three times daily.

Chapter 8
Estrus and Estrous Behavior

1. Estrus dates: Numbers ashore peak between 7 and 14 July (fig. 3.3), and observed copulations peak between 9 and 14 July (fig. 3.6).

2. Copulations: We recorded date, hour, identity of the male and female, vibrissae color of the female (an age index, Scheffer, 1962; Vladimirov and Nikulin, 1991), initiator of the interaction, number of different mounts, duration of the interaction (with a stopwatch), duration of the final mount, duration of rhythmic pelvic thrusting (RPT), representative samples of thrust rate (number of thrusts in 15 sec ∞ 4), occurrence of female vocalizations, and terminator of the copulation.

3. Rhythmic pelvic thrusting: The RPT phase was considered to have begun after fifty pelvic thrusts were made without any pauses. The duration of the RPT phase averaged 1.3 minute (n = 158 copulations, SD = 0.08).

4. Ejaculation: Typically the thrusting rate was low (90/min) early in the RPT phase, reached a peak a few seconds before the end of the phase, and slowed somewhat just before dismount. Thrusting never changed abruptly, and facial expression and posture gave no hint that ejaculation had occurred.

5. Sex ratio: Females arrive and enter estrus over at least a month's time, and female foraging trips remove two-thirds of the total females from land on a given day (fig. 3.4). Thus, staggered mating and foraging produce an "in-group" ratio on shore (chapter 2) that is substantially lower than the overall sex ratio of the population.

6. Exceptional female: The female had been released only 24 hours after the observed onset of estrus and may not have been completely anestrous when we released her.

7. Behavioral categories: Categories 7, 16, 17, 18, 19, and 20 from table 8.1 were scored for either the male or the female. All others were scored as dual components because male actions and female responses occurred as highly stereotyped, predictable units.

8. Definition of estrus: ♂ *Mount*/ ♀ *Lordosis* (table 8.1) always occurred just before males attempted intromission. As soon as we saw it, we pulled the female out of the test arena by the rope to prevent the male from effecting intromission around the perineal shield (which he had learned to do in the initial trials).

9. Scoring: Components 18, 19, and 20 (table 8.1) occurred continuously through some sessions so their presence was scored only once in each one-minute interval. All other components were scored once each time they occurred. One person (RLG) observed the interactions as they occurred and called out the components to the scorer (JRH or CBH) who tallied them on the protocol.

10. Estrus definitions: The duration of estrus was calculated only for females that

were nonreceptive (i.e., aggressive) during at least one session before and one session after tests in which ♂ *Mount*/♀ *Lordosis* occurred. The time of estrus onset was taken as the first test session in which ♂ *Mount*/♀ *Lordosis* was seen. The time of estrus' end was taken as halfway between the last session when ♂ *Mount*/♀ *Lordosis* was scored and the next session.

11. Receptivity definitions: Females were considered Fully Receptive if ♂ *Mount*/♀ *Lordosis* occurred, Nonreceptive if their behavior was mostly aggressive or evasive, and Partly Receptive if we saw the component ♀ *No Response* but not the component ♂ *Mount*/♀ *Lordosis*.

12. RQ: The number of "Receptive" components (those that occurred concurrently with ♂ *Mount*/♀ *Lordosis* at least 75% of the time) divided by the total components scored in that session.

13. Estrus look: Scored as the components ♀ *Sleepy* and ♀ *Pant/Flex* in table 8.1.

14. Behavior: Behavioral components differed in the degree to which they corresponded with the main indicator of estrus, ♂ *Mount*/♀ *Lordosis* (fig. 8.7, Prop.), and in the frequency with which they were seen (fig. 8.7, Percent components). The names of the components in figure 8.7 that indicated the various stages of receptivity were Fully Receptive = ♂ *Touch Neck*/♀ *No Response* followed immediately by a mount (seen in thirty-two test sessions); Partly Receptive = ♂ *Touch Neck*/♀ *No Response* not followed immediately by a mount; and Nonreceptive = ♂ *Aggression* (seen in twenty-one test sessions).

15. Nose touches: Females were more tolerant of being touched by the male during Full Receptivity than at other times. Touches per minute were much greater when females were receptive ("Rate" columns, table 8.3) than when they were partially or nonreceptive. Most of this increase was from the male touching the female's neck and back. Female aggression waned as the number of components scored per minute increased.

16. Mean RQ scores: Nonreceptive = 0.020, Fully Receptive = 0.702, Partly Receptive = 0.244 (reject the null hypothesis of no difference, ANOVA, $p < 0.001$).

17. Exclusivity of behavior: When female responses were listed in descending order of their co-occurrence with ♂ *Mount*/♀ *Lordosis* ("Fully Receptive" column, table 8.2), the list that resulted showed ascending concurrence with Nonreceptive behavior ("Not Receptive" column, table 8.2). Partial receptivity was a mix of behavioral elements seen in the other two stages.

18. Estrus onset: 4/18 females shifted to Fully Receptive and 9/18 shifted to Partly Receptive within a test session (examples A and B, table 8.4). Test sessions may have been too brief to allow the full change to occur (Burley, 1980). We extended the test session for one of the latter females and observed a mount in minute 20 (example C, table 8.4).

19. End of estrus: In these two cases, the male used aggressive components to suppress aggression in the formerly receptive females, with the result that these females began to show receptivity again.

20. Female mate choice: According to theory, a species should show female mate choice if male reproductive success is much more variable than for females (Fisher, 1930; Bateman, 1948; Trivers, 1972; Clutton-Brock, 1988; Boness, 1991). Females should not mate with young or peripheral males that have lower fitness than adult, territorial males (the "marginal male hypothesis"). Female choice is

related to the form of polygyny maintained (Emlen and Oring, 1977; Clutton-Brock, 1989a), leks (Boness, 1991), and delayed implantation (Sandell, 1990). Measures of actual reproductive choices that females make are generally not available but are badly needed (Wittenberger, 1983).

21. Sample sizes: Nine females were used to address questions 1 and 2, four were used to address questions 3–5, and the remainder were used to test the stimuli associated with the onset and termination of estrus, described in part 5 of this chapter.

22. Male sizes: The male classes were juvenile (ca. 45 kg), peripheral (ca. 135 kg), territorial (ca. 200 kg), and senescent (ca. 200 kg). The latter were identifiable by their gaunt appearance, worn canine teeth, scars, and worn pelage.

23. End of estrus: We allowed coitus to occur in eight pairings with juvenile males, seven with peripheral males, and two with senescent males. Postcoital receptivity could only be assessed in fifteen of these seventeen cases due to nonresponsiveness of the test male late in the season.

24. Ketones: We tested for the presence of acetoacetate (using clinical Ketosticks) but found none (D. Costa, unpub. data).

25. Vaginal code: A memory for the species-typical code is stored in the central nervous system, which when triggered causes prolactin release that activates corpora lutea (Terkel et al., 1990). The vaginal code mechanism separates the hormonal from the gametic requirements for pregnancy.

26. Suppression of aggression: Male W, a 135 kg test male, would approach females face to face and perform "♂ *Nose*" (definitions in table 8.1) until the female responded with "Open Mouth Lunge," whereupon he would withdraw about 2 meters and wicker for 4–5 minutes. He would then move behind the female and rest his chest upon her rear flippers, which elicited more lunges to which he would respond by recoiling his head and making a guttural snoring sound. The male appeared to be establishing proximity. He would then suddenly strike the female flat with a blow of his muzzle (scored as ♂ *Aggression*). Immediately thereafter he would perform ♂ *Nose* again, to which the female usually responded not with lunging as before, but with ♀ *Nose*. That is, aggression from the male suppressed further aggression from the female. The male would next attempt Touch Neck, Touch Back, or AGI, thus escalating his investigation. The female usually responded with ♀ *Soft Bite*, to which the male would respond by striking as before or threatening with an open mouth. In response to male aggression, the female would stop biting and only try to block the male by interposing her head between his head and her body. With female resistance almost nonexistent, more ♂ *AGI* would be followed by ♂ *Touch Back*/♀ *No Response*, and mounting would occur soon thereafter.

Chapter 9
Ontogeny of Male Territorial Behavior

1. Visits: Daily movements were studied for 180 branded juveniles on St. George Island for three seasons (Gentry, 1981b). "Visit" was defined as a series of daily sightings, no two of which were separated by more than 10 days of absence. The mean length of the season in which the average 3.1 visits occurred was 80 days, and the mean number of days spent on shore per year was 20 (± 10.6). Males sometimes visited two landing areas or both islands in the same day. Inter-island move-

ments of these animals were reported to me by M. Griben, who was conducting a similar study (Griben, 1979).

2. Size categories: Class 1, vibrissae all black (age 1–2 years); Class 2, vibrissae mixed black and white (age 2–4 years); Class 3, vibrissae white, no secondary sexual characteristics (age 4–5 years); Class 4, vibrissae white, some secondary sexual characteristics (long hair on head and neck) (age 5–8 years); Class 5, vibrissae white, neck and chest massive, adult male size and configuration (ages 8+ years). Ages are only approximate (Scheffer, 1962). These definitions differ from, and are not intended to replace, the size classes used in the annual count of adult males (Standing Scientific Committee, 1963).

3. Definition of interaction: Interactions were defined as obvious, mutual, coordinated movements of animals in close proximity. They usually occurred between dyads and often included stereotyped components reminiscent of adult behavior patterns.

4. Play: Play in northern fur seals consisted of many components of adult behavior mixed in atypical combinations, as in other otariids (Gentry, 1974). Play in this species was accompanied by a metacommunication signal (Bateson, 1955) indicating nonserious intent (a languid, sideways "S" movement of the head and neck with the mouth held open). However, because these animals often shifted within a bout from playful to serious intent, "play" was an inconsistent category.

5. Forays: For each foray, observers recorded the date, identity (if known), size class, grid subsectors (5 ∞ 5 m units) passed through, biting or chasing by other males, and whether a territory was established.

6. Rest sites: Of 103 focal males recorded in 1977, 42 selected a rest site during the observation period, 17 of these by displacing another male.

7. Size preference: Opponent size preference was analyzed for 1,015 interactions recorded in 1977. The analysis included all fifteen pairwise combinations of the five size classes. A χ^2 analysis was done to test the null hypothesis that for each size class the number of interactions was equally distributed among opponent size classes.

Chapter 10
Female Attendance Behavior

1. Strategy: To repeat from chapter 1, the term "strategy" here refers to an adaptive response of animals that is shaped by natural selection and does not imply cognition.

2. Comparative trips to sea: The temperate New Zealand fur seal averaged 2.6 days at sea in one year, and 7.7 days the following season at the same site (Goldsworthy, 1992). Females late in the season made trips of 11.2 days, much longer than for more southerly subantarctic fur seals (2 days at Macquarie Island). However, subantarctic fur seals at Amsterdam Island are reported to make trips of 20–30 days (C. Guinet, unpub. data discussed at the International Symposium and Workshop on Otariid Reproductive Strategies and Conservation, National Zoo, Washington, D.C., 12–16 April 1996). The temperate Juan Fernandéz fur seal makes trips to sea an average of 16.5 days (Francis et al., 1995). The temperate Guadalupe fur seal makes trips of 9–13.5 days, depending on the year (Figueroa Carranza, 1994).

3. Shore visits: Shore visits are shorter than trips to sea (1.7 days long) in New Zealand fur seals, and are said to be stable (Goldsworthy, 1992; Miller, 1971).

4. Sexual selection theory: It predicts that in highly polygynous species in which reproductive success varies more widely in one sex than in the other, mothers should differentially invest in the more variable sex (Bateman, 1948; Trivers and Willard, 1973; Maynard Smith, 1980; Clutton-Brock et al., 1981). The northern fur seal is highly polygynous, has extreme sexual dimorphism, and male reproductive success is much more variable than for females (see fig. 4.4). Therefore, the northern fur seal is an excellent species on which to test this theoretical prediction.

5. Dive patterns: "Deep" divers, with maximum depths to 200 meters and average depths more than 100 meters, made up about 30% of the females instrumented at St. George Island in one study (Gentry et al., 1986c). "Shallow" divers, with no dives greater than 75 meters and average dives less than 50 meters, comprised another 30%, and the other 40% were "mixed" divers that used the deep pattern on the first and last days at sea (when they were crossing the shelf), and the shallow pattern on all other days (when they were in deep water beyond the shelf).

6. Short trips: Another reason why deep divers make shorter trips than shallow divers is that they appear to be more efficient (receive more energy per dive), perhaps by taking larger or more energy-rich prey (from isotope dilution studies; Costa and Gentry, 1986).

7. Young females: The average duration of trips to sea in summer by females 5 years and ≤ 4 years old averaged 4.5 and 6.7 days, respectively (Goebel, 1988). This difference persists into October, although it diminishes.

8. Attendance behavior, sample size: The records of 737 individual mothers (East Reef = 289, Zapadni = 311, Kitovi = 137) fit the restrictions. Some were present for several years, resulting in a data set of 943 female-years of data that included 4,471 trips to sea, or on average about five trips to sea per female per year. Five years of data (1976–78, 1983, and 1985) were available for comparing East Reef, Zapadni, and Kitovi. Twelve years of data (1974–85) were available for comparing East Reef and Zapadni.

9. Feeding trip length: Annual mean trips for both sites for all years were ranked and two-way Kruskal-Wallis tests were run for site, year group (1974–78 and 1979–85), and site-by-year group. The group means for the two time periods were 5.6 and 4.4 days, respectively. The difference between groups, and the interaction of site-by-year group were significant, while the difference between sites was not (H = 13.61 for year group, 3.89 for site-by-year group, and H = .046 for site, all vs. tabled $\chi^2_{(0.05,1)} = 3.84$). A Wilcoxon-Mann-Whitney two-sample test showed that the sites were significantly different before 1979 but not afterward (1974–78, U = 25 vs. tabled $U_{(0.05,5,5)} = 23$; 1979–85, U = 40 vs. tabled $U_{(0.05,7,7)} = 41$).

10. Shore visit length: An annual mean was calculated for shore visits at each site (excluding the first by each female). Differences in the year groups 1974–78 and 1979–85 (same groups as for trips to sea) were tested using a Mann-Whitney two-sample test. The two groups were significantly different at East Reef (2.3 decreasing to 1.8 days, U = 30 vs. tabled $U_{(0.05,5,7)} = 30$), but not at Zapadni (1.7 days in both groups, U = 25 vs. tabled $U_{(0.05,5,7)} = 30$).

11. Age effects on foraging trip length: Using 3,177 trips for multiyear females

recorded between 19 June and 28 August, a mean duration of trips to sea was calculated for tenure years 1, 2, 3, 4, and 5–13 combined (due to small sample size in those years) for East Reef and Zapadni, and were compared by a Kruskal-Wallis one-way ANOVA. Trip length did not differ by tenure year at East Reef ($H = 2.46$ vs. tabled $H_{(0.05,4)} = 9.49$). But at Zapadni, trips in the first 2 years were significantly different (longer) than in subsequent years ($H = 57.05$ vs. tabled $H(0.05,4) = 9.49$).

Chapter 11
Neonatal Growth and Behavior

1. Newborn definition: Newborns were those that had a fresh, bloody, umbilical stump or were dragging moist, pink placental membranes. Most were born at night and we captured them before noon.

2. Pup marking: We marked them using tags, bleach, and pelage clipping.

3. Data recording: For all captures we recorded the location along the catwalk, date, weight, and presence of the mother.

4. Daily relative weight gain: We use the formula (*weight at time t − birth weight*)/(*birth weight*) ∞ 100. Only pups born early in the season would have been 50 days of age when sampling ended.

5. Sine function: We used only data prior to 30 days of age when the oscillation appeared to break down. For each sex-by-island combination we fit the following multiple linear regression:

$$Y = a + bt + c \, [\sin (57.2958 \times (2\pi t / L^2)],$$

where Y = relative weight gain, a = y-intercept, b = growth rate, t = pup age in days, c = coefficient of the sine wave amplitude, and L = period length of the sine wave in days. Each mean value of relative weight gain was weighted by the reciprocal of its variance. We ran the weighted regression trying different values of L until r^2 was maximized. We used a Fourier transform analysis to supplement this empirical approach to defining the oscillation pattern.

6. Absolute growth: For each pup weighed three or more times we linearly regressed absolute weight on age using the model $w_t = a + bt$, where w_t = absolute weight, a = y-intercept, b = growth rate, and t = pup age in days. We then ranked the slopes (absolute growth rate) and tested the ranks in a Kruskal-Wallis two-way ANOVA.

7. Relative growth: Using the same ranking and testing procedure as in note 6, we also evaluated pup growth with two other measures of weight. In the first, we calculated the slope based on a model using the pup weight relative to birth weight as a function of age. The regression model was $y = a + bt$, where y = (*weight at time t/birth weight*) ∞ 100, a = y-intercept, b = growth rate, and t = pup age in days. The model implies a constant percentage of weight increase over time. In the second measure, we calculated the slope based on the logarithm of absolute weight regressed against pup age. The regression model was $\log (w_t) = a + bt$, where w_t + absolute weight at time t, a = y-intercept, b = growth rate, and t = pup age in days. In this model, the weight on any day depended on the weight of the previous day.

8. Sample size: We weighed and marked 815 pups total (405 at Staraya Artil,

410 at Reef) and excluded 110 of them from analysis because they were weighed only once (63 at Staraya Artil, 47 at Reef). These comprised the "excluded" sample, leaving the "included" sample as 362 and 350 for Reef and Staraya Artil, respectively.

9. Excluded vs. included for sex ratio: Staraya Artil χ^2 = 3.12, df = 1, p = .08; Reef χ^2 = 7.31, df = 1, p = .01.

10. Excluded vs. included for mean date of birth: Mann-Whitney test U values, Staraya Artil, p = .70; Reef, p = .00.

11. Excluded vs. included for birth weight: Staraya Artil females (means 5.19 vs. 5.04 kg, ANOVA, p = .15); Staraya Artil males (means 5.64 vs. 5.83, ANOVA, p = .21): Reef females (mean 5.32 vs. 4.81, ANOVA, p = .00); Reef males (mean 5.87 vs. 5.42, ANOVA, p = .02).

12. Inter-island differences: Two-way ANOVA; sex, p = .0001; island, p = .0004.

13. Runs: For example, males comprised only 29% of the sample on 2 July at Staraya Artil (n = 14), and the next day comprised 67% of the sample (n = 15).

14. Sample size for sex ratio: Sample made up of the "excluded" and "included" categories, plus the 1979 sample and a few others.

15. Location along catwalk and birth weight: Reef, p = .20; Staraya Artil, p = .07; sex ratio, Reef, p = .70; Staraya Artil, p = .34.

16. Sample size for growth rate: Estimates were based on 1,480 measurements at Reef (for 162 females and 199 males) and 1,035 measurements at Staraya Artil (for 184 females and 169 males). The range in number of weights per pup was 1–8 at Staraya Artil and 1–16 at Reef, but 72–100% of the pups, depending on sex and island, were weighed only 2–5 times.

17. Male pups: A 17–20-day cycle would imply that mothers spent 11 days at sea after the perinatal period. In fact, first trips were unusually short (1–3 days). Furthermore, mothers of female and male pups were not different in the length of their trips to sea.

18. Growth rate differences: Kruskal-Wallis test, by sex (H = 16.7 vs. tabled $\chi^2_{(.05,1)}$ = 3.8); by island (H = 0.95).

19. Site differences over time: We tested the data in 5-day increments of age, using only one weight per pup per increment, and compared the sites by two-way ANOVA using sex and island as the major variables.

20. Relative growth: The analyses of growth relative to birth weight and growth relative to the previous day showed no significant differences between sexes or islands (all H values < tabled $\chi_{(.05,1)}$ = 3.8).

21. Factors contributing to pup weight: Done by stepwise multiple linear regression from birth to 50 days. Age, sex, island, age by sex, and age by island were all significant (r^2 = .48, p ≤ .001 for all factors). Age and sex contributed the most (r^2 = 0.400 for age, and r^2 = 0.47 with sex added). Other factors contributed progressively smaller increments. To eliminate repeated measures errors, the model was run several times using only one randomly selected weight per pup. The r^2 values differed slightly on different runs but consistently showed that age and sex contributed most to the fit. Island of origin contributed significantly in some runs but not others (therefore it had a marginal effect on growth rates).

Chapter 12
Female Foraging Behavior: Inter- and Intra-Annual Variation in Individuals

1. Sample size: The intra-annual sample includes twelve foraging records from five females that were instrumented in July or August (less than one month after parturition) and again in October (approximately one month before weaning and the start of migration). The sample included two females from East Reef (St. George Island) in 1983 and three females from Zapadni Reef (St. Paul Island) in 1985. The inter-annual sample includes nine foraging records from four females instrumented in July or August of different years.

2. Equipment: Females were captured using a noose pole, removed from the central breeding area, and placed on a restraint board (Gentry and Holt, 1982). Each female was tagged on the fore-flippers (Allflex sheep ear tag) and equipped with a photomechanical TDR (Meer Instruments, Solana Beach, CA) that stored data on a strip of film (Kooyman et al., 1976). TDRs were attached by a harness (Gentry and Kooyman, 1986a). On their next visit ashore, females were recaptured with a hoop net and the TDR was removed by cutting the harness. All October captures were made with a one-meter hoop net. When possible, the mass for each female was measured at each capture using a platform scale accurate to 0.1 pound.

3. Techniques: Film was developed in either Agfa Rodinol or Kodak D-19, and the negatives were reproduced on paper with a 7 ∞ enlargement using copy-flow xerography. At least three points—the start, end, and maximum depth of each dive—were digitized on an electronic digitizing pad. These points allowed computation of dive duration, depth, and interdive surface interval.

4. Weight gain: July weights = 39.5, 39.5, and 40.4 kg; October weights = 43.5, 44, and 41 kg.

5. Trip length: Mean trip duration increased (p = .041) from 149 h (n = 3, SD = 81) in July to 296 h (n = 3, SD = 8.5) in October.

6. Mean depth of dive: It increased from 45 m (n = 1,575, SD = 14) in July to 93 m (n = 2,325, SD = 31) in October (p < .001). Also, most July dives occurred at depths of less than 50 m and in October they were normally distributed about 90 m (fig. 12.1b).

7. Dive pattern: Female no. 308 shifted from diving around 90 m in July to 90.5 m in October (SD = 14.8, n = 660); female no. 767 shifted from diving around 70 m in July to 63.6 m in October (SD = 20.7, n = 1075); and female no. 854 had few dives less than 100 m in July but averaged 124.9 m in October (SD = 32.6, n = 590).

8. Dive rate per trip: The rate of diving (dives per hour for the entire trip) did not change between July and October (p = .16).

9. Foraging bouts per hour: Foraging bouts per hour did not change (p = 0.98) from July (mean = .09, n = 3, SD = .02) to October (mean = .09, n = 3, SD = .02).

10. Duration of foraging bouts: Duration of bouts did not change (p = .68) between July (mean = 3.4, n = 41, SD = 0.7) and October (mean = 3.7, n = 79, SD = 1.6).

11. Number of dives within bouts: The mean number of dives within foraging bouts decreased in October (p = .01; July mean = 42, n = 41, SD = 16; October mean = 29, n = 79, SD = 17).

12. Dive rate within bouts: The mean rate of dives (dives/h) within bouts decreased (p = .05, July mean = 12, n = 41, SD = 2; October mean = 8, n = 79, SD = 2).

13. Dives by hour: The temporal distribution of dives did not change between July and October when considering an earlier dusk and a later dawn.

14. Shallow dives: For female P9, the mean depth of dive increased (p < .001) from 127 m in July (SD = 46.6, n = 175) to 170 in October (SD = 72.4, n = 301). For female no. 1789 it decreased (p < .001) from 69 m in July (SD = 23.3, n = 148) to 56 m in October (SD = 25.6, n = 1221).

15. Dive bouts per trip: The number of diving bouts increased (p = .028) between July (mean = 9, SD = 2.8) and October (mean = 20.5, SD = 2.1).

16. Percentage of time in foraging bouts: It doubled from 15.5 to 30.2 between July and October.

17. Dive rate (dive/h): For female no. 1789 the number of dives per hour increased considerably from July to October but not by a significant amount (p = .448) overall.

18. Number of bouts per hour: The number of bouts per hour remained unchanged (p = .500) in July (mean = .07, SD = .03) compared to October (mean = .09, SD = .01).

19. Duration of dive bouts: They did not change (p = .154) between July (mean = 2.3, SD = 1.6, n = 18) and October (mean = 3.2, SD = 2.1, n = 41).

20. Number of dives per bout: They did not change (p = .872) between July (mean = 15, SD = 12, n = 18) and October (mean = 36, SD = 57, n = 41).

21. Dive rate (dives/h) within bouts: The rate did not change (p = .068) between July (mean = 8, SD = 5, n = 18) and October (mean = 10, SD = 7, n = 41).

22. Nonbout dives: Decreased greatly from 18 (July) to 1.7 (October).

23. Sample: One of the four females, J8, was instrumented for two successive trips in her second year of being instrumented.

24. Trip duration: No difference in first and second years (p = .34; first-year mean = 152.2, SD = 35.7, n = 4; second-year mean = 142.2, SD = 29.5, n = 5; range, 106–192 h). See table 12.4.

25. Dives per trip: No difference in first and second years (p = .96; first-year mean = 242.2, SD = 90.7, n = 4; second-year mean = 226.6, SD = 143.6, n = 5).

26. Dive rate: No difference in first and second years (p = .93; first-year mean = 1.6, SD = 0.6, n = 4; second-year mean = 1.5, SD = 0.6, n = 5).

27. Number of dive bouts: No difference in first and second years (p = .77; first-year mean = 9.2, SD = 5.9, n = 4; second-year mean = 11.4, SD = 4.6, n = 5).

28. Dive bouts per hour: No difference in first and second years (p = .46; first-year mean = 0.06, SD = 0.03, n = 4; second-year mean = 0.08, SD = 0.02, n = 5).

29. Number of dives per bout: They did not change for any female (♀ J8, p = .50; ♀ 1789, p = .16; ♀ 2775, p = .33; ♀ 592, p = .81).

30. Mean depth of dive: It changed for three females (♀ J8, p < .001; ♀ 1789, p < .001; ♀ 592, p < .001) but not the fourth (♀ 2775, p = .38).

31. Duration of dive bouts: It changed for female no. 1789 (p = .03) but not for others (♀ J8, p = .78; ♀ 2775, p = .053; ♀ 592, p = .054).

32. Dive rate (dives/h) within bouts: it changed for female no. 2775, p = .04 but not the others (♀ J8, p = .10; ♀ 1789, p = .47; ♀ 592, p = .26)

Chapter 13
Female Foraging Behavior: Effects of Continental Shelf Width

1. Field methods: We captured animals with hoop nets or noose poles and physically restrained them while instruments were attached with a nylon harness (Gentry and Holt, 1982). The instruments were Wildlife Computers Mark 3+ microprocessor-controlled TDRs set for sampling rates for depth and water temperature every 10 seconds, with resolutions of ± 2 m and ± 0.1 C, respectively. The "geolocation" feature of TDRs used at Medny Island sampled light level every 15 minutes to compute approximate location at sea (DeLong et al., 1992; Hill, 1994). Data reduction was accomplished with PC software available from Wildlife Computers.

2. Methods of analysis: We analyzed dives for descent and ascent rates, time spent at maximum depth (bottom time), and shape. Dive shape was assigned to one of thirteen categories depending on the number and phase of the dive in which reversals in depth occurred (see fig. 13.2). Using special software, all depth reversals on the descent and ascent that exceeded 2 meters in extent were analyzed for the depth and water temperature at which they began.

3. Dive pattern: This term refers to a combination of the maximum depths in the record, and the way depth changes with time. They are best seen by plotting maximum depth against time for every day of the record, or by three-dimensional plots of time, depth, and frequency.

4. Dive-bout criteria: Probability plots were constructed using the natural log of interdive intervals for each female.

5. Vertical distance moved: We selected twelve of the deepest sets of dives available from each island (St. George Island, 100–143 m; Medny Island, 20–63 m). Sets were selected that had the briefest and most regular interdive intervals, and that were less than 2.75 hours in duration. These sets sometimes occurred within longer diving sequences.

6. Instrument effect: Trips to sea for uninstrumented and instrumented females were, for St. George, 6.9 and 7.5 days, respectively (Gentry et al., 1986b); for Medny, 3.4 (Vladimirov, 1983) and 11.1 days (present result), respectively.

7. Traveling dives: Mann-Whitney U test, p = .313.

8. Transit: Outbound transit = the interval from leaving shore to the first dive; inbound transit = the interval from the last dive to landing on shore. Mann-Whitney U test, outbound, p = .138; inbound, p = .917.

9. Inbound sample: We did not get a measure of inbound transits on all records because the TDR memory often had filled during the trip such that instruments were not recording at trip's end.

10. Foraging time: Mann-Whitney U test, U = 140 vs. tabled $U_{.05,10,17}$ = 125, p = .006.

11. Dive rate (dives/hr from first to last dive on the record): Mann-Whitney U test, U = 128 vs. tabled $U_{.05,10,17}$ = 125, p = .031.

12. Time spent under water: Traveling dives included, Mann-Whitney U test, U = 97 vs. tabled $U_{.05,9,14}$ = 95; p = .032.

13. Total dives per trip: Mann-Whitney U test, traveling dives included, p = .482; traveling dives excluded, p = .16.

14. Vertical distance moved per hour: Mann-Whitney U test for entire trip,

p = .95, for selected dive bouts, p = .214.

15. Vertical distance moved on a foraging trip: St. George = 1,392 m, Medny = 1,455 m; Mann-Whitney U test, p = .078.

16. Mean depth and duration: Mann-Whitney U test, depth, p = .175; duration, p = .428.

17. Single dive depth and duration: Mann-Whitney U test, depth, U = 126 vs. tabled $U_{.05,9,14}$ = 95, p = .000; duration, U = 113 vs. tabled $U_{.05,9,14}$ = 95, p = .002.

18. Depth vs. duration: Both the correlation coefficients and the slopes for the regression of dive duration on dive depth were significantly different between the two islands (see table 13.2); Mann-Whitney U test, correlation, U = 119, p = .000; slope, U = 125, p = .000, both vs. tabled $U_{.05,9,14}$ = 95. Long, shallow dives at Medny Island decreased the correlation coefficient (all but one $r^2 < .65$) and increased the regression slopes.

19. Bottom times: Mann-Whitney U test, U = 95 vs. tabled $U_{.05,9,14}$ = 95, p = .044.

20. Ascent vs. descent rates: One-sample t-test, t = 4.55, df = 22, p = .00.

21. Sample size for shape: We categorized 3,599 dives ≥ 20 m as to shape, an average of 185 and 138 dives per female at St. George and Medny Islands, respectively.

22. Spike dives: The older conclusion was based on photo-mechanical TDRs which had poorer resolution than newer electronic TDRs.

23. Island differences in dive shapes 0 and 4: Mann-Whitney U test, shape 0, U = 97 vs. tabled $U_{.05,9,13}$ = 89, p = .01; shape 4, U = 83 vs. tabled $U_{.05,9,12}$ = 82, p = .039.

24. Island differences in dive shapes 1, 12, and 13: Mann-Whitney U test, shape 1, U = 97 vs. tabled U.05,9,13 = 89, p = .01; shape 12, U = 85.5 vs. tabled $U_{.05,8,12}$ = 74, p = .006; shape 13, U = 55 vs. tabled $U_{.05,6,10}$ = 49, p = .007.

25. Depth by dive shape: Mann-Whitney U test, shape 0, U = 115 vs. tabled $U_{.05,9,13}$ = 89, p = .000; shape 1, U = 104.5 vs. tabled $U_{.05,9,13}$ = 89, p = .002; shape 2, U = 88 vs. tabled $U_{.05,9,12}$ = 82, p = .016; shape 4, U = 105 vs. tabled $U_{.05,9,12}$ = 82, p = .000; shape 5, U = 101 vs. tabled $U_{.05,9,13}$ = 89, p = .004; shape 14, U = 70 vs. tabled $U_{.05,7,10}$ = 56, p = .001.

26. Depth of reversals: Mann-Whitney U test, U = 118 vs. tabled $U_{.05,9,14}$ = 95, p = .001.

27. Inter-island differences in water temperature: Mann-Whitney U test, surface, U = 80 vs. tabled $U_{.05,7,12}$ = 66 , p = .001; descent reversal, U = 89.5, vs. tabled $U_{.05,7,13}$ = 71, p = .000; ascent reversal, U = 90.5 vs. tabled $U_{.05,7,13}$ = 71, p = .000; maximum depth, U = 75 vs. tabled $U_{.05,7,12}$ = 66, p = .005.

28. Location at sea: The calculated positions are not shown graphically because we failed to calculate the TDR's time drift accurately, which resulted in poor accuracy (Hill, 1994).

Chapter 14
Synthesis

1. Entanglement rates: Entanglement rate at the Commander Islands from 1978 to 1985, n = 6 yr of data, mean = 0.77% of harvest entangled, SD = 0.16 (data from VNIRO, 1978–85). Entanglement rate at St. Paul Island from 1969 to 1980, n = 12 yr

of data, mean = .43% of harvest entangled, SD = 0.13; data from Scordino and Fisher, 1983; Fowler, 1982, 1987). The data were collected by dissimilar methods, so close comparison is probably not justified.

2. Salmon: They should be considered in this analysis because landing records extend back to 1925 (Beamish and Bouillon, 1993), salmon landings reflect major climate changes (Beamish, 1993; Beamish and Bouillon, 1993, 1995), and they reflect landings of other commercial species (Bakkala, 1993; Quinn et al., 1995). The abundance of many prey items taken by northern fur seals is never measured, but may in general follow the trend in commercially valuable fish.

3. Pressures, 1940–70s: Mean atmospheric pressures in this period were generally high, except for some intense lows from about 1959 to 1964 (Beamish and Bouillon, 1993; see fig. 7 in Trenberth and Hurrell, 1994).

4. Timing: A suggested mechanism is that fur seal survival may be enhanced by the onset of intense atmospheric low pressures, possibly due to wind effects enhancing primary productivty and other aspects of the food web. This notion receives some support from the finding that the only time fur seals increased between 1956 and 1983 was in the period 1970–76 (fig. 1.8), six years after the intense atmospheric lows of 1959–64 reported in note 3 above.

5. Potential biological removal (PBR): Defined as "the maximum number of animals, not including natural mortalities, that may be removed from a marine mammal stock while allowing that stock to reach or maintain its optimum sustainable population" (Barlow et al., 1995). It is calculated as the product of a minimum population estimate, half the maximum net productivity rate, and a recovery factor. The minimum population level (N_{min}) is a conservative estimate of the population size in that it compensates for variability in the original abundance estimates. The maximum net productivity rate (R_{max}) reflects the balance between recruitment and natural losses, and can be estimated or derived empirically. The recovery factor (F_r) ranges from 1.0 to 0.1, depending on several factors, including status relative to the carrying capacity, and status relative to the Endangered Species Act.

6. Kill of juvenile males, 1956–68: The loss of juvenile males during the reduction of females may have had no effect on the population trend. For example, more males were killed from 1943 to 1955 (n = 888,139, average of 68,318 per year; data from Engle et al., 1980) than from 1956 to 1968 (n = 676,965, average of 52,074 per year; data from Engle et al., 1980), yet the total population increased from 1943 to 1955, and decreased from 1955 to 1968 (fig. 1.5).

Literature Cited

Allen, J. A. 1880. *History of North American Pinnipeds: A Monograph of the Walruses, Sea-Lions, Sea-Bears and Seals of North America.* Washington, D.C.: U.S. Government Printing Office.

Anderson, S. S., R. W. Burton, and C. F. Summers. 1975. Behavior of grey seals (*Halichoerus grypus*) during a breeding season at North Rona. *J. Zool. Lond.* 177:179–195.

Anonymous. 1969. Fur seal investigations, 1966. *Spec. Sci. Rep.* no. 584. U.S. Bureau of Commercial Fisheries. Available from National Marine Mammal Laboratory, Seattle.

Anonymous. 1971. Some observations on the behavior of tagged pups on Seal Island. *Izvestiya Tikhookeanskogo nauchno-issledovatel'skogo inst. rybnogokhozyaistva i oceanografia* 76:168–169. In Russian. English translation available from NOAA, National Marine Mammal Laboratory, Seattle.

Anonymous. 1973. Appendix D. Coordinated Pribilof Islands–Bering Sea research proposal. *Fed. Regist.* 38(147):20599–20601.

Antonelis, G. A., Jr. 1976. A comparison of the diurnal and nocturnal behavior of the northern fur seal on San Miguel Island. M.S. thesis, San Diego State University, San Diego.

Antonelis, G. A. 1990. Migration studies of northern fur seals. *NWAFSC Quart. Rep.* (April–June), 1–4. Alaska Fisheries Science Center, Seattle.

Antonelis, G. A., B. S. Stewart, and W. F. Perryman. 1990a. Foraging characteristics of female northern fur seals (*Callorhinus ursinus*) and California sea lions (*Zalophus californianus*). *Can. J. Zool.* 68:150–158.

Antonelis, G. A., C. W. Fowler, E. S. Sinclair, and A. E. York. 1990b. Population assessment, Pribilof Islands, Alaska. In H. Kajimura, ed., *Fur Seal Investigations, 1989.* NOAA Tech. Mem. NMFS F/NWC-190:8–21, Alaska Fisheries Science Center, Seattle.

Antonelis, G. A., E. H. Sinclair, R. R. Ream, and B. W. Robson. 1993. Inter-island variation in the diet of female northern fur seals (*Callorhinus ursinus*) in the Bering Sea. Abstract for the Tenth Biennial Conference on the Biology of Marine Mammals, Galveston, Texas, November 1993.

Antonelis, G. A., T. J. Ragen, and N. J. Rooks. 1994a. Male-biased secondary sex ratios of northern fur seals on the Pribilof Islands, Alaska, 1989 and 1992. In E. H. Sinclair, ed., *Fur Seal Investigations, 1992.* NOAA Tech. Mem. NMFS-AFSC-45:84–89.

Antonelis, G. A., A. E. York, B. W. Robson, R. G. Towell, and C. W. Fowler. 1996. Population assessment, Pribilof Islands, Alaska. In E. H. Sinclair ed., *Fur Seal Investigations, 1994.* NOAA Tech. Mem. NMFS-AFSC-69:9–29. NOAA, National Marine Mammal Laboratory, Seattle.

Antonelis, G. A., E. H. Sinclair, R. R. Ream, and B. W. Robson. 1997. Inter-island variation in the diet of female northern fur seals (*Callorhinus ursinus*) in the Bering Sea. *J. Zool. Lond.* (in press).

Asaga, T., Y. Naito, B. J. Le Boeuf, and H. Sakurai. 1994. Functional analysis of dive types of female northern elephant seals. In B. J. Le Boeuf and R. M. Laws eds. *Elephant Seals: Population Ecology, Behavior, and Physiology*, 310–327. Berkeley: University of California Press.

Baker, J. D. 1989. Aquatic copulation in the northern fur seal, *Callorhinus ursinus. Northw. Nat.* 70:33–36.

Baker, J. D. 1991. Trends in female northern fur seal, *Callorhinus ursinus*, feeding cycles indicated by nursing lines in juvenile male teeth. M.S. thesis, University of Washington, Seattle.

Baker, J. D., and C. W. Fowler. 1990. Tooth weights of juvenile male northern fur seals, *Callorhinus ursinus. Mar. Mamm. Sci.* 6(1):32–47.

Baker, J. K., and C. W. Fowler. 1992. Pup weight and survival of northern fur seals, *Callorhinus ursinus. J. Zool. Lond.* 227:231–238.

Baker, J. D., C. W. Fowler, and G. A. Antonelis. 1994. Mass change in fasting immature male northern fur seals. *Can. J. Zool.* 72:326–329.

Baker, J. D., G. A. Antonelis, C. W. Fowler, and A. E. York. 1995. Natal site fidelity in northern fur seals, *Callorhinus ursinus. Anim. Behav.* 50:23–247.

Bakkala, R. G. 1993. *Structure and Historical Changes in the Groundfish Complex of the Eastern Bering Sea.* NOAA Tech. Rep. NMFS 114:1–91, Alaska Fisheries Science Center, Seattle.

Bakkala, R. G., and Wakabayashi, K., eds. 1985. Results of cooperative U.S.-Japan groundfish investigations in the Bering Sea during May–August 1979. *Int. North Pac. Fish. Comm. Bull.* 44.

Bakkala, R. G., V. G. Wespestad, and J. J. Traynor. 1987. Walleye pollock. In R. Bakkala and J. Balsiger, eds. *Condition of Groundfish Resources of the Eastern Bering Sea and Aleutian Islands Region in 1986*, 11–29. NOAA Tech. Mem. NMFS F/NWC-117, NOAA Alaska Fisheries Science Center, Seattle.

Barlow, J., S. L. Swartz, T. C. Eagle, and P. R. Wade. 1995. *U.S. Marine Mammal Stock Assessments: Guidelines for Preparation, Background, and a Summary of the 1995 Assessments.* NOAA Tech. Mem. NMFS-OPR-6:1–73.

Barry, R. G. 1989. The present climate of the Arctic Ocean and possible past and future states. In Y. Herman, ed., *The Arctic Seas: Climatology, Oeanography, Geology, and Biology*, 1–46. New York: Van Nostrand Reinhold.

Bartholomew, G. A. 1953. Behavioral factors effecting social structure in the Alaska fur seal. *Trans. 18th North Amer. Wildl. Conf.*, 481–502.

Bartholomew, G. A. 1959. Mother-young relations and the maturation of pup behavior in the Aslaka fur seal. *Anim. Behav.* 7:163–171.

Bartholomew, G. A. 1970. A model for the evolution of pinniped polygyny. *Evolution* 24:546–559.

Bartholomew, G. A., and P. G. Hoel. 1953. Reproductive behavior of the Alaska fur seal, *Callorhinus ursinus. J. Mamm.* 34:417–436.

Bateman, A. J. 1948. Intrasexual selection in Drosophila. *Heredity* 2:349–368.

Bateson, G. 1955. A theory of play and fantasy. *Psychiat. Res. Rep.* 2:39–51.

Baumgartner, T. R., A. Soutar, and V. Ferreira-Bartrina. 1992. Reconstruction of the history of Pacific sardine and northern anchovy populations over the past two millennia from sediments of the Santa Barbara Basin, California. *CalCOFI Rep.* 33:24–40.

Beach, F. A. 1976. Sexual attractivity, proceptivity, and receptivity in female mammals. *Horm. Beh.* 7:105–138.

Beamish, R. J. 1993. Climate and exceptional fish production off the west coast of North America. *Can. J. Fish. Aquat. Sci.* 50:2270–2291.

Beamish, R. J., and D. R. Bouillon. 1993. Pacific salmon production trends in relation to climate. *Can. J. Fish. Aquat. Sci.* 50:1002–1016.

Beamish, R. J., and D. R. Bouillon. 1995. Marine fish production trends off the Pacific coast of Canada and the United States. In R. J. Beamish, ed., Climate change and northern fish populations. *Can. Spec. Publ. Aquat. Sci.* 121:585–591.

Bengtson, J. L., L. M. Ferm, T. J. Härkönen, and B. S. Stewart. 1990. Abundance of Antarctic fur seals in the South Shetland Islands, Antarctica, during the 1986/87 austral summer. In K. R. Kerry and G. Hempel, eds., *Antarctic Ecosystems*, 265–270. Berlin: Springer-Verlag.

Berta, A., and T. A. Deméré. 1986. *Callorhinus gilmorei* n. sp. (Caranivora: Otariidae) from the San Diego Formation (Blancan) and its implications for otariid phylogeny. *Trans. San Diego Soc. Nat. Hist.* 21(7):111–126.

Berta, A., C. E. Ray, and A. R. Wyss. 1989. Skeleton of the oldest known pinniped, *Enaliarctos mealsi. Science* 244:60–62.

Bigg, M. A. 1973. Adaptations in the breeding of the harbour seal, *Phoca vitulina. J. Reprod. Fert.* (suppl.) 19:131–142.

Bigg, M. A. 1979. Studies on captive fur seals, progress report no. 3. Unpublished report submitted to North Pacific Fur Seal Commission, 1979. Available from Pacific Biological Station, Nanaimo, British Columbia.

Bigg, M. A. 1986. Arrival of northern fur seals, *Callorhinus ursinus*, on St. Paul Island, Alaska. *Fish. Bull.* 84:383–394.

Bigg, M. A. 1990. Migration of northern fur seals (*Callorhinus ursinus*) off western North America. *Can. Tech. Rep. Fish. Aquat. Sci.* no. 1764, Pacific Biological Station, Nanaimo, British Columbia.

Bignami, G., and F. A. Beach. 1968. Mating behavior in the chinchilla. *Anim. Behav.* 16:45–53.

Blandau, R. J., J. L. Boling, and W. C. Young. 1941. The length of heat in albino rat as determined by the copulatory response. *Anat. Rec.* 79:453–463.

Blix, A. S., L. K. Miller, M. C. Keyes, H. J. Grav, and R. Elsner. 1979. Newborn northern fur seals (*Callorhinus ursinus*)—do they suffer from cold? *Amer. J. Physiol. Reg. Integrat. Comp. Physiol.* 5(3):R322–R27.

Block, M. S., L. C. Volpe, and M. J. Hayes. 1981. Saliva as a chemical cue in the development of social behavior. *Science* 211:1062–1064.

Boltnev, A. I. 1991. Information on the postnatal development of the northern fur seal. In USSR Ministry of Fisheries, Kamchatka Division of KOTINRO, *Research into Biology and Population Dynamics of commercial Fish Species of the Kamchatka Shelf.* Collected Research Papers, Edition 1, Part 2, 137–149. Petropavlovsk-Kamchatskii, 1991.

Boltnev, A. I., A. E. York, and G. A. Antonelis. 1997. Growth of northern fur seal (*Callorhinus ursinus*) neonates at Bering Island. Manuscript available from NOAA, National Marine Mammal Laboratory, Seattle. (in prep).

Boness, D. J. 1991. Determinants of mating systems in the Otariidae (Pinniedia). In D. Renouf, ed., *The Behaviour of Pinnipeds*, 1–44. London: Chapman and Hall.

Boness, D. J., and P. Majluf. (In prep.) Abstract for the International Symposium and Workshop on Otariid Reproductive Strategies and Conservation, Smithsonian Institution, Washington, D.C., April 1996.

Bowen, W. D., O. T. Oftedal, and D. J. Boness. 1985. Birth to weaning in 4 days: Remarkable growth in the hooded seal, *Cystophora cristata. Can. J. Zool.* 63:2841–2846.

Bowen, W. D, O. T. Oftedal, and D. J. Boness. 1992. Mass and energy transfer during lactation in a small phocid, the harbor seal (*Phoca vitulina*). *Physiol. Zool.* 65(4):844–866.

Boyd, I. L. 1993. Pup production and distribution of breeding antarctic fur seals (*Arctocephalus gazella*) at South Georgia. *Antarctic Sci.* 5(1):17–24.

Boyd, I. L. 1996. Individual variation in the duration of pregnancy and birth date in Antarctic fur seals: The role of environment, age, and sex of fetus. *J. Mamm.* 77(1):124–133.

Boyd, I.L., and J.P. Croxall. 1992. Diving behaviour of lactating female antarctic fur seals. *Can. J. Zool.* 70: 919–928.

Boyd, I. L., and T. S. McCann. 1989. Pre-natal investment in reproduction by female Antarctic fur seals. *Behav. Ecol. Sociobiol.* 24:377–385.

Boyd, I. L., J.P.Y. Arnould, T. Barton, and J. P. Croxall. 1994. Foraging behaviour of Antarctic fur seals during periods of contrasting prey abundance. *J. Anim. Ecol.* 63 N3:703–713.

Bradbury, J. W. 1981. The evolution of leks. In R. D. Alexander and D. Tinkle, eds., *Natural Selection and Social Behavior*, 138–169. New York: Chiron Press.

Braddock, J. C. 1949. The effect of prior residence upon dominance in the fish *Platypoecilus maculatus. Physiol. Zool.* 22:161–169.

Brodeur, R. D., and D. M. Ware. 1992. Long-term variability in zooplankton biomass in the subarctic Pacific Ocean. *Fish. Oceanogr.* 1:32–38.

Bryant, C. 1870–71. On the habits of the Northern fur seal (*Callorhinus ursinus Gray*), with a description of the Pribyloff group of islands. *Bull. Mus. Comp. Zool.* 2(1):89–108.

Buckley, B. M., R. D. D'Arrigo, and G. C. Jacoby. 1992. Tree-ring records as indicators of air-sea interaction in the Northeast Pacific sector. In K. T. Redmond, ed., *Proceedings of the Eighth Annual Pacific Climate (PACLIM) Workshop*, Asilomar, Calif., March 1991.

Budd, G. M., and M. C. Downes. 1969. Population increase and breeding in the Kerguelen fur seal, *Arctocephalus gazella*, at Heard Island. *Mammalia* 33(1):58–67.

Burley, R. A. 1980. Precopulatory and copulatory behavior in relation to states of the oestrus cycle in the female Mongolian gerbil. *Behaviour* 72:211–241.

Busch, B. C. 1985. *The War against the Seals*. Montreal: McGill-Queen's University Press.

Bychkov, V.A., and S. V. Dorofeev. 1962. On the behavior of fur seal bulls during the harem period of their life. *Zool. Zh.* 41(9):1433–1435. In Russian. English translation available from NOAA, National Marine Mammal Laboratory, Seattle.

Cahill, C. F. Jr., E. B. Marliss, and T. T. Aoki. 1970. Fat and nitrogen metabolism in fasting man. In *Adipose Tissue: Regulation and Function*, 181–85. New York: Academic Press.

Calambokidis, J., and R. L. Gentry. 1985. Mortality of northern fur seal pups in relation to growth and birth weights. *J. Wildl. Dis.* 21(3):327–330.

Caldwell, J. D., G. F. Jirikowski, E. R. Greer, and C. A. Pederson. 1989. Medial preoptic area oxytocin and female sexual receptivity. *Behav. Neur.* 103(3):655–662.

Campagña, C., B. J. Le Boeuf, and H. L. Cappozzo. 1988. Group raids: A mating strategy of male southern sea lions. *Behaviour* 105(3–4):224–249.

Cane, M. A. 1983. Oceanographic events during El Niño. *Science* 222:1189–1195.

Carter, C. S. 1973. Stimuli contributing to the decrement in sexual receptivity of female golden hamsters (*Mesocricetus auratus*). *Anim. Behav.* 21:827–834.

Carter, C. S., D. M. Witt, S. R. Manoch, K. A. Adams, J. M. Bahr, and K. Carlstead. 1989. Hormonal correlates of sexual behavior and ovulation in male-induced and postpartum estrus in female prairie voles. *Physiol. Behav.* 46(6):941–948.

Carter, L. D., J. Brigham-Grette, L. Marincovich, Jr., V. L. Pease, and J. W. Hillhouse. 1986. Late cenozoic Arctic Ocean sea ice and terrestrial paleoclimate. *Geology* 14:675–678.

Chapman, D. G. 1961. Population dynamics of the Alaska fur seal herd. *Trans. 26th N. Amer. Wildl. Conf.*, 356–369.

Chapman, D. G. 1964. A critical study of Pribilof fur seal population estimates. *U.S. Fish Wildl. Serv. Fish. Bull.* 63:657–669.

Chelnokov, F. G. 1982. Homing and distribution of fur seals in the harem territories of Southeast rookery, Medny Island. In The study, preservation and rational exploitation of marine mammals, *Rep. Eighth All-Union Conf. Study Mar. Mamm*, 401–403. In Russian. English translation available from NOAA, National Marine Mammal Laboratory, Seattle.

Clutton-Brock, T. H. 1988. Reproductive success. In T. H. Clutton-Brock, ed., *Reproductive Success: Studies of Individual Variation in Contrasting Breeding Systems*, 472–485. Chicago: University of Chicago Press.

Clutton-Brock, T. H. 1989a. Mammalian mating systems. *Proc. R. Soc. Lond. B* 236:339–372.

Clutton-Brock, T. H. 1989b. Female transfer and inbreeding avoidance in social mammals. *Nature* 337:70–72.

Clutton-Brock, T., and K. McComb. 1993. Experimental tests of copying and mate choice in fallow deer (*Dama dama*). *Behav. Ecol.* 4(3):191–193.

Clutton-Brock, T. H., S. D. Albon, and F. E. Guinness. 1981. Parental investment in male and female offspring in polygynous mammals. *Nature* 289:487–489.

Conaway, C. H. 1971. Ecological adaptation and mammalian reproduction. *Biol. Repro.* 1:239–247.

Costa, D. P. 1991a. Reproductive and foraging energetics of pinnipeds: Implications for life history patterns. In D. Renouf, ed., *Behavioural Biology of the Pinnipeds*, 300–344. London: Chapman and Hall.

Costa, D. P. 1991b. Reproductive and foraging energetics of high latitude penguins, albatrosses, and pinnipeds: Implications for life history patterns. *Amer. Zool.* 31:111–130.

Costa, D. P. 1993. The relationship between reproductive and foraging energetics and the evolution of the Pinnipedia. In I. L. Boyd, ed., Marine mammals: Advances in behavioural and population biology. *Symp. Zool. Soc. Lond.* 66:293–314.

Costa, D. P., and R. L. Gentry. 1986. Free-ranging energetics of northern fur seals. In R. L. Gentry and G. L. Kooyman, eds., *Fur Seals: Maternal Strategies on Land and at Sea*, 79–101. Princeton, N.J.: Princeton University Press.

Costa, D. P., F. Trillmich, and J. P. Croxall. 1988. Intraspecific allometry of neonatal size in the Antarctic fur seal (*Arctocephalus gazella*). *Behav. Ecol. Sociobiol.* 22:361–364.

Costa, D. P., J. P. Croxall, and C. D. Duck. 1989. Foraging energetics of antarctic fur seals in relation to changes in prey availability. *Ecology* 70(3):596–606.

Coulson, J. C., and G. Hickling. 1961. Variations in the secondary sex-ratio of the grey seal *Halichoerus grypus* (Fab.) during the breeding season. *Nature* 190:281.

Cox, A., D. M. Hopkins, and G. B. Dalrymple. 1966. Geomagnetic polarity epochs: Pribilof Islands, Alaska. *Geol. Soc. Amer. Bull.* 77:883–910.

Cox, C. R., and B. J. Le Boeuf. 1977. Female incitation of male competition: A mechanism in sexual selection. *Amer. Nat.* 11:317–335.

Craig, A. M. 1964. Histology of reproduction and the estrus cycle in the female fur seal, *Callorhinus ursinus*. *J. Fish. Bd. Canada* 21(4):773–811.

Crawley, M. C. 1975. Growth of New Zealand fur seal pups. *N.Z. J. Mar. Freshw. Res.* 9(4):539–545.

Crawley, M. C., and D. B. Cameron. 1972. New Zealand sea lions, *Phocarctos hookeri*, on the Snares Islands. *N.Z. J. Mar. Freshw. Res.* 6:127–132.

Crocker, D. E., B. J. Le Boeuf, Y. Naito, T. Asaga, and D. P. Costa. 1994. Swim speed and dive function in a female northern elephant seal. In B. J. Le Boeuf and R. M. Laws, eds., *Elephant Seals: Population Ecology, Behavior, and Physiology*, 328–329. Berkeley: University of California Press.

Croxall, J. P., and R. L. Gentry, eds. 1987. *Status, Biology, and Ecology of Fur Seals*. Proceedings of an international symposium and workshop, Cambridge, England, April 1984. NOAA Tech. Rep. NMFS 51.

Croxall, J. P., I. Everson, G. L. Kooyman, C. Ricketts, and R. W. Davis. 1985. Fur seal diving behavior in relation to vertical distribution of krill. *J. Anim. Ecol.* 54:1–8.

Daniel, J. C., Jr. 1974. Circulating levels of estradiol 17-beta during early pregnancy in the Alaska fur seal showing an estrogen surge preceding implantation. *J. Reprod. Fert.* 37:425–428.

Daniel, J. C., Jr. 1975. Concentrations of circulating progesterone during early pregnancy in the northern fur seal, *Callorhinus ursinus*. *J. Fish. Res. Bd. Can.* 32:65–66.

Daniel, J. C., Jr. 1981. Delayed implantation in the northern fur seals (*Callorhinus ursinus*) and other pinnipeds. *J. Reprod. Fert.* suppl. 29:35–50.

Daniel, J. C., Jr., and R. S. Krishnan. 1969. Studies on the relationship between uterine fluid components and the diapausing state of blastocycts from mammals having delayed implantation. *J. Exp. Zool.* 172:267–281.

D'Arrigo, R. D., and G. C. Jacoby. 1991. A 1000-year record of winter precipitation from northwestern New Mexico, USA: A reconstruction from tree-rings and its relation to El Niño and the southern oscillation. *Holocene* 1,2:95–101.

David, J.H.M. 1987a. South African fur seal, *Arctocephalus pusillus pusillus*. In J. P. Croxall and R. L. Gentry, eds., *Status, Biology, and Ecology of Fur Seals*. NOAA Tech. Rep. NMFS 51:65–77.

David, J.H.M. 1987b. Diet of the South African fur seal (1974–1985) and an assessment of competition with fisheries in Southern Africa. *S. Afr. Mar. Sci.* 5:693–713.

Davies, J. L. 1958. Pleistocene geography and the distribution of northern pinnipeds. *Ecology* 39(1):96–113.

Decker, M. B., G. L. Hunt, Jr., and G. V. Byrd, Jr. 1995. Relations among sea surface temperature, the abundance of juvenile walleye pollock (*Theragra calcogramma*), and the reproductive performance and diets of sea birds at the Pribilof Islands, S.E. Bering Sea. In R. J. Beamish, ed., *Climate Change and Northern Fish Populations. Can. Spec. Publ. Aquat. Sci.* 121:425–437.

DeLong, R. L. 1982. Population biology of northern fur seals at San Miguel Island, California. Ph.D. diss., University of California, Berkeley.

DeLong, R. L. 1990. Population and behavioral studies, San Miguel Island, California. In H. Kajimura, ed., *Fur Seal Investigations, 1989.* NOAA Tech. Mem. NMFS F/NWC-190. NOAA, National Marine Mammal Laboratory, Seattle.

DeLong, R. L., and G. A. Antonelis. 1985. 3. Population growth and behavior, San Miguel Island (Adams Cove and Castle Rock). In *Fur Seal Investigations, 1982.* NOAA Tech. Mem. NMFS-F/NWC-71:54–58, NOAA, National Marine Mammal Laboratory, Seattle.

DeLong, R. L., and G. A. Antonelis. 1991. Impact of the 1982–1983 El Niño on the northern fur seal population at San Miguel Island, California. In F. Trillmich and K. A. Ono, eds., *Pinnipeds and El Niño*, 74–83. Berlin: Springer-Verlag.

DeLong, R. L., B. S. Stewart, and R. D. Hill. 1992. Documenting migrations of northern elephant seals using day length. *Mar. Mamm. Sci.* 8(2):155–159.

Deutsch, J. C. and R.J.C. Nefdt. 1992. Olfactory cues influence female choice in two lek-breeding antelopes. *Nature* 356:596–598.

de Vries, J., and M. R. van Eerden. 1995. Thermal conductance in aquatic birds in relation to the degree of water contact, body mass, and body fat: Energetic implications of living in a strong cooling environment. *Physiol. Zool.* 68(6):1143–1163.

Dewsbury, D. A. 1972. Patterns of copulatory behavior in male animals. *Quart. Rev. Biol.* 47:1–33.

Diamond, M. 1970. Intromission pattern and species vaginal code in relation to induction of pseudopregnancy. *Science* 169:995–997.

Doidge, D. W. 1987. Rearing of twin offspring to weaning in *Arctocephalus gazella*. In J. P. Croxall and R. L. Gentry, eds., *Status, Biology, and Ecology of Fur Seals.* NOAA Tech. Rep. NMFS 51:107–111.

Doidge, D. W., and J. P. Croxall. 1989. Factors affecting weaning weight in antarctic fur seals. *Polar Biol.* 9:155–160.

Doidge, D. W., J. P. Croxall, and C. Ricketts. 1984. Growth rates of Antarctic fur seal, *Arcotcephalus gazella*, pups at South Georgia. *J. Zool. Lond.* 203:87–93.

Doidge, D. W., T. S. McCann, and J. P. Croxall. 1986. Attendance behavior of Antarctic fur seals. In R. L. Gentry and G. L. Kooyman, eds., *Fur Seals: Maternal Strategies on Land and at Sea*, 102–114. Princeton, N.J.: Princeton University Press.

Duck, C. D. 1990. Annual variation in the timing of reproduction in Antarctic fur seals, *Arctocephalus gazella*, at Bird Island, South Georgia. *J. Zool. Lond.* 222:103–116.

Dunbar, R.I.M. 1988. *Primate Social Systems*. Ithaca, N.Y.: Comstock Associates, Cornell University Press.

Elliott, H. W. 1882. A monograph of the seal-islands of Alaska. *U.S. Comm. Fish Fish. Spec. Bull.* Reprinted from *Report on Fishery Industry* (10th Census) 176:1–176. Washington, D.C.: U.S. Government Printing Office.

Emlen, S. T., and L. W. Oring. 1977. Ecology, sexual selection, and the evolution of mating systems. *Science* 197:215–223.

Enders, R. K., O. P. Pearson, and A. K. Pearson. 1946. Certain aspects of reproduction in the fur seal. *Anat. Rec.* 94:213–227.

Engle, R. M., R. H. Lander, A. Y. Roppel, P. Kozloff, J. R. Hartley, and M. C. Keyes. 1980. Population data, collection procedures, and management of the northern fur seal, *Callorhinus ursinus*, of the Pribilof Islands, Alaska. NWAFC Proc. Rep. 80–11, Northwest and Alaska Fisheries Center, Seattle.

Erskine, M. S., E. Kornberg, and J. A. Cherry. 1989. Paced copulation in rats: Effects of intromission frequency and duration on luteal activation and estrus length. *Phys. Behav.* 45:33–39.

Fahrbach, E., F. Trillmich, and W. Arntz. 1991. 1. The time sequence and magnitude of physical effects of El Niño in the Eastern Pacific. In F. Trillmich and K. A. Ono, eds., *Pinnipeds and El Niño: Responses to Environmental Stress*, 8–21. Berlin: Springer-Verlag.

Feldkamp, S. D., R. L. DeLong, and G. A. Antonelis. 1989. Diving patterns of California sea lions, *Zalophus californianus*. *Can. J. Zool.* 67:872–883.

Figueroa Carranza, A. L. 1994. Early lactation and attendance behavior of the Guadalupe fur seal females (*Arctocephalus townsendi*). M.S. thesis, University of California, Santa Cruz.

Fisher, R. A. 1930. *The Genetical Theory of Natural Selection*. Oxford: Clarendon Press.

Fowler, C. W. 1982. Interactions of northern fur seals and commercial fisheries. *Trans. 47th N. Amer. Wildl. Nat. Res. Conf.*, 278–292.

Fowler, C. W. 1987. Marine debris and northern fur seals: A case study. *Mar. Pollut. Bull.* 18(6b):326–335.

Fowler, C. W. 1990. Density dependence in northern fur seals (*Callorhinus ursinus*). *Mar. Mamm. Sci.* 6(3):171–195.

Fowler, C. W. 1997. Northern fur seals of the Pribilof Islands. In *The Northern Fur Seal: Species of the USSR and Contiguous Countries*. Nauka, Moscow: Russian Academy of Sciences (in press).

Fowler, C. W., G. A. Antonelis, and J. D. Baker. 1993. Studies of juvenile males tagged as pups and resighted during roundups in 1992. In E. H. Sinclair, ed., *Fur Seal Investigations, 1992*, 48–70 NOAA Tech. Mem. NMFS-AFSC-45.

Francis, J. M. 1987. Interfemale aggression and spacing in the northern fur seal *Callorhinus ursinus* and the California sea lion *Zalophus californianus*. Ph.D. diss., University of California, Santa Cruz.

Francis, J. M., and D. J. Boness. 1991. The effect of thermoregulatory behavior on the mating system of the Juan Fernandéz fur seal, *Arctocephalus philipi*. *Behaviour* 119(1–2):114–126.

Francis, J., D. Boness, and H. Ochoa. 1997. A protracted foraging and attendance cycle in female Juan Fernández fur seals. *Mar. Mamm. Sci.* (accepted Feb. 1997).

Fritz, L. W., R. C. Ferrero, and R. J. Berg. 1995. The threatened status of Steller sea lions, *Eumetopias jubatus*, under the Endangered Species Act: Effects on Alaska groundfish fisheries management. *Mar. Fish. Rev.* 57(2)14–27.

Frost, O. W., and M. A. Engel. 1988. *Georg Wilhelm Steller: Journal of a Voyage with Bering, 1741–1742.* Stanford, Calif.: Stanford University Press.

Gales, N. J., A. J. Cheal, G. J. Pobar, and P. Williamson. 1992. Breeding biology and movements of Australian sea-lions, *Neophoca cinerea*, off the west coast of Western Australia. *Wildl. Res.* 19(4):405–416.

Gales, N. J., P. D. Shaughnessy, and T. E. Dennis. 1994. Distribution, abundance and breeding cycle of the Australian sea lion *Neophoca cinerea* (Mammalia: Pinnipedia). *J. Zool. Lond.* 234:353–370.

Gallo, J. P. 1994. Factors affecting the population of Guadalupe fur seal, *Arctrocephalus townsendi* (Merriam 1897), at Isla de Guadalupe, Baja California, Mexico. Ph.D. diss., University of California, Santa Cruz.

Gemmell, J. J. 1996. Molecular ecology of fur seals: Fuzzy bands from fuzzy beasts. Abstract for the International Symposium and Workshop on Otariid Reproductive Strategies and Conservation, Smithsonian Institution, Washington, D.C., April 1996.

Gentry, R. L. 1970. Social behavior of the Steller sea lion. Ph.D. diss., University of California, Santa Cruz.

Gentry, R. L. 1973. Thermoregulatory behavior of eared seals. *Behaviour* 46:73–93.

Gentry, R. L. 1974. The development of social behavior through play in the Steller sea lion. *Amer. Zool.* 14:391–403.

Gentry, R. L. 1975. Comparative social behavior of eared seals. *Rapp. P.-v. Réun. Cons. Int. Explor. Mer* 169:188–194.

Gentry, R. L. 1981a. Seawater drinking in eared seals. *Comp. Biochem. Physiol.* 68A:81–86.

Gentry, R. L. 1981b. Land-sea movements of northern fur seals relative to commercial harvesting. In J. A. Chapman and D. Pursley, eds., *Worldwide Furbearer Conference Proceedings*, vol. 2, 1328–1359, August 1980, Frostburg, Md.

Gentry, R. L. 1991. El Niño effects on adult northern fur seals at the Pribilof Islands. In F. Trillmich and K. Ono, eds., *Pinnipeds and El Niño*, 84–93. Berlin: Springer-Verlag.

Gentry, R. L., and C. A. Goebel-Diaz. 1987–88. Behavior and biology of northern fur seals, Pribilof Islands, Alaska, 1987–88. In *Fur Seal Investigations, 1987 and 1988*, 56–61. NOAA Tech. Mem. NMFS F/NWC-180. Available from NOAA, National Marine Mammal Laboratory, Seattle.

Gentry R. L., and J. R. Holt. 1982. *Equipment and Techniques for Handling Northern Fur Seals.* NOAA Tech. Rep. NMFS SSRF-758, Alaska Fisheries Research Center, Seattle.

Gentry, R. L., and J. R. Holt. 1986. Attendance behavior of northern fur seals. In R. L. Gentry and G. L. Kooyman, eds., *Fur Seals: Maternal Strategies on Land and at Sea*, 41–60. Princeton, N.J.: Princeton University Press.

Gentry, R. L., and J. H. Johnson. 1975. Part 3. Behavior—St. George Island. In Marine Mammal Division, *Fur Seal Investigations, 1974*, 25–34. Alaska Fisheries Research Center, Seattle.

Gentry, R. L., and J. H. Johnson. 1978. Physical restraint for immobilizing fur seals. *J. Wildl. Manag.* 42:944–946.

Gentry, R. L., and J. H. Johnson. 1981. Predation by sea lions on northern fur seal neonates. *Mammalia* 45(4):423–430.

Gentry, R. L., and G. L. Kooyman. 1986a. Methods of dive analysis. In R. L. Gentry and G. L. Kooyman, eds., *Fur Seals: Maternal Strategies on Land and at Sea*, 28–40. Princeton, N.J.: Princeton University Press.

Gentry, R. L., and G. L. Kooyman, eds. 1986b. *Fur Seals: Maternal Strategies on Land and at Sea*. Princeton, N.J.: Princeton University Press.

Gentry, R. L., and W. E. Roberts. In prep. Mating system of the Hooker's sea lion, *Phocarctos hookeri*.

Gentry, R. L., D. P. Costa, J. P. Croxall, J.H.M. David, R. W. Davis, G. L. Kooyman, P. Majluf, T. S. McCann, and F. Trillmich. 1986a. Synthesis and Conclusions. In R. L. Gentry and G. L. Kooyman, eds., *Fur Seals: Maternal Strategies on Land and at Sea*, 220–264. Princeton, N.J.: Princeton University Press.

Gentry, R. L., G. L. Kooyman, and M. E. Goebel. 1986b. Feeding and diving behavior of northern fur seals. In R. L. Gentry and G. L. Kooyman, eds., *Fur Seals: Maternal Strategies on Land and at Sea*, 61–78. Princeton, N.J.: Princeton University Press.

Gentry, R. L., M. E. Goebel, and W. E. Roberts. 1986c. Behavior and biology, Pribilof Islands, Alaska. In P. Kozloff, ed., *Fur Seal Investigations, 1984*. NOAA Tech. Mem. NMFS F/NWC-97:29–40, Alaska Fisheries Science Center, Seattle.

Gisiner, R. C. 1985. Male territorial and reproductive behavior in the Steller sea lion, *Eumetopias jubatus*. Ph.D. diss., University of California, Santa Cruz.

Gloersen, P. 1995. Modulation of hemispheric sea-ice cover by ENSO events. *Nature* 373:503–506.

Goebel, M. E. 1988. Duration of feeding trips and age-related reproductive success of lactating females, St. Paul Island, Alaska. In P. Kozloff and H. Kajimura, eds. *Fur Seal Investigations, 1985*. NOAA Tech. Mem. NMFS F/NWC:28–33, Alaska Fisheries Science Center, Seattle.

Goebel, M. E., J. L. Bengtson, R. L. DeLong, R. L. Gentry, and T. R. Loughlin. 1991. Diving patterns and foraging locations of female northern fur seals. *Fish. Bull.* 89:171–179.

Goldfoot, D. A., and R. W. Goy. 1970. Abbreviation of behavioral estrus in guinea pigs by coital and vagino-cervical stimulation. *J. Comp. Physiol. Psychol.* 72:426–434.

Goldsworthy, S. D. 1992. Maternal care in three species of southern fur seal (*Arctocephalus spp.*). Ph.D. diss. Monash University, Australia.

Graham, N. E. 1995. Simulation of recent global temperature trends. *Science* 267:666–671.

Griben, M. R. 1979. A study of the intermixture of subadult male fur seals, *Callorhinus ursinus* (Linnaeus 1758) between the Pribilof Islands of St. George and St. Paul, Alaska. M.A. thesis, University of Washington, Seattle.

Gustafson, C. E. 1968. Prehistoric use of fur seals: Evidence from the Olympic coast of Washington. *Science* 161:49–51.

Hamilton, J. E. 1934. The southern sea lion, *Otaria byronia* (de Blainville). *Disc. Rep.* 8:269–318.

Hamilton, J. E. 1939. A second report on the southern sea lion, *Otaria byronia* (de Blainville). *Disc. Rep.* 19:121–164.

Hamilton, J. E. 1956. Scent of otariids. *Nature* 177:900.

Harcourt, R. 1992. Factors affecting early mortality in the South American fur seal (*Arctocephalus australis*) in Peru: Density-related effects and predation. *J. Zool. Lond.* 226:259–270.

Harcourt, R. 1993. Individual variation in predation on fur seals by sea lions (*Otaris byronia*) in Peru. *Can. J. Zool.* 71:1908–1911.

Harrison, R. J. 1963. A comparison of factors involved in delayed implantation in badgers and seals in Great Britain. In A. C. Enders, ed., *Delayed Implantation*, 440–482. Chicago: University of Chicago Press.

Harrison, R. J. 1969. Reproduction and reproductive organs. In H. T. Andersen, ed., *The Biology of Marine Mammals*, 253–348. New York: Academic Press.

Heath, C. B. 1990. The behavioral ecology of the California sea lion, *Zalophus californianus*. Ph.D. diss., University of California, Santa Cruz.

Higgins, L. V., and L. Grass. 1993. Birth to weaning: Parturition, duration of lactation, and attendance cycles of Australian sea lions (*Neophoca cinerea*). *Can. J. Zool.* 71:2047–2055.

Hill, R. D. 1994. Theory of geolocation by light levels. In B. J. Le Boeuf and R. M. Laws, eds., *Elephant Seals: Population Ecology, Behavior, and Physiology*, 227–236. Berkeley: University of California Press.

Hinde, R. A. 1976. Interactions, relationships and social structure. *Man* 11:1–17.

Hoover, A. A. 1988a. Harbor seal. In J. Lentfer, ed., *Selected Marine Mammals of Alaska*, 125–157. Marine Mammal Commission, 1988, no. PB88–178462. Springfield, Va.: National Technical Information Service.

Hoover, A. A. 1988b. Steller sea lion. In J. Lentfer, ed., *Selected Marine Mammals of Alaska*, 159–193. Marine Mammal Commission, 1988, no. PB88–178462. Springfield, Va.: National Technical Information Service.

Hopkins, D. M. 1973. Sea level history in Beringia during the past 250,000 years. *Quat. Res.* 3:520–540.

Hopkins, D. M., and T. Einarsson. 1966. Pleistocene glaciation on St. George, Pribilof Island. *Science* 152(3720):343–345.

Horning, M. 1996. Development of diving in Galapagos fur seals. Abstract for the International Symposium and Workshop on Otariid Reproductive Strategies and Conservation, Smithsonian Institution, Washington, D.C., April 1996.

Huck, U. W., R. D. Lisk, S. Kim, and A. B. Evans. 1989. Olfactory discrimination of estrous condition by the male golden hamster (*Mesocrecetus auratus*). *Behav. Neur. Biol.* 51:1–10.

Huelsbeck, D. R. 1983. Mammals and fish in the subsistence economy of Ozette. Ph.D. diss., Washington State University, Pullman.

Ichihara, T., and K. Yoshida. 1972. Diving depth of northern fur seals in the feeding time. *Whales Res. Inst. Sci. Rep.* 24:145–148.

Ignell, S. 1990. Diagnostic indices of oceanographic and atmospheric conditions in the central north Pacific Ocean. INPFC document, Auke Bay Laboratory, Alaska Fisheries Science Center, Auke Bay, Alaska.

Insley, S. J. 1992. Mother-offspring separation and acoustic stereotypy: A comparison of call morphology in two species of pinnipeds. *Behaviour* 120(1–2):103–122.

Irving, S. M., R. W. Goy, and G. A. Davis. 1980. Copulation-induced abbreviation of estrus in the female guinea pig: Block by clonidine. *Pharmacol. Biochem. Behav.* 12:755–759.

Iverson, S. J. 1993. Milk secretion in marine mammals in relation to foraging: Can milk fatty acids predict diet? *Symp. Zool. Soc. Lond.* 66:263–291.

Jacobs, G. A., H. E. Hurlburt, J. C. Kindle, E. J. Metzger, J. L. Mitchell, W. J. Teague, and A. J. Wallcraft. 1994. Decade-scale trans-Pacific propagation and warming effects of an El Niño anomaly. *Nature* 370:360–363.

Jameson, G. S. 1971. The functional significance of corneal distortion in marine mammals. *Can. J. Zool.* 49:421–423.

Johnson, A. M. 1968. Annual mortality of territorial male fur seals and its management significance. *J Wildl. Manag.* 32(1):94–99.

Jordan, D. S. 1898. *Fur Seals and Fur-seal Islands of the North Pacific Ocean.* Part 4. Washington, D.C.: U.S. Government Printing Office.

Jouventin, P., and A. Cornet. 1980. The sociobiology of pinnipeds. In J. S. Rosenblatt, R. A. Hinde, C. Beer, and M.-C. Busnel, eds., *Advances in the Study of Behaviour*, 11:121–141. New York: Academic Press.

Kajimura, H. 1980. Distribution and migration of northern fur seals (*Callorhinus ursinus*) in the eastern Pacific. In H. Kajimura, R. A. Lander, M. A. Perez, and A. E. York, eds., Further analysis of pelagic fur seal data collected by the United States and Canada during 1958–74, 4–19. Unpublished report, NOAA, National Marine Mammal Laboratory, Seattle.

Kajimura, H. 1984. *Opportunistic Feeding of the Northern Fur Seal*, Callorhinus ursinus, *in the Eastern North Pacific Ocean and Eastern Bering Sea.* NOAA Tech. Rep. NMFS SSRF-779. Alaska Fisheries Research Center, Seattle.

Kajimura, H. 1990. *Fur Seal Investigations, 1989.* NOAA Tech. Rep. NMFS F/NWC-190, Alaska Fisheries Research Center, Seattle.

Kajimura, H., and T. R. Loughlin. 1988. Marine mammals in the oceanic food web of the eastern subarctic Pacific. *Bull. Ocean Res. Inst. Univ. Tokyo* 26(I):187–223.

Kajimura, H., R. H. Lander, M. A. Perez, and A. E. York. 1979. Preliminary analysis of pelagic fur seal data collected by the United States and Canada during 1958–74. Unpublished report, submitted to 23rd annual meeting of the North Pacific Fur Seal Commission, April 1980, Moscow. NOAA, National Marine Mammal Laboratory, Seattle.

Kellogg, R. 1925. New pinnipeds from the Miocene diatomaceous earth near Lompoc, California. Part 4 *in* Additions to the tertiary history of the pelagic mammals of the Pacific coast of North America. *Carnegie Inst. Wash. Contr. Palaeontol.* 348:71–96.

Kennett, J. P. 1982. *Marine Geology.* Englewood Cliffs, N.J.: Prentice Hall.

Kenyon, K. W. 1960. Territorial behavior and homing in the Alaska fur seal. *Mammalia* 24(2):431–444.

Kenyon, K. W., and F. Wilke. 1953. Migration of the northern fur seal, *Callorhinus ursinus. J. Mamm.* 34(1):86–98.

Kerley, G.I.H. 1985. Pup growth in the fur seals *Arctocephalus tropicalis* and *A. gazella* on Marion Island. *J. Zool. Lond.* 205:315–324.

Keyes, M. C. 1965. Pathology of the northern fur seal. *J. Am. Vet. Med. Assoc.* 147:1090–1095.

Keyes, M. C., A. W. Smith, R. J. Brown, N. A. Vedros, E. J. Izquierdo, and M. A. Stedham. 1975. Management considerations. In *Fur Seal Investigations, 1974*, 66–67. NOAA, National Marine Mammal Laboratory, Seattle.

Khlebnikova, K. T. 1830–31. Russkaya Amerika Neopublikovannykh Zapiskakh. Unpublished notes. In V. A. Aleksandrova, ed., Izdatel'stvo "Nauka," Leningrad, 1979. In Russian. English translation by A. Y. Roppel. Available from NOAA, National Marine Mammal Laboratory, Seattle.

Kogan, V. M. 1970. Resul'taty iskus stvennogo rasshireniya garemnogo lezhbishcha na O. Tyulen'em (Results of the artificial spread of the harem rookery on Robben Island). Izvestiya Tikhookeanskogo Hauchno-issledovatel'skogo Instituta Rybnogo Khozyaistva i Okeanografii, T. 70:235–237. In Russian. English translation available from NOAA, National Marine Mammal Laboratory, Seattle.

Kooyman, G. L. 1972. Deep diving behaviour and effects of pressure in reptiles, birds, and mammals. *In* Society for Experimental Biology Symposium (London) 26, *The Effects of Pressure on Organisms*, vol. 26, 295–311.

Kooyman, G. L. 1988. Pressure and the diver. *Can. J. Zool.* 66:84–88.

Kooyman, G. L. 1989. *Diverse Divers*. Berlin: Springer-Verlag.

Kooyman, G. L., and R. L. Gentry. 1986. Diving behavior of South African fur seals. In R. L. Gentry and G. L. Kooyman, eds., *Fur Seals: Maternal Strategies on Land and at Sea*, 142–152. Princeton, N.J.: Princeton University Press.

Kooyman, G. L., and F. Trillmich. 1986. Diving behavior of Galapagos fur seals. In R. L. Gentry and G. L. Kooyman, eds., *Fur Seals: Maternal Strategies on Land and at Sea*, 168–185. Princeton, N.J.: Princeton University Press.

Kooyman, G. L., R. L. Gentry, and D. L. Urquhart. 1976. Northern fur seal diving behavior: A new approach to its study. *Science* 193:411–412.

Kooyman, G. L., J. O. Billups, and W. D. Farwell. 1983a. Two recently developed recorders for monitoring diving activity of marine birds and mammals. In A. G. Macdonald and I. G. Priede, eds., *Experimental Biology at Sea*, 197–214. New York: Academic Press.

Kooyman, G. L., M. A. Castellini, R. W. Davis, and R. A. Maue. 1983b. Aerobic dive limits in immature Weddell seals. *J. Comp. Physiol.* 151:171–174.

Kow, L. M., and D. W. Pfaff. 1973–74. Effects of estrogen treatment on the size of the receptive field and response threshold of pudendal nerve in the female rat. *Neuroendocrinology* 13:299–313.

Kuzin, A. E., M. K. Maminov, and A. S. Perlov. 1973. Population dynamics and prospects for growth in the fur-seal stock of the Kurile Islands. *Ekologiya* 4:63–67. In Russian. UDC translation number 591.524.

Landau, D., and W. W. Dawson. 1970. The histology of retinas from the Pinnipedia. *Vision Res.* 10:691–702.

Lander, R. H. 1979. Role of land and ocean mortality in yield of male Alaskan fur seal, *Callorhinus ursinus. Fish. Bull.* 77(1):311–314.

Lander, R. H. 1980a. *Summary of Northern Fur Seal Data and Collection Procedures*, vol. 1: *Land Data of the United States and Soviet Union (Excluding Tag and Recovery Records)*. NOAA Tech. Mem. NMFS F/NWC-3, Alaska Fisheries Research Center, Seattle.

Lander, R. H. 1980b. *Summary of Northern Fur Seal Data and Collection Procedures*, vol. 2: *Eastern Pacific Pelagic Data of the United States and Canada*

(Excluding Fur Seals Sighted). NOAA Tech. Mem. NMFS F/NWC-4, Alaska Fisheries Research Center, Seattle.

Lander, R. H. 1981–82. A life table and biomass estimate for Alaskan fur seals. *Fish. Res.* 1(1981/82):55–70.

Lander, R. H., and H. Kajimura. 1982. Status of northern fur seals. In, Mammals in the Seas. *FAO Fish. Ser.* 4(5):319–345.

Lavigne, D. M., and K. Ronald. 1975. Evidence of duplicity in the retina of the California sea lion (*Zalophus californianus*). *Comp. Biochem. Physiol.* 50A:65–70.

Laws, R. M. 1956. The elephant seal (*Mirounga leonina* Linn.). 3. The physiology of reproduction. *Falk. Isl. Depend. Surv. Sci. Rep.* 15:1–66.

Le Boeuf, B. J. 1967. Interindividual associations in dogs. *Behaviour* 29:268–295.

Le Boeuf, B. J. 1972. Sexual behavior in the northern elephant seal *Mirounga angustirostris*. *Behaviour* 41:1–26.

Lento, G. M., C. S. Baker, and D. Penny. 1996. Molecular systematic relationships among the otariids: A challenge to the traditional view. Abstract for the International Symposium and Workshop on Otariid Reproductive Strategies and Conservation, Smithsonian Institution, Washington, D.C., April 1996.

Lisitsyna, T. Yu. 1973. Povedenie i zvukovaya signalizatsiya severnogo morskogo kotika (*Callorhinus ursinus*) na lezhbishchaka (The behavior and acoustical signals of the northern fur seal [*Callorhinus ursinus*] on the rookery). *Zool. Zh.* 52(8):1220–1228. In Russian. Translation available from NOAA, National Marine Mammal Laboratory, Seattle.

Lluch-Belda, D., R.J.M. Crawford, T. Kawasaki, A. D. MacCall, R. H. Parrish, R. A. Schwartzlose, and P. E. Smith. 1989. Worldwide fluctuations of sardine and anchovy stocks: The regime problem. *S. Afr. J. Mar. Sci.* 8:195–205.

Lodder, J., and G. H. Zeilmaker. 1976. Role of pelvic nerves in the postcopulatory abbreviation of behavioral estrus in female rats. *J. Comp. Physiol. Psych.* 90:925–929.

Loughlin, T. R., and R. V. Miller. 1989. Growth of the northern fur seal colony on Bogoslof Island, Alaska. *Arctic* 42(4):368–372.

Loughlin, T. R., J. L. Bengtson, and R. L. Merrick. 1987. Characteristics of feeding trips of female northern fur seals. *Can. J. Zool.* 65:2079–2084.

Loughlin, T. R., G. A. Antonelis, B. W. Robson, and E. H. Sinclair. 1993. Summer and fall foraging characteristics of post-parturient female northern fur seals (*Callorhinus ursinus*) in the Bering Sea. Abstract for the Tenth Biennial Conference on the Biology of Marine Mammals, Galveston, Texas, November 1993.

Loughlin, T. R., G. A. Antonelis, J. D. Baker, A. E. York, C. W. Fowler, R. L. DeLong, and H. W. Braham. 1994. Status of the northern fur seal population in the United States during 1992. In E. H. Sinclair, ed., *Fur Seal; Investigations, 1992*. NOAA Tech. Mem. NMFS-AFSC-45:9–28. Available from NOAA, National Marine Mammal Laboratory, Seattle.

Lunn, N. J., and I. L. Boyd. 1991. Pupping-site fidelity of Antarctic fur seals at Bird Island, South Georgia. *J. Mammal.* 72(1):202–206.

Macy, S. K. 1982. Mother-pup interactions in the northern fur seal. Ph.D. diss. University of Washington, Seattle.

Mattlin, R. H. 1981. Pup growth of the New Zealand fur seal *Arctocephalus forsteri* on the Open Bay Islands, New Zealand. *J. Zool. Lond.* 193:305–314.

Mattlin, R. H. 1993. Seasonal diving behaviour of the New Zealand fur seal, *Arctocephalus forsteri*. Abstract for the Tenth Biennial Conference on the Biology of Marine Mammals, Galveston, Texas, November 1993.

Maynard Smith, J. 1980. A new theory of sexual investment. *Behav. Ecol. Sociobiol.* 7:247–251.

McCann, T. S. 1980. Territoriality and breeding behavior of adult male Antarctic fur seal, *Arctocephalus gazella*. *J. Zool. Lond.* 192:295–310.

McCann, T. S., and J. P. Croxall. 1986. Attendance behavior of antarctic fur seals. In R. L. Gentry and G. L. Kooyman, eds., *Fur Seals: Maternal Strategies on Land and at Sea*, 102–114. Princeton, N.J.: Princeton University Press.

McCann, T. S., and D. W. Doidge. 1987. Antarctic fur seals, *Arctocephalus gazella*. In J. P. Croxall and R. L. Gentry, eds., *Status, Biology, and Ecology of Fur Seals*. NOAA Tech. Rep. NMFS 51:5–8, British Antarctic Survey, Cambridge, England.

McLaren, I. A. 1967. Seals and group selection. *Ecology* 18(1):104–110.

Melin, S. R. 1995. Winter and spring attendance patterns of California sea lion (*Zalophus californianus*) females and pups at San Miguel Island, California, 1991–94. M.S. thesis, University of Washington, Seattle.

Merrick, R. 1990. Census of northern fur seals on Bogoslof Island. In H. Kajimura, ed., *Fur Seals: Investigations, 1989*. NOAA Tech. Mem. NMFS F/NWC-190:40, Alaska Fisheries Research Center, Seattle.

Merrick, R. L., T. R. Loughlin, and D. G. Calkins. 1987. Decline in the abundance of the northern sea lion (*Eumetopias jubatus*) in Alaska, 1956–86, *Fish. Bull.* (U.S.) 85:351–365.

Merrick, R. L., G. A. Antonelis, and I. L. Boyd. 1996. Population status of sea lions and fur seals worldwide—a review and suggested directions for future conservation related study. Abstract for the International Symposium and Workshop on Otariid Reproductive Strategies and Conservation, Smithsonian Institution, Washington, D.C., April 1996.

Millar, J. S. 1977. Adaptive features of mammalian reproduction. *Evolution* 31:370–386.

Miller, E. H. 1971. Social and thermoregulatory behaviour of the New Zealand fur seal. M.S. thesis, University of Canterbury, Christ Church, N.Z.

Miller, E. H. 1974. Social behavior between adult male and female New Zealand fur seals, *Arctocephalus forsteri* (Lesson) during the breeding season. *Aust. J. Zool.* 22:155–173.

Miller, E. H. 1975. Social and evolutionary implications of territoriality in adult male New Zealand fur seals, *Arctocephalus forsteri* (Lesson 1828), during the breeding season. *Rapp. P.-v. Réun. Cons. Int. Explor. Mer* 169:170–187.

Miller, L. K. 1978. Energetics of the northern fur seal in relation to climate and food resources of the Bering Sea. Unpublished report no. MMC-75-/08, U.S. Marine Mammal Commission, Washington, D.C.

Murphy, E. C., A. M. Springer, and D. G. Roseneau. 1986. Population status of common guillemots *Uria aalge* at a colony in western Alaska: Results and simulations. *Ibis* 128:348–363.

Murphy, H. D. 1970. Microscopic studies on the testis of the northern fur seal, *Callorhinus ursinus*. In T. C. Poulter, ed., *Proceedings of the Seventh Annual Conference on Biological Sonar and Diving Mammals*, 97–103, SRI International, Menlo Park, Calif., October 1970.

National Marine Fisheries Service. 1963–91. Fur seal investigations. Unpublished report series (one year per volume), NOAA, National Marine Mammal Laboratory, Seattle.

National Marine Fisheries Service. 1993. Conservation plan for the northern fur seal *Callorhinus ursinus*. Unpublished report, NOAA, Office of Protected Resources, National Marine Fisheries Service, Silver Spring, Md.

National Research Council. 1996. *The Bering Sea Ecosystem*. Washington, D.C.: National Academy Press.

Niebauer, H. J., and R. H. Day. 1989. Causes of interannual variability in the sea ice cover of the eastern Bering Sea. *Geojournal* 18(1):45–59.

NOAA. 1991. NOAA monthly summary of sea surface temperatures for August 1990. NOAA, Alaska Fisheries Science Center, Seattle.

Noble, R. G. 1979a. Limited coital stimulation facilitates sexual responses of the female hamster. *Physiol. Behav.* 23:1007–1010.

Noble, R. G. 1979b. The sexual responses of the female hamster. *Physiol. Behav.* 23:1001–1005.

North Pacific Fur Seal Commission. 1958–85. Proceedings of the first through 28th annual meetings, respectively. Unpublished reports. NOAA, National Marine Mammal Laboratory, Seattle.

North Pacific Fur Seal Commission. 1980. Report on investigations during 1973–76. Unpublished report, NOAA, National Marine Mammal Laboratory, Seattle.

North Pacific Fur Seal Commission. 1984. Report on investigations during 1977–80. Unpublished report, NOAA, National Marine Mammal Laboratory, Seattle.

Nutting, C. C. 1891. Some of the causes and results of polygamy among the pinnipedia. *Amer. Nat.* 25:103–112.

Odell, D. K. 1981. California sea lion *Zalophus californianus* (Lesson 1828). In S. H. Ridgway and R. J. Harrison, eds., *Handbook of Marine Mammals*, vol. 1, 67–97. New York: Academic Press.

Oftedal, O. T., D. J. Boness, and R. A. Tedman. 1987. The behavior, physiology and anatomy of lactation in the Pinnipedia. In H. H. Genoways, ed., *Current Mammalogy*, vol. 1, 175–245. New York: Plenum Press.

Oliver, J. R. 1913. The spermiogenesis of the Pribilof fur seal (*Callorhinus alascanus* J and C). *Amer. J. Anat.* 14:473–499.

Ono, K. A. 1991. Introductory remarks and the natural history of the California sea lion. In F. Trillmich and K. A. Ono, eds., *Pinnipeds and El Niño: Responses to Environmental Stress*, 109–111. Berlin: Springer-Verlag.

Ono, K. A., and D. J. Boness. 1996. Sexual dimorphism in sea lion pups: Differential maternal investment, or sex-specific differences in energy allocation? Abstract for the International Symposium and Workshop on Otariid Reproductive Strategies and Conservation, April 1996, Smithsonian Institution, Washington, D.C.

Orians, G. H., and N. E. Pearson. 1977. On the theory of central place foraging. In D. J. Horn, G. R. Stairs, and R. D. Mitchell, eds., *Analysis of Ecological Systems*, 153–177. Columbus: Ohio State University Press.

Osgood, W. H., E. A. Preble, and G. H. Parker. 1915. The fur seals and other life of the Pribilof Islands, Alaska, in 1914. *U.S. Bur. Fish. Bull.* 34:1–172.

Parkes, A. S. 1960. The role of odorous substances in mammalian reproduction. *J. Reprod. Fert.* 1:312–314.

Parsons, S. D., and G. L. Hunter. 1967. Effect of the ram on duration of oestrus in the ewe. *J. Reprod. Fert.* 14:61–70.

Paulian, P. 1964. Contribution a l'étude de l'otarie de l'île Amsterdam. *Mammalia* 28 (suppl.) 1:3–146.

Payne, M. R. 1977. Growth of a fur seal population. *Phil. Trans. R. Soc. Lond.* B 279:67–79.

Payne, M. R. 1979. Growth in the Antarctic fur seal *Arctocephalus gazella*. *J. Zool. Lond.* 187:1–20.

Pearcy, W. G., E. E. Krygier, R. Mesecar, and F. Ramsey. 1977. Vertical distribution and migration of oceanic micronekton off Oregon. *Deep Sea Res.* 24:223–245.

Pearson, A. K., and R. K. Enders. 1951. Further observation on the reproduction of the Alaskan fur seal. *Anat. Rec.* 111:695–711.

Perez, M. A., and M. A. Bigg. 1986. Diet of northern fur seals, *Callorhinus ursinus*, off western North America. *Fish. Bull.* 84(4):957–971.

Perez, M. A., and E. E. Mooney. 1985. Increased food and energy consumption of lactating northern fur seals, *Callorhinus ursinus*. *Fish. Bull.* 84:371–381.

Peters, R. H. 1976. Tautology in evolution and ecology. *Amer. Nat.* 110(971):1–12.

Peterson, R. S. 1961. Behavior of fur-seal pups during autumn. Unpublished report, NOAA, National Marine Mammal Labaoratory, Alaska Fisheries Research Center, Seattle.

Peterson, R. S. 1965. Behavior of the northern fur seal. Ph.D. diss., Johns Hopkins University, Baltimore, Md.

Peterson, R. S. 1968. Social behavior in pinnipeds with particular reference to the northern fur seal. In R. J. Harrison, R. C. Hubbard, R. S. Peterson, C. E. Rice, and R. J. Schusterman, eds., *The Behavior and Physiology of Pinnipeds*, 3–53. New York: Appleton-Century-Crofts.

Peterson, R. S., and G. A. Bartholomew. 1967. The natural history and behavior of the California sea lion. *Amer. Soc. Mamm. Spec. Publ.* 1:1–79.

Peterson, R. S., and B. J. Le Boeuf. 1969. Fur seals in California. *Pac. Discov.* 22:12–15.

Peterson, R. S., C. L. Hubbs, R. L. Gentry, and R. L. DeLong. 1968a. The Guadalupe fur seal: Habitat, behavior, population size, and field identification. *J. Mamm.* 49(4):665–675.

Peterson, R. S., B. J. Le Boeuf, and R. L. Delong. 1968b. Fur seals from the Bering Sea breeding in California. *Nature* 219(5157):899–901.

Pierson, M. O. 1978. A study of the population dynamics and breeding behavior of the Guadalupe fur seal, *Arctocephalus townsendii*. Ph.D. diss., University of California, Santa Cruz.

Pike, G. C., and B. E. Maxwell. 1958. The abundance and distribution of the northern sea lion (*Eumetopias jubata*) on the coast of British Columbia. *J. Fish. Res. Bd. Can.* 15(1):5–17.

Pitcher, K. W. 1990. Major decline in number of harbor seals, *Phoca vitulina richardsi*, on Tugidak Island, Gulf of Alaska. *Mar. Mamm. Sci.* 6(2):121–134.

Polovina, J. J. 1996. Decadal variation in the trans-Pacific migration of northern bluefin tuna (*Thunnus thynnus*) coherent with climate-induced change in prey abundance. *Fish. Oceanogr.* 5(2):114–119.

Polovina, J. J., G. T. Mitchum, N. E. Graham, M. P. Craig, E. E. DeMartini, and

E. N. Flint. 1994. Physical and biological consequences of a climate event in the central North Pacific. *Fish. Oceanogr.* 3(1):15–21.

Ponganis, P. J., R. L. Gentry, E. P. Ponganis, and K. V. Ponganis. 1992. Analysis of swim velocities during deep and shallow dives of two northern fur seals, *Callorhinus ursinus. Mar. Mamm. Sci.* 8(1):69–75.

Queller, D. C. 1987. The evolution of leks through female choice. *Anim. Behav.* 35:1424–1432.

Quinn, T. J., II, and H. J. Niebauer. 1995. Relation of eastern Bering Sea walleye pollock (*Theragra calcogramma*) recruitment to environmental and oceanographic variables. In R. J. Beamish, ed., *Climate Change and Northern Fish Populations. Can. Spec. Publ. Aquat. Sci.* 121:497–507.

Ragen, T. J., G. A. Antonelis, and M. Kyota. 1995. Early migration of northern fur seal pups from St. Paul Island, Alaska. *J. Mammal.* 76(4):1137–1148.

Ralls, K. 1976. Mammals in which females are larger than males. *Quart. Rev. Biol.* 51:245–276.

Ralls, K. 1977. Sexual dimorphism in mammals: Avian models and unanswered questions. *Amer. Nat.* 111:917–938.

Rand, R. W. 1955. Reproduction in the female Cape fur seal, *Arctocephalus pusillus. J. Zool. Soc. Lond.* 24:717–741.

Rand, R. W. 1956. The Cape fur seal (*Arctocephalus pusillus*): Its general characteristics and moult. *Invest. Rep. Div. Un. S. Afr.* 21:1–52.

Rand, R. W. 1959. The Cape fur seal (*Arctocephalus pusillus*): Distribution, abundance and feeding habits off the south western coast of the Cape Province. *Invest. Rep.. Div. Un. S. Afr.* 34:1–75.

Rand, R. W. 1967. The Cape fur seal (*Arctocephalus pusillus*), 3: General behavior on land and at sea. *Invest. Rep. Div. Sea Fish. S. Afr.* 60:1–39.

Rasmusson, E. M., and J. M. Wallace. 1983. Meteorological aspects of the El Niño/ Southern Oscillation. *Science* 222:1195–1202.

Ream, R. R., G. A. Antonelis, and J. D. Baker. 1994. Trends in pup production of rookeries on St. George Island, Alaska. *Fur Seal Investigations, 1992.* NOAA Tech. Mem. NMFS-AFSC-45:76–83.

Repenning, C. A., and R. H. Tedford. 1977. Otarioid seals of the neogene. *Geol. Surv. Prof. Papers* no. 992 (catalogue number 024–011–02988–0). Washington, D.C.: U.S. Government Printing Office.

Repenning, C. A., C. E. Ray, and D. Grigorescu. 1979. Pinniped biogeography. In J. Gray and A. J. Boucot, eds., *Historical Biogeography, Plate Tectonics, and the Changing Environment*, 357–369. Corvallis: Oregon State University Press.

Richardson, W. J. 1995. Chapter 8, Marine mammal hearing. In W. J. Richardson, C. R. Greene, Jr., C. I. Malme, and D. H. Thomson, eds., *Marine Mammals and Noise*. San Diego: Academic Press.

Ridgway, S. H., and R. J. Harrison. 1981. *Handbook of Marine Mammals*, vol. 1: *The Walrus, Sea Lions, Fur Seals and Sea Otters*. London: Academic Press.

Roby, D. D., and K. L. Brink. 1986. Decline of breeding least auklets on St. George Island, Alaska. *J. Field Ornithol.* 57:57–59.

Roemmich, D. 1992. Ocean warming and sea level rise along the Southwest U.S. Coast. *Science* 257:373–375.

Roemmich, D., and J. McGowan. 1995. Climatic warming and the decline of zooplankton in the California current. *Science* 267:1324–1326.

Ronald, K., and A. W. Mansfield, eds. 1975. Biology of the seal. *Rapp. P.-v. Réun. Cons. Int. Explor. Mer* 169:1–557.

Roper, C.F.E., and R. E. Young. 1975. Vertical distribution of pelagic cephalopods. *Smithson. Contrib. Zool.* 209:1–51.

Roppel, A. Y. 1984. Management of northern fur seals on the Pribilof Islands, Alaska, 1786–1981. NOAA Tech. Rep. NMFS 4, Alaska Fisheries Research Center, Seattle.

Roppel, A. Y., and S. P. Davey. 1965. Evolution of fur seal management on the Pribilof Islands. *J. Wildl. Manag.* 29(3)448–463.

Roux, J.-P. 1987. Recolonization processes in the Subantarctic fur seal, *Arctocephalus tropicalis*, on Amsterdam Island. In J. P. Croxall and R. L. Gentry, eds., *Status, Biology, and Ecology of Fur Seals*, 189–194. NOAA Tech. Rep. NMFS 51:9–21, British Antarctic Survey, Cambridge, England.

Royer, T. C. 1989. Upper ocean temperature variability in the northeast Pacific Ocean: Is it an indicator of global warming? *J. Geophys. Res.* 94(C12):18,175–18,183.

Royer, T. C. 1993. High-latitude oceanic variability associated with the 18.6-year nodal tide. *J. Geophys. Res.* 98:4639–4644.

Rubenstein, D. I., and R. W. Wrangham. 1986. Socioecology: Origins and trends. In D. I. Rubenstein and R. M. Wrangham, eds., *Ecological Aspects of Social Evolution*. Princeton, N.J.: Princeton University Press.

Sadlier, R.M.F.S. 1969. *The Ecology of Reproduction in Wild and Domestic Mammals*. London: Methuen.

Sandegren, F. E. 1976. Courtship display, agonistic behavior and social dynamics in the Steller sea lion (*Eumetopias jubatus*). *Behaviour* 55:159–172.

Sandell, M. 1990. The evolution of seasonal delayed implantation. *Quart. Rev. Biol.* 65(1):23–42.

Scammon, C. M. 1869. Fur seals. *Overland Monthly* 3(5):393–399.

Scammon, C. M. 1874. *The Marine Mammals of the Northwestern Coast of North America Together with an Account of the American Whale Fishery*, 141–163. San Francisco: John H. Carmany. (Dover edition, 1968. New York: Dover Publications.)

Scheffer, V. B. 1950. Growth of the testes and baculum in the fur seal, *Callorhinus ursinus*. *J. Mammal.* 31:384–394.

Scheffer, V. B. 1958. *Seals, Sea Lions, and Walruses: A Review of the Pinnipedia*. Stanford, Calif.: Stanford University Press.

Scheffer, V. B. 1962. Pelage and surface topography of the northern fur seal. *U.S. Fish Wildl. Serv. North Am. Fauna* 64:206.

Scheffer, V. B., and F. Wilke. 1953. Relative growth in the northern fur seal. *Growth* 17:129–145.

Scheffer, V. B., C. H. Fiscus, and E. I. Todd. 1984. History of scientific study and management of the Alaskan fur seal, *Callorhinus ursinus*, 1786–1964. NOAA Tech. Rep. NMFS SSRF-780, Alaska Fisheries Research Center, Seattle.

Schinckel, P. G. 1954. The effect of the ram on the incidence and occurrence of oestrus in ewes. *Aust. Vet. J.* 30:189–195.

Schusterman, R. J. 1966. Serial discrimination—reversal learning with and without errors by the California sea lion. *J. Exp. Anal. Beh.* 9:593–600.

Schusterman, R. J. 1972. Visual acuity in Pinnipeds. In H. E. Winn and B. L. Olla, eds., *Behavior of Marine Animals*, 469–492. New York: Plenum Press.

Schusterman, R. J. 1981a. Steller sea lion *Eumetopias jubatus* (Schreber 1776). In S. H. Ridgway and R. J. Harrison, eds., *Handbook of Marine Mammals*, vol. 1, 119–141. New York: Academic Press.

Schusterman, R. J. 1981b. Behavioral capabilities of seals and sea lions: A review of their hearing, visual, learning and diving skills. *Psych. Rec.* 31:125–143.

Schusterman, R. J., and R. L. Gentry. 1970. Development of a fatted male phenomenon in California sea lions. *Dev. Psychobiol.* 4(4):333–338.

Schusterman, R. J., and D. Kastak. 1993. A California sea lion (*Zalophus californianus*) is capable of forming equivalence relations. *Psych. Rec.* 43:823–839.

Schusterman, R. J., and K. Krieger. 1986. Artificial language comprehension and size transposition by a California sea lion (*Zalophus clifornianus*). *J. Comp. Psych.* 100:348–355.

Schusterman, R. J., and T. Thomas. 1966. Shape discrimination and transfer in the California sea lion. *Psychon. Sci.* 5(1):21–22.

Schusterman, R. J., R. Gisiner, B. K. Grimm, and E. B. Hanggi. 1992. Behavioral control by exclusion and attempts at establishing semanticity in marine mammals using Match-to-Sample paradigms. In, H. L. Roitblat, L. M. Herman, and P. E. Nachtigall, eds., *Language and Communication: Comparative Perspectives*, 249–274. Hillsdale, N.J.: Erlbaum.

Scordino, J., and R. Fisher. 1983. Investigations on fur seal entanglement in net fragments, plastic bands and other debris in 1981 and 1982, St. Paul Island, Alaska. Unpublished background paper presented to the 26th Annual Meeting of the North Pacific Fur Seal Commission, Washington, D.C., March/April 1983. Available from NOAA, National Marine Mammal Laboratory, Seattle.

Shaughnessy, P. D. 1984. Historical population levels of seals and seabirds on islands off southern Africa, with special reference to Seal Island, False Bay. *Invest. Rep. Sea Fish. Res. Inst. S. Afr.* 127:1–61.

Shaughnessy, P. D., and L. Fletcher. 1987. Fur seals, *Arctocephals* spp., at Macquarie Island. In J. P. Croxall and R. L. Gentry, eds., *Status, Biology, and Ecology of Fur Seals*. NOAA Tech. Rep. NMFS 51:177–188, British Antarctic Survey, Cambridge, England.

Shaughnessy, P. D., and S. D. Goldsworthy. 1990. Population size and breeding season of the Antarctic fur seal *Arctocephalus gazella* at Heard Island—1987/88. *Mar. Mamm. Sci.* 6(4):292–304.

Shelton, M. 1960. Influence of the presence of a male goat on the initiation of estrus cycling and ovulation of angora does. *J. Anim. Sci.* 19:368–375.

Short, R. V. 1984. Species differences in reproductive mechanisms. In R. Austin and R. V. Short, eds., *Reproduction in Mammals*, vol. 4: *Reproductive Fitness*, 24–61. Cambridge, England: Cambridge University Press.

Sinclair, E., T. Loughlin, and W. Pearcy. 1994. Prey selection by northern fur seals (*Callorhinus ursinus*) in the eastern Bering Sea. *Fish. Bull.* 92:144–156

Sivertsen, E. 1954. A survey of the eared seals (family Otariidae) with remarks on the antarctic seals collected by M/K "Norvegica" in 1928–1929. *Det Norske Vi-*

denskaps-Academi i Oslo, Sci. Results Norweg. Antarct. Exped., 1927–29 et sqq . . . Lars Christensen, 36:1–76.

Slimp, J. E. 1977. Reduction of genital sensory input and sexual behavior of female guinea pigs. *Physiol. Behav.* 18:1027–1031.

Small, R. J., and D. P. DeMaster. 1995. Alaska marine mammal stock assessments, 1995. NOAA Tech. Mem. NMFS-AFSC-57:9–11.

Smith, G. B. 1981. The biology of walleye pollock. In D. W. Hood and J. A. Calder, eds., *The Eastern Bering Sea Shelf: Oceanography and Resources*, vol. 1, 527–551. Washington, D.C.: U.S. Government Printing Office.

Spotte, S., and G. Adams. 1981. Photoperiod and reproduction in captive female northern fur seals. *Mamm. Rev.* 11:31–35.

Standing Scientific Committee of the North Pacific Fur Seal Commission. 1963. Glossary of terms used in fur seal research and management. *U.S. Fish Wildl. Serv., Bur. Comm. Fish. Leaf.* 546:1–10.

Stearns, S. C. 1977. The evolution of life history traits: A critique of the theory and a review of the data. *Ann. Rev. Ecol. Syst.* 8:145–171.

Stejneger, L. 1896. *The Russian Fur-Seal Islands*. Washington, D.C.: U.S. Government Printing Office.

Stejneger, L. 1925. Fur-seal industry of the Commander Islands, 1897–1922. *Bull. Bur. Fish.* 41(986):289–332.

Steller, G. W. 1749. The beasts of the sea. *Mem. Imp. Acad. Sci. St. Petersburg.* In Latin. Translation by W. Miller, 1899. In D. S. Jordan, ed., *The Fur Seals and Fur-Seal Islands of the North Pacific Ocean*, part 3, 179–218. Washington, D.C.: U.S. Government Printing Office.

Stirling, I. A. 1971. Variations in sex ratio of newborn Weddell seals during the pupping season. *J. Mamm..* 52:842–844.

Stirling, I. 1983. The evolution of mating systems in pinnipeds. In J. A. Eisenberg and D. G. Kleimann, eds., *Advances in the Study of Mammalian Behavior. Amer. Soc. Mamm. Spec. Publ.* no. 7:489–527.

Stirling, I. A., and N. J. Lunn. 1997. Environmental fluctuations in arctic marine ecosystems as reflected by variability in reproduction of polar bears and ringed seals. *In* M. Young and S. Woodin, eds., *Ecology of Arctic Ecosystems*, 167–181. Oxford: Blackwell Scientific.

Taylor, F.H.C., M. Fujinaga, and F. Wilke. 1955. Distribution and food habits of the fur seals of the North Pacific Ocean. Report of cooperative investigations by the Governments of Canada, Japan, and the United States, February–July 1952. U.S. Fish and Wildlife Service, Washington, D.C. Files of NOAA, National Marine Mammal Laboratory, Seattle.

Temte, J. L. 1985. Photoperiod and delayed implantation in the northern fur seal (*Callorhinus ursinus*). *J. Reprod. Fert.* 73:127–131.

Terkel, J., J. A. Witcher, and N. T. Adler. 1990. Evidence for memory of cervical stimulation for the promotion of pregnancy in rats. *Horm. Behav.* 24:40–49.

Townsend, C. H. 1899. Pelagic sealing with notes on the fur seals of Guadalupe, the Galapagos, and Lobos Islands. In D. S. Jordan, ed., *The Fur Seals and Fur-Seal Islands of the North Pacific Ocean*, part 3, 223–274. Washington, D.C.: U.S. Government Printing Office.

Traynor, J., and D. Smith. 1996. Summer distribution and abundance of age-0 walleye pollock (*Theragra calcogramma*) in the Bering Sea. In *Ecology of Juvenile Walleye Pollock* (*Theragra calcogramma*). Papers from the Workshop on the Importance of Prerecruit Walleye Pollock to the Bering Sea and North Pacific Ecosystems, October 1993, Seattle, Wash. NOAA Tech. Rep. 126:57–59.

Trenberth, K. E. 1990. Recent observed interdecadal climate changes in the northern hemisphere. *Bull. Amer. Meteor. Soc.* 71(7):988–993.

Trenberth, K. E., and J. W. Hurrell. 1994. Decadal atmosphere-ocean variations in the Pacific. *Clim. Dyn.* 9:303–319.

Trillmich, F. 1986. Attendance behavior of Galapagos sea lions. In R. L. Gentry and G. L. Kooyman, eds., *Fur Seals: Maternal Strategies on Land and at Sea*, 196–208. Princeton, N.J.: Princeton University Press.

Trillmich, F. 1990. The behavioral ecology of maternal effort in fur seals and sea lions. *Behaviour* 114(1–4):3–20.

Trillmich, F., G. L. Kooyman, P. Majluf, and M. Sanchez-Grinan. 1986. Attendance and diving behavior of South American fur seals during El Niño in 1983. In R. L. Gentry and G. L. Kooyman, eds., *Fur Seals: Maternal Strategies on Land and at Sea*, 153–167. Princeton, N.J.: Princeton University Press.

Trillmich, T., K. A. Ono, D. P. Costa, R. L. De Long, S. D. Feldkamp, J. M. Francis, R. L. Gentry, C. B. Heath, B. J. Le Boeuf, P. Majluf, and A. E. York. 1991. The effects of El Niño on pinniped populations in the Eastern Pacific. In F. Trillmich and K. A. Ono, eds., *Pinnipeds and El Niño*, 247–270. Berlin: Springer-Verlag.

Trites, A. W. 1991. Fetal growth of northern fur seals: Life-history strategy and sources of variation. *Can. J. Zool.* 69:2608–2617.

Trites, A. W. 1992a. Fetal growth and the condition of pregnant northern fur seals off western North America from 1958 to 1972. *Can. J. Zool.* 70:2125–2131.

Trites, A. W. 1992b. Northern fur seals: Why have they declined? *Aquat. Mamm.* 18.1:3–18.

Trites, A. W., and G. A. Antonelis. 1994. The influence of climatic seasonality on the life cycle of the Pribilof northern fur seal. *Mar. Mamm. Sci.* 10(3):311–324.

Trites, A. W., and A. E. York. 1993. Unexpected changes in reproductive rates and mean age at first birth during the decline of the Pribilof northern fur seal (*Callorhinus ursinus*). *Can. J. Fish. Aquat. Sci.* 50(4):858–864.

Trivers, R. L. 1972. Parental investment and sexual selection. In B. Campbell, ed., *Sexual Selection and the Descent of Man*, 136–179. Chicago: Aldine.

Trivers, R. L. 1974. Parent-offspring conflict. *Am. Zool.* 14:249–264.

Trivers, R. L., and D. E. Willard. 1973. Natural selection of parental ability to vary the sex ratio of offspring. *Science* 179:90–92.

U.S. Congress, Senate. 1957. Interim convention on conservation of North Pacific fur seals. *Message from the President of the United States*. 85th Congress, 1st sess., Exec. J. Washington, D.C.: U.S. Government Printing Office.

Vaz-Ferreira, R. 1950. Observaciones sobre la Isla de Lobos. *Rev. Fac. Hum. y Cien.* 5:145–176.

Veniaminov, I. 1839. The sea bear. *Phoca ursina*. Morskoit ko. In German. In K. E. von Baer, Information regarding the Russian Provinces on the Northwest Coast of America, *Acad. Sci. St. Petersburg*. Partial translation by L. Stejneger, 1899. In D. S. Jordan, ed., *The Fur Seals and Fur-Seal Islands of the North Pacific Ocean*, part 3, 219–222. Washington, D.C.: U.S. Government Printing Office.

Venrick, E. L., J. A. McGowan, D. R. Cayan, and T. L. Hayward. 1987. Climate and chlorophyll a: Long-term trends in the central North Pacific Ocean. *Science* 238:70–72.

Versaggi, C. S. 1981. Studies on the internal structure, development, and function of bone in pinnipeds and the sea otter. Ph.D. diss., University of California, Santa Barbara.

Vladimirov, V. A. 1983. On reproductive biology of sea bear females on the Urilye rookery (Mednyi Island). *Bull. Mosk. O-va Ispyt. Prir. Otd. Biol.* 88(4)52–61. In Russian, with English summary. NOAA, National Marine Mammal Laboratory, Seattle.

Vladimirov, V. A. 1987. Age-specific reproductive behavior in northern fur seals on the Commander Islands. In J. P. Croxall and R. L. Gentry, eds., *Status, Biology, and Ecology of Fur Seals.* NOAA Tech. Rep. NMFS 51:113–120, British Antarctic Survey, Cambridge, England.

Vladimirov, V. A. 1993. Information report on the results of monitoring fur seal populations in Russia, 1992. *Russ. Fed. Res. Inst. Fish. Oceanogr.* (VNIRO). Unpublished report, NOAA, National Marine Mammal Laboratory, Seattle.

Vladimirov, V. A., and V. S. Nikulin. 1991. Preliminary investigation of age-sex structure of northern fur seals on the Pribilof Islands, 1991. In E. H. Sinclair, ed., *Fur Seal Investigations, 1991.* NOAA Tech. Mem. NMFS-AFSC-24:61–73.

VNIRO (All-Union Research Institute of Marine Fisheries and Oceanography). 1978–85. Annual reports on USSR fur seal investigations. Unpublished reports submitted to the 22nd through 28th annual meetings of the Standing Scientific Committee of the North Pacific Fur Seal Commission, NOAA, National Marine Mammal Laboratory, Seattle.

Wada, K. 1971. Food and feeding habits of northern fur seals along the coast of Sanriku. *Bull. Tokai Reg. Fish. Res. Lab.* 64:1–37.

Walker, P. L. 1979. Archaeological evidence concerning the prehistoric occurrence of sea mammals at Point Bennett, San Miguel Island. *Calif. Fish Game* 65:(1):50–54.

Walker, W. A., and L. L. Jones. 1990. Food habits of northern right whale dolphin, Pacific white-sided dolphin, and northern fur seal caught in the high seas driftnet fishery of the North Pacific Ocean, 1990. *Internat. N. Pac. Fish. Comm. Bull.* 53(2):285–295.

Wallace, J. M., and D. S. Gutzler. 1981. Teleconnections in the geopotential height field during the northern hemisphere winter. *Mon. Weather Rev.* 109:784–812.

Walls, G. L. 1942. The vertebrate eye and its adaptive radiation. *Cranbrook Inst. Sci. Bull.* 19, Bloomfield Hills, Michigan. (Reprinted 1963 by Hafner, New York.)

Wespestad, V. G. 1991. Pacific herring population dynamics, early life history, and recruitment variation relative to eastern Bering Sea oceanographic factors. Ph.D. diss., School of Fisheries, University of Washington, Seattle.

Wespestad, V. G., and J. J. Traynor. 1988. Walleye pollock. In R. G. Bakkala, ed., *Condition of Groundfish Resources of the Eastern Bering Sea and Aleutian Region in 1987.* NOAA Tech. Mem. NMFS-F/NWC-139:11–32.

Whalen, R. E. 1963. Sexual behavior of cats. *Behaviour* 20:321–343.

Whitten, W. K. 1956a. Modifications of the oestrous cycle of the mouse by external stimuli associated with the male. *J. Endocrinol.* 13:399–404.

Whitten, W. K. 1956b. The effect of the olfactory bulbs on the gonads of mice. *J. Endocrinol.* 14:160–163.

Wilke, F., and K. W. Kenyon. 1954. Migration and food of the northern fur seal. *Trans. 19th N. Amer. Wildl. Conf.* 430–440.

Wilson, E. O. 1975. *Sociobiology: The New Synthesis.* Cambridge, Mass.: Belknap Press of Harvard University Press.

Wilson, R. P., W. S. Grant, and D. C. Duffy. 1986. Recording devices on free-ranging marine animals: Does measurement affect foraging performance? *Ecology* 67(4):1091–1093.

Wittenberger, J. F. 1983. Tactics of mate choice. In P. Bateson, ed., *Mate Choice,* 435–448. Cambridge, England: Cambridge University Press.

Wolfson, A. 1958. Role of light and darkness in the regulation of the annual stimulus for spring migration and reproductive cycles. *Internat. Ornithol. Congr. Proc.* 12:758–789.

York, A. E. 1983. Average age at first reproduction of the northern fur seal (*Callorhinus ursinus*). *Can. J. Fish. Aquat. Sci.* 40(2):121–127.

York, A. E. 1985. Juvenile survival of fur seals. In P. Kozlof, ed., *Fur Seal Investigations, 1982.* NOAA Tech. Mem. NMFS-F/NWC-71:34–45, Alaska Fisheries Science, Seattle.

York, A. E. 1987a. Northern fur seal, *Callorhinus ursinus*, Eastern Pacific population (Pribilof Islands, Alaska, and San Miguel Island, California). In J. P. Croxall and R. L. Gentry, eds., *Status, Biology, and Ecology of Fur Seals.* NOAA Tech. Rep. NMFS 51:133–140, British Antarctic Survey, Cambridge, England.

York, A. E. 1987b. Growth of northern fur seal fetuses. Abstract for the Seventh Biennial Conference on the Biology of Marine Mammals, December 1987, Miami, Florida.

York, A. E. 1990. Trends in numbers of pups born on St. Paul and St. George Islands 1973–88. In H. Kajimura, ed., *Fur Seal Investigations, 1988 and 1988.* NOAA Tech. Mem. NMFS-F/NWC-180:31–37, Alaska Fisheries Science, Seattle.

York, A. E. 1991a. An analysis of tagging studies done on northern fur seals 1947–68 emphasizing their applicability for future studies. Unpublished manuscript, NOAA, National Marine Mammal Laboratory, Seattle.

York, A. E. 1991b. Sea surface temperatures and their relationship to the survival of juvenile male northern fur seals from the Pribilof Islands. In F. Trillmich and K. A. Ono eds., *Pinnipeds and El Niño,* 94–106. Berlin: Springer-Verlag.

York, A. E. 1995. The relationship of several environmental indices to the survival of juvenile male northern fur seals (*Callorhinus ursinus*) from the Pribilof Islands. *Can. Spec. Publ. Fish. Aquat. Sci.* 21:317–327.

York, A. E., and J. R. Hartley. 1981. Pup production following harvest of female northern fur seals. *Can. J. Fish. Aquat. Sci.* 38(1):84–90.

York, A. E., and P. Kozloff. 1987. On the estimation of numbers of northern fur seal, *Callorhinus ursinus*, pups born on St. Paul Island, 1980–86. *Fish. Bull.* 85:367–375.

York, A. E., and V. B. Scheffer. 1997. Implantation date of northern fur seals (*Callorhinus ursinus*) from the Pribilof Islands. *J. Mammal.* 78(2):674–683.

Yoshida, K., and N. Baba. 1982. A study of behavior and ecology of female fur seals on a breeding island by radio wave telemetry and visual observation. *Bull. Far Seas Res. Lab.* 20:37–61.

Yoshida, K., N. Baba, M. Oya, and K. Mizue. 1977. On the formation and regression of curpus luteum in the northern fur seal overies. *Sci. Rep. Whales Res. Inst.* 29:121–128.

Yoshida, K., N. Baba, M. Oya, and K. Mizue. 1978. Seasonal changes in the ovary of the northern fur seal. *Bull. Far Seas Fish. Res. Lab.* 16:93–101.

Zar, J. H. 1984. *Biostatistical Analysis.* 2d ed. Englewood Cliffs, N.J.: Prentice Hall.

Zeusler, F. A. 1936. Report of the oceanographic cruise of the United States Coast Guard Cutter *Chelan*, Bering Sea and Bering Strait, 1934, and other related data. Unpublished report. Washington, D.C.: U.S. Coast Guard Headquarters.

Author Index